Introduction to Machine Learning and Bioinformatics

Chapman & Hall/CRC
Computer Science and Data Analysis Series

The interface between the computer and statistical sciences is increasing, as each discipline seeks to harness the power and resources of the other. This series aims to foster the integration between the computer sciences and statistical, numerical, and probabilistic methods by publishing a broad range of reference works, textbooks, and handbooks.

SERIES EDITORS
David Madigan, Rutgers University
Fionn Murtagh, Royal Holloway, University of London
Padhraic Smyth, University of California, Irvine

Proposals for the series should be sent directly to one of the series editors above, or submitted to:

Chapman & Hall/CRC
23-25 Blades Court
London SW15 2NU
UK

Published Titles

Bayesian Artificial Intelligence
Kevin B. Korb and Ann E. Nicholson

Computational Statistics Handbook with MATLAB®, Second Edition
Wendy L. Martinez and Angel R. Martinez

Pattern Recognition Algorithms for Data Mining
Sankar K. Pal and Pabitra Mitra

Exploratory Data Analysis with MATLAB®
Wendy L. Martinez and Angel R. Martinez

Clustering for Data Mining: A Data Recovery Approach
Boris Mirkin

Correspondence Analysis and Data Coding with Java and R
Fionn Murtagh

Design and Modeling for Computer Experiments
Kai-Tai Fang, Runze Li, and Agus Sudjianto

Introduction to Machine Learning and Bioinformatics
Sushmita Mitra, Sujay Datta, Theodore Perkins, and George Michailidis

R Graphics
Paul Murrell

Semisupervised Learning for Computational Linguistics
Steven Abney

Statistical Computing with R
Maria L. Rizzo

CH Computer Science and Data Analysis Series

Introduction to Machine Learning and Bioinformatics

Sushmita Mitra

Indian Statistical Institute

Kolkata, India

Sujay Datta

Texax A&M University

College Station, TX, U.S.A.

Theodore Perkins

McGill Centre for Bioinformatics

Montreal, Quebec, Canada

George Michailidis

University of Michigan

Ann Arbor, MI, U.S.A.

CRC Press
Taylor & Francis Group
Boca Raton London New York

CRC Press is an imprint of the
Taylor & Francis Group, an **informa** business

A CHAPMAN & HALL BOOK

CRC Press
Taylor & Francis Group
6000 Broken Sound Parkway NW, Suite 300
Boca Raton, FL 33487-2742

First issued in paperback 2019

© 2008 by Taylor & Francis Group, LLC
CRC Press is an imprint of Taylor & Francis Group, an Informa business

No claim to original U.S. Government works

ISBN-13: 978-1-58488-682-2 (hbk)
ISBN-13: 978-0-367-38723-5 (pbk)

Library of Congress Cataloging-in-Publication Data

Mitra, Sushmita, 1962-
    Introduction to machine learning and bioinformatics / Sushmita Mitra, Sujay Datta.
        p. cm. -- (Computer science and data analysis series)
    Includes bibliographical references and index.
    ISBN 978-1-58488-682-2 (hardback : alk. paper)
    1. Bioinformatics. 2. Machine learning. I. Datta, Sujay. II. Title. III. Series.

QH324.2.M58 2008
572.80285--dc22   2008008649

Visit the Taylor & Francis Web site at
http://www.taylorandfrancis.com

and the CRC Press Web site at
http://www.crcpress.com

To my precious daughter *Moni*, and my beloved late *Ma*.

S. Mitra

To my ever-inspiring mother, beloved wife and sweet little daughter and to all those who, even in this pragmatic world, love knowledge for the sake of knowledge.

S. Datta

To my wife, Doina, for her constant encouragement.

T. Perkins

To the memory of my father Constantine.

G. Michailidis

Preface

Our motivation behind proposing this book was the desire to summarize under one umbrella the latest developments at the interface between two areas of tremendous contemporary interest—bioinformatics and machine learning. In addition, we wanted to provide a thorough introduction to the basic ideas of each individual area. At present, there is no other book offering such an informative yet accessible overview of the ways in which these two increasingly intertwined areas borrow strength or motivation from each other.

There are many different definitions of bioinformatics. Generally speaking, it is an emerging field of science growing from an application of mathematics, statistics and information science to the study and analysis of very large biological datasets with the help of powerful computers. No matter what precise definition is being used, there seems to be a consensus that bioinformatics is all about extracting knowledge from the deluge of information (in the form of huge datasets) produced by modern-day high-throughput biological experiments. And this is precisely the task that machine learning is meant for—it is supposed to provide innovative tools and techniques enabling us to handle the information overload that the scientific community is currently experiencing. So these two areas are destined to be married. From the numerous books that have been published in these areas in the recent past, most of which tend to focus on specific aspects of this marriage, one would get a piecemeal impression of how these two fields have been nourishing each other. There are a couple of notable exceptions, but even those books suffer from certain drawbacks, such as being inaccessible to people without a technically sophisticated background or focusing solely on classical statistical and combinatorial approaches. As a result, it is not uncommon to find bioinformatics and/or machine learning courses that are forced to use multiple textbooks or none at all (in which case, the course material might be a customized collection of research papers scattered over the internet).

Keeping this in mind, our intention was to attempt a coherent and seamless integration of the flurry of activities and the plethora of new developments that had taken place at the interface between these two areas in the last few years. As more and more people choose to get involved in these fields, the demand is growing for quality training programs and good textbooks/references. Our book aims to satisfy this demand. We believe that it will be comprehensive and self-explanatory enough to be used alone in a course. More and more graduate and advanced undergraduate courses in bioinformatics or machine learning these days tend to spend significant amounts of time cross-training students in the other discipline before even talking about the interface. To facilitate this process, our book provides sufficient background material in each of the first five chapters. It describes the main problems in bioinformatics, explains the fundamental concepts and algorithms of machine learning and provides many real or realistic examples to demonstrate the capabilities of the key machine learning techniques. Also, in order to facilitate proper digestion of the subject matter, each chapter (except chapter 2 and the special applications chapters 7-11) ends with problem sets. Once the groundwork has been done, chapters 7-11 will expose students to quite a few interesting examples where state-of-the-art machine learning techniques are being applied to bioinformatics problems. Although the subject matter is highly technical at times, a great deal of care has been taken to maintain lucidity in the style of exposition throughout the book, so that it is accessible to a wide audience—not just a handful of advanced researchers. The abundance of examples will help in this process.

Bioinformatics and machine learning being two of the hottest areas of contemporary research, there are quite a few books on them currently in the market. They can be broadly classified into the following six categories.

(A) Books that introduce the basic concepts of bioinformatics to readers with technical or non-technical backgrounds. Some examples are:

1. *Basic Bioinformatics* by Ignacimuthu, S. (2004, Alpha Science International, Ltd.)
2. *Bioinformatics (The Instant Notes Series)* by Westhead, D.R., Parish, J.H. and Twyman, R.M. (2002, BIOS Scientific publishers)
3. *Fundamental Concepts of Bioinformatics* by Krane, D.E. and Raymer, M.L. (2002, Benjamin Cummings)

(B) Books that introduce the basic concepts of machine learning to readers with different levels of technical background. Examples are:

1. *Machine Learning* by Mitchell, T.M. (1997, McGraw-Hill)
2. *Introduction to Machine Learning (Adaptive Computation and Machine Learning)* by Alpaydin, E. (2004, MIT Press)
3. *The Elements of Statistical Learning: Data Mining, Inference and Prediction* by Hastie, T., Tibshirani, R. and Friedman, J. (2001, Springer)
4. *Computational Learning Theory* by Kearns, M. and Vazirani, U. (1994, MIT Press)

(C) Those that focus on some special aspect of bioinformatics (e.g., microarrays or protein structures) or the application of a special modeling/computational technique (e.g., evolutionary computation or hidden Markov models) in bioinformatics. Examples are:

1. *Microarray Bioinformatics* by Stekel, D. (2003, Cambridge University Press)
2. *Bioinformatics: Sequence and Genome Analysis* by Mount, D.W. (2001, Cold Spring Harbor Laboratory Press)
3. *Protein Bioinformatics: An Algorithmic Approach to Sequence and Structure Analysis* by Eidhammer, I., Jonassen, I. and Taylor, W.R. (2004, John Wiley and Sons)
4. *Evolutionary Computation in Bioinformatics* by Fogel, G.B. and Corne, D.W. (2002, Morgan Kaufmann)
5. *Statistical Methods in Bioinformatics* by Ewens, W.J. and Grant, G.R. (2001, Springer-Verlag)
6. *Neural Networks and Genome Informatics* by Wu, C.H. and McLarty, J.W. (2000, Elsevier)
7. *Hidden Markov Models of Bioinformatics* by Koski, T. (2002, Kluwer Academic Publishers)

(**D**) Those that focus on some special aspect of machine learning (such as Bayesian neural networks or support vector machines) or introduce machine learning as part of a bigger agendum (such as artificial intelligence). Some examples are:

1. *Artificial Intelligence: A Modern Approach* by Russel, S. and Norvig, P. (2003, Prentice-Hall)
2. *Pattern Classification* by Duda, R., Hart, P. and Stork, D. (2001, Wiley)
3. *Neural Networks for Pattern Recognition* by Bishop, C.M. (1995, Oxford University Press)
4. *An Introduction to Support Vector Machines and Other Kernel-Based Learning Methods* by Christianini, N. & Shawe-Taylor, J. (2000, Cambridge University Press)
5. *Bayesian Learning for Neural Networks* by Neal, R.M. (1996, Springer)

(**E**) Edited volumes consisting of articles contributed by a number of authors that either provide a general introduction to or deal with advanced topics in bioinformatics or machine learning. Some examples are:

1. *Machine Learning and Its Applications: Advanced Lectures* edited by Paliouras, G. et al. (2001, Springer)
2. *Introduction to Bioinformatics: A Theoretical and Practical Approach* edited by Krawetz, S.A. and Womble, D.D. (Humana Press, 2003)
3. *Bioinformatics: Sequence, Structure and Databanks: A Practical Approach* edited by Higgins, D. and Taylor, W. (2000, Oxford University Press)

(**F**) Books that focus primarily on data mining. Examples are:

1. *Data Mining: Concepts and Techniques* by Han, J. and Kamber, M. (2000, Morgan Kaufmann)

2. *Data Mining: Multimedia, Soft Computing and Bioinformatics* by Mitra, S. and Acharya, T. (2003, John Wiley and Sons)

3. *Machine Learning and Data Mining: Methods and Applications* edited by Michalski. R.S., Bratko, R. and Kubat, M. (1998, John Wiley and Sons)

Each of these books has its own strengths and limitations. The ones in category (A) deal primarily with the basics of bioinformatics and offer very little on machine learning. Some of them (such as (A) (II) and (A) (III)) are essentially guide-books for internet resources. Those in category (B) deal primarily with machine learning, not necessarily in the context of bioinformatics. An exception is (B) (III) (i.e., Hastie, Tibshirani and Friedman), which has plenty of examples from bioinformatics and is an excellent textbook. However, its level of mathematical sophistication makes it inaccessible to a large part of our target audience. Others (such as B(I) and B(IV)) are a bit outdated since many years have passed since their publication. Books in category (C) are clearly narrow in their coverage, each specializing in a particular aspect of bioinformatics that may or may not have something to do with machine learning. A similar comment applies to the books in category (D), most of which specialize in some particular aspect of machine learning that may or may not be motivated by bioinformatics applications. Some in this category (such as (D)(I)) have a wider agendum (e.g., artificial intelligence) and devote several introductory chapters to machine learning in that context. But we believe this is not the kind of introduction that a beginner will find most helpful when he/she is trying to get into a new and challenging area. The books in category (E) are actually edited volumes of articles, dealing with bioinformatics or machine learning (but not both) at various levels. There are fundamental differences between a book and an edited volume. The latter often suffers from the cut-and-paste syndrome, that is, thematic incoherence among the constituent articles or lack of proper organization preventing a seamless integration of those articles. In our case, we chose not to follow this path, especially when one of our main objectives was to make our product usable as a primary or supplementary textbook. Finally, the books in category (F) are thematically centered on data mining, which is quite different from our focus.

So it should be clear that in the long list of books above, there is hardly a single one that succeeds in achieving exactly what our book tries to accomplish. While many of them address some needs that students and researchers may have, none is entirely satisfactory on its own. As a result, none of them can be projected as a direct substitute for our book. One could use a combination of several of them in a course (as, we believe, most instructors are forced to do at present). But as we mentioned earlier, a primary motivation behind writing this book was to provide instructors and researchers with a better option—-that of using a single book to learn all about the latest developments at the interface of bioinformatics and machine learning. To our knowledge, right now there is only one other book that has a comparable approach: *Bioinformatics: The Machine Learning Approach* by Baldi, P. and Brunak, S. (2001, MIT Press). However, with its abstract mathematical style, it does not enjoy the reputation of being particularly user-friendly—especially to students with a bioscience-related background. Also, Baldi and Brunak seem to emphasize the Bayesian paradigm and thus downplay many other approaches to machine learning that do not subscribe to this philosophy. In comparison, our book has a more balanced approach, so that it is less likely to turn off readers who are interested in the non-Bayesian avenue.

We strongly believe that our book will serve as a valuable source of up-to-date information for the benefit of Ph.D. students and advanced Masters students in bioinformatics, machine intelligence, applied statistics, biostatistics, computer science and related areas. Any university or research institution with a graduate-level program in any of these areas will find it useful. Advanced undergraduate-level courses offered by reputed universities may also find it suitable for adoption. In addition to its potential popularity as an informative reference, the material covered in it has been carefully chosen so that it can serve as the main textbook (or at least a supplementary textbook) in a variety of bioinformatics and machine learning courses.

A project like this would never be successful without the unselfish and silent contributions made by a number of distinguished colleagues who kindly accepted our invitations to contribute materials for the special applications chapters 7, 8, 10 and 11. We offer them our sincerest gratitude and consider it an honor to acknowledge all of them by name. They are Frank DiMaio, Ameet Soni and Jude Shavlik of the University of Wisconsin (Madison), Haider Banka of the Centre for Soft Computing Research, (India), Zhaohui S. Qin, Peter J. Ulintz, Ji Zhu and Philip Andrews of the University of Michigan (Ann Arbor), Bani K. Mallick of the Texas A&M University (College Station), Debashis Ghosh of the University of Michigan (Ann Arbor) and Malay Ghosh of the University of Florida (Gainesville).

It has been a real pleasure to work with the editorial and production staff at Taylor & Francis—right from the initial planning stage to the completion of the project. Robert Stern was patient, cooperative and encouraging throughout the duration of the project. Without his editorial experience and the technical advice of Marsha Pronin, it would be a much more daunting task for us and we truly appreciate their help. S. Datta would like to thankfully acknowledge the support provided by the Department of Statistics, Texas A&M University, through a National Cancer Institute grant (# CA90301). G. Michailidis acknowledges support from NIH grant P41-18627.

Collectively and individually, we express our indebtedness to the colleagues, students and staff at our home institutions. Last but not least, we hereby lovingly acknowledge the never-ending support and encouragement from our family members that gave us the strength to go all the way to the finish line.

S. Mitra

S. Datta

T. J. Perkins

G. Michailidis

Appreciation

To those colleagues who most kindly offered to contribute the material for Chapters 8 through 12 and shared their expertise and vision at various junctures of writing this book, the authors express their sincerest gratitude and appreciation. The authors consider it a privilege on their part to mention each contributor by name:

Philip Andrews	Bani K. Mallick
Haider Banka	Jude Shavlik
Frank DiMaio	Ameet Soni
Debashis Ghosh	Zhaohui S. Qin
Malay Ghosh	Peter J. Ulintz
Ji Zhu	

Contents

1 Introduction   1

2 The Biology of a Living Organism   5

   2.1 Cells   5

   2.2 DNA and Genes   8

   2.3 Proteins   12

   2.4 Metabolism   15

   2.5 Biological Regulation Systems: When They Go Awry   17

   2.6 Measurement Technologies   19

References   24

3 Probabilistic and Model-Based Learning   25

   3.1 Introduction: Probabilistic Learning   25

   3.2 Basics of Probability   27

   3.3 Random Variables and Probability Distributions   40

   3.4 Basics of Information Theory   56

   3.5 Basics of Stochastic Processes   58

   3.6 Hidden Markov Models   62

   3.7 Frequentist Statistical Inference   66

   3.8 Some Computational Issues   86

   3.9 Bayesian Inference   89

   3.10 Exercises   97

References   100

4 Classification Techniques   **101**

4.1   Introduction and Problem Formulation   101

4.2   The Framework   103

4.3   Classification Methods   108

4.4   Applications of Classification Techniques to Bioinformatics Problems   124

4.5   Exercises   124

References   **125**

5 Unsupervised Learning Techniques   **129**

5.1   Introduction   129

5.2   Principal Components Analysis   129

5.3   Multidimensional Scaling   136

5.4   Other Dimension Reduction Techniques   139

5.5   Cluster Analysis Techniques   141

5.6   Exercises   151

References   **153**

6 Computational Intelligence in Bioinformatics   **155**

6.1   Introduction   155

6.2   Fuzzy Sets (FS)   156

6.3   Artificial Neural Networks (ANN)   161

6.4   Evolutionary Computing (EC)   167

6.5   Rough Sets (RS)   171

6.6   Hybridization   173

6.7   Application to Bioinformatics   175

6.8   Conclusion   199

6.9   Exercises   200

References   **201**

7 Connections between Machine Learning and Bioinformatics   **211**

   7.1   Sequence Analysis   211

   7.2   Analysis of High-Throughput Gene Expression Data   218

   7.3   Network Inference   223

   7.4   Exercises   230

References   **231**

8 Machine Learning in Structural Biology: Interpreting 3D Protein Images   **237**

   8.1   Introduction   237

   8.2   Background   237

   8.3   ARP/WARP   247

   8.4   RESOLVE   252

   8.5   TEXTAL   258

   8.6   ACMI   264

   8.7   Conclusion   273

   8.8   Acknowledgments   275

References   **275**

9 Soft Computing in Biclustering   **277**

   9.1   Introduction   277

   9.2   Biclustering   278

   9.3   Multi-Objective Biclustering   283

   9.4   Fuzzy Possibilistic Biclustering   287

   9.5   Experimental Results   291

   9.6   Conclusions and Discussion   297

References   **298**

10 Bayesian Machine-Learning Methods for Tumor Classification Using Gene Expression Data   **303**

   10.1 Introduction   303

   10.2 Classification Using RKHS   306

   10.3 Hierarchical Classification Model   308

   10.4 Likelihoods of RKHS Models   310

   10.5 The Bayesian Analysis   312

   10.6 Prediction and Model Choice   314

   10.7 Some Examples   315

   10.8 Concluding Remarks   321

   10.9 Acknowledgments   322

References   **322**

11 Modeling and Analysis of Quantitative Proteomics Data Obtained from iTRAQ Experiments   **327**

   11.1 Introduction   327

   11.2 Statistical Modeling of iTRAQ Data   328

   11.3 Data Illustration   330

   11.4 Discussion and Concluding Remarks   332

   11.5 Acknowledgments   334

References   **334**

12 Statistical Methods for Classifying Mass Spectrometry Database Search Results   **339**

   12.1 Introduction   339

   12.2 Background on Proteomics   341

   12.3 Classification Methods   342

   12.4 Data and Implementation   347

   12.5 Results and Discussion   350

   12.6 Conclusions   356

   12.7 Acknowledgments   357

References   **357**

Index   **361**

CHAPTER 1

Introduction

This book is meant to provide an informative yet accessible overview of the ways in which the two increasingly intertwined areas of *bioinformatics* and *machine learning* borrow strength or motivation from each other. There are many different definitions of bioinformatics. Generally speaking, it is an emerging field of science growing from an application of mathematics, statistics and information science to the study and analysis of very large biological datasets with the help of powerful computers. No matter what precise definition is being used, there seems to be a consensus that bioinformatics is all about extracting knowledge from the deluge of information (in the form of huge datasets) produced by modern-day high-throughput biological experiments. And this is precisely the task that machine learning is meant for—it is supposed to provide innovative tools and techniques enabling us to handle the information overload that the scientific community is currently experiencing. So these two areas were destined to be married. In this book, our goal is to shed enough light on the key aspects of both areas for our readers to understand why this 'couple' is 'made for each other.'

Among the natural sciences, biology is the one that studies highly complex systems known as living organisms. Before the era of modern technology, it was primarily a descriptive science that involved careful observation and detailed documentation of various aspects of a living being (e.g., its appearance, behavior, interaction with the surrounding environment, etc.). These led to a reasonably accurate classification of all visible forms of life under the sun (the *binomial nomenclature* by Carolas Linnaeus) and to a theory of how the life-forms we see all around us came into being over billions of years (the theory of *evolution* by Charles Darwin). However, the technology available in those days was not good enough for probing the internal mechanisms that made life possible and sustained it—the complex biochemistry of metabolism, growth and self-replication. So the biological datasets that were available for statistical analysis in those days (including clinical and epidemiological data) were relatively small and manageable, and standard classical procedures (such as two-sample t-tests, ANOVA and linear regression) were adequate to handle them. But starting from the middle of the twentieth century, key breakthroughs in the biomedical sciences and rapid technological development changed everything. Not only did they enable us to probe the inner sanctum of a living organism at the molecular and genetic levels, but also brought a sea-change in our concept of medicine. A series of new discoveries gave us unprecedented insight into the *modus operandi* of living systems (the most famous one being the Franklin-Watson-Creek *double helix* model for DNA) and the ultimate goal of medical scientists became personalized medicine. It is a concept diametrically opposite to the way people used to view medicine earlier—a bunch of so-called experts taking their chances with chemicals they know little about to cure diseases they know nothing about. All these, together with the advent of modern computers, forced statisticians and information scientists to come out of their comfort zone and adapt to the new reality. New experiments powered by advanced technology were now generating enormous amounts of data on various features of life and increasingly efficient computing machines were making it possible to create and query gigantic databases. So statisticians and information scientists were now inundated with an incredible amount of data and demand for methods that could handle huge datasets violating some basic assumptions of classical statistics was high. Enhanced data-storage capability meant that there was an abundance of prior information on the unknown parameters in many applications and the lack of nice analytical solutions mattered little as our number-crunching machines churned out approximate numerical solutions much more quickly. This is the perfect scenario for Bayesians and, as a result, the Bayesian paradigm is now firmly in the driver's seat. It allows much more flexibility for modeling complex phenomena than the classical (or frequentist) paradigm and the inference that is drawn subsequently is almost always based on computationally intensive methods known as Markov chain Monte Carlo. To sum it up all, a modern-day statistician almost invariably finds himself/herself navigating an ocean of information where traditional statistical methods and their underlying assumptions often seem like flimsy boats, and yet, the race is on for the quickest way to 'fish out' relevant pieces of information. Unlike the old days, mathematical elegance and technical puritanism are no longer at the centerstage—speed and efficiency are much higher in the priority list. Perhaps the poet would say: "Data, data everywhere, not much time to think."

It is easy to see the similarity between this situation and the circumstances that motivated the industrial revolution that once occurred during the course of human civilization. The pressing need for rapid mass-production of commodities provided the driving force behind almost all scientific and technological innovations for a long time thereafter. Gone were the days of handcrafted products that boasted of an artiste's personal touch, because they were 'mountains of inefficiency' and not amenable to mechanized production in an assembly line. Speed and volume were the buzzwords. Of course, none of these would be possible without the *machines* that made mass-manufacturing a reality. The situation today is quite similar except that it is taking place in the realm of information. The processes of data-mining and recognizing patterns in search of information, converting the information to knowledge and, above all, repeating these tasks very rapidly a large number of times are ideally suited for machines—not humans. Our learning machines are computers that are getting smarter and faster at an astonishing rate. The phrase 'machine learning' can be interpreted either as building sophisticated machines that are capable of 'learning' (i.e., capable of being programmed to perform the tasks mentioned above repeatedly with pre-specified accuracy) or as mechanizing the process of 'learning.' The first interpretation leads to disciplines such as computer engineering and artificial intelligence, but here we mostly confine ourselves to the second interpretation. For example, two important tasks that researchers have tried to mechanize are *clustering* and *classification*. The first one is a form of pattern recognition that falls under the category of *unsupervised learning*, whereas the second one belongs to the category of *supervised learning*. Details can be found in Chapter 4.

Anyone who has watched a human baby grow up will agree that 'learning' as we know it is not exactly a mechanical process. It requires the perceptual and cognitive capabilities that human beings (and, to a lesser extent, the great apes) have acquired over millions of years through evolution. A computer, such as a von Neumann machine, is far behind human beings in this respect. It can outperform us only in tasks that involve substantial amounts of repetitive computation, such as inverting a large matrix. But when it comes to recognizing shapes of different sizes and orientations, even in an occluded environment, we are clearly the winner. In other words, a von Neumann machine is good for well-structured problems, whereas the human brain is typically better in solving ill-defined and imprecisely formulated problems of the real world. To overcome the limitations of the traditional computing paradigm, scientists have been searching for new computational approaches that can be used to model—at least partially—the human thought process and the functioning of our brains. As a result, in the recent past, several novel modes of computation have emerged. Some of them are collectively known as *soft computing*. Information processing in a biological system is a complex phenomenon, which enables a living organism to survive by recognizing its surroundings, making predictions and planning its future activities. This kind of information processing has both a logical and an intuitive aspect. Conventional computing systems are good for the former but way behind human capabilities in the latter. As a first step toward accomplishing human-like information processing, a computing system should be flexible enough to support the following three characteristics: *openness*, *robustness* and *real-time processing*. Openness of a system is its ability to adapt or extend itself on its own to cope with changes encountered in the real world. Robustness of a system means its stability and tolerance when confronted with distorted, incomplete or imprecise information. The real-time processing capability of a system implies that it can react to an event within a reasonably small amount of time.

The primary concerns of traditional computing have always been precision, rigor and certainty. We call this *hard* computing. In contrast, the main principle in soft computing is that precision and certainty carry a cost, and that the processes of reasoning and decision-making should exploit (wherever possible) the tolerance for imprecision, uncertainty, approximate reasoning and partial truth in order to obtain low-cost solutions. After all, these are what enable a human being to understand distorted speech, decipher sloppy handwriting, comprehend the nuances of natural languages, summarize text, recognize/classify images and drive safely in congested traffic. Our ability to make rational decisions in an environment of uncertainty and imprecision has distinguished us from other life-forms. So in soft computing, the challenge is to devise methods of computation that lead to an *acceptable solution at low cost*. A good example is the problem of parking a car. Generally, a car can be parked rather easily because, although the boundaries of a parking spot are marked,the final position of the car is not specified with extreme precision. If it were specified to within, say, a fraction of a millimeter and a few seconds of angular arc, it would take hours (if not days) of maneuvering to satisfy those specifications. This simple example points to the important fact that in general, high precision carries a high cost and, in many real-life situations, high precision is unnecessary.

The major components of soft computing, at this juncture, are *fuzzy sets* (FS), *artificial neural networks* (ANN), *evolutionary computing* (EC), *rough sets* (RS) and various hybrids of these. Fuzzy sets primarily deal with imprecision and approximate reasoning. Artificial neural networks provide a kind of machinery for learning. Searching and optimization in large datasets can be carried out via evolutionary computing. Knowledge extraction and dimensionality reduction are possible using rough sets. The different hybrids of these involve a synergistic integration of the individual paradigms so as to incorporate their generic and application-specific merits for the purpose of designing intelligent systems. Soft computing—an emergent technology—currently shows a lot of promise in bioinformatics mainly because of the imprecise nature of biological data. The power of soft computing lies in its ability to (i) handle subjectivity, imprecision and uncertainty in queries, (ii) model relevance as a *gradual* (as opposed to a *crisp*) property, (iii) provide deduction capability to the search engines, (iv) provide learning capability and (v) deal with the dynamism, scale and heterogeneity of information.

With this background, we now invite the reader to begin the fascinating journey.

CHAPTER 2

The Biology of a Living Organism

Understanding the intricate chemistry by which life survives, propagates, responds and communicates is the overarching goal of molecular biology—a field which relies increasingly on automated technologies for performing biological assays and computers for analyzing the results. While the complexity and diversity of life on earth is a source of neverending awe, at the molecular level there are many commonalities between different forms of life. In this chapter, we briefly review some essential concepts in cellular and molecular biology, in order to provide the necessary background and motivation for the bioinformatics problems discussed throughout the book. First, we discuss cells and organelles. Then we consider DNA, the carrier of genes, and protein. Biochemical networks are the topic of the following two sections, focussing in particular on cell metabolism and the control of proliferation. For more details on these topics, we refer the reader to any modern textbook on molecular and cellular biology, such as Alberts et al. [1]. We conclude the chapter with a discussion of some classical laboratory techniques of molecular biological as well as the new, automated techniques that create an ever-growing need for bioinformatics and machine learning.

2.1 Cells

The basic structural and functional unit of a living organism is a *cell*. Robert Hooke, a British scientist of the 17th century, made a chance discovery of them when he looked at a piece of cork under the type of microscope that was available at that time. Soon it was hypothesized, and ultimately verified, that every living organism is made of cells. In the three and a half centuries since their discovery, thanks to increasingly powerful microscopes and other sophisticated technologies, astonishing details are available about all aspects of cells, including their internal structure, day-to-day function, proliferation and death. Primitive organisms such as *bacteria* (e.g., *Proteobacteria* that include dangerous organisms like *Escherichia coli* and *Salmonella*, *Cyanobacteria* that are capable of converting the Sun's electromagnetic energy to chemical energy through photosynthesis or *Archaebacteria* that survive in extreme environments, including the deep-ocean hydrothermal vents) are single-cell organisms known as *prokaryotes*. They lack an organelle, called a nucleus, that is present

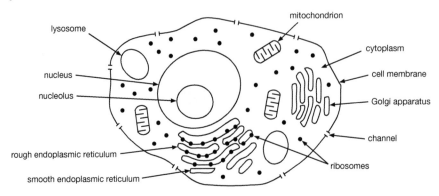

Figure 2.1 *Schematic of a Eukaryotic cell, with its various organelles.*

in the cells of all higher organisms known as *eukaryotes*. Eukaryotes, which include everything from a bath sponge to an elephant, may be uni-cellular or multi-cellular. In multi-cellular organisms, the cells are organized into *tissues* or groups of cells that collectively perform specific tasks, such as transport of oxygen or absorption of nutrients. Several such tissues collectively form an *organ*, such as the leaves of a plant or the lungs in our body. All the different types of cells in a multi-cellular organism originate from the same "mother cells" in the embryo, known as *stem cells*, which are capable of transforming into various types of specialized cells through a process called *differentiation*. This is why embryonic stem cells are considered so promising in the medical sciences as a possible cure for certain debilitating and degenerative diseases. The process of differentiation is still not well understood and bioinformatics may play a key role in helping us understand it.

The human body consists of trillions of eukaryotic cells. There are at least two hundred different types of cells in our body and they may be as little as a few thousandths of a millimeter in diameter and up to a meter in length (for neurons with long axons). Each cell has an outer membrane which encloses the *cytoplasm* and various components called *organelles* (see Figure 2.1). They include the *nucleus*, the *endoplasmic reticulum*, the *Golgi apparatus*, the *mitochondria*, the *ribosomes* and the *lysosomes*. A plant cell also has large cavities called *vacuoles* and different types of *plastids* (*chloroplasts* containing chlorophyll for photosynthesis, *leucoplasts* and *chromoplasts*). Each of these organelles has some specific role to play in the life of a cell. Because mitochondria and chloroplasts can multiply within a cell and bear structural similarity to certain unicellular prokaryotic organisms, it is hypothesized that long ago some prokaryotic cells started living inside others, forming a *symbiotic* or mutually beneficial relationship with their hosts. This process, called *endosymbiosis*, is believed to have led ultimately to the development of eukaryotic cells.

Before discussing the roles of the various organelles further, we introduce the main types of molecules present within a cell. There are four major classes of small organic (carbon-containing) molecules in cells: *sugars*, *lipids*, *amino acids*, and *nucleotides*. These molecules may exist on their own in the cell, but may also be bound into *macromolecules* or *polymers*, large molecules comprising specific patterns of different types of small organic molecules. For example, simple sugars, such as glucose, act as a source of energy for the cell and as a means for storing energy. Glucose can also be assembled into *cellulose*, a polymer which plays a structural role by forming the cell walls of plant cells. Lipids are a crucial component of cellular membranes, including the outer cell membrane and the membranes of organelles. They are also used for the storage of energy and act as *hormones*, transmitting signals between different cells. Amino acids are the building blocks of *proteins*, which perform numerous functions, and are discussed at greater length in Section 2.3. The nucleotide adenosine triphosphate (ATP) is used to transfer energy for use in a variety of reactions, while long chains of nucleotides form the *deoxyribonucleic acids* (DNA) and *ribonucleic acids* (RNA) that act as the basis of heredity and cellular control. DNA and RNA are discussed further in Section 2.2. In addition to these organic molecules, there are a variety of inorganic molecules within the cell. By mass, the majority of a cell is water. A few other important inorganic compounds include oxygen and carbon monoxide (the latter is usually considered inorganic, despite containing carbon), various ions (the medium for transmitting electrical signals in neurons and heart cells), and phosphate groups (crucial in intracellular signaling).

Returning to the organization of the cell, let us continue our discussion with organelles other than the nucleus. The cell membrane is primarily made of *phospholipid*, with various *membrane proteins* and other substances embedded in it. A phospholipid molecule has a *hydrophobic* (water-repelling) end and a *hydrophilic* (water-attracting) end. A cell membrane consists of a phospholipid bi-layer, with the hydrophobic "heads" of the molecules tucked inside and the hydrophilic "tails" forming surfaces that touch water. In addition to serving as an enclosure for the organelles, the membrane controls the passage of substances into and out of the cell and various receptors embedded in it play a crucial role in *signal transduction*. The endoplasmic reticulum primarily serves as a transport network and storage area for various cellular substances. It is continuous with the membrane of the nucleus. Various nuclear products (such as the messenger RNA produced by transcription—see Section 2.2) are transported to places where they are needed through the endoplasmic network. This reticulum has smooth surfaces and rough surfaces, the rough appearance being the result of numerous ribosomes attached to it. The ribosomes are the cell's protein-manufacturing plants where the macromolecules of proteins are synthesized component by component according to the instructions in the messenger RNA. The proteins produced by the ribosomes are transported to the Golgi apparatus (named after the Italian scientist Camillo Golgi who first noticed them) where they are further modified if necessary and then sent off in *vesicles* (little bubble-like sacks) to other organelles or to the cell membrane for secretion outside the cell. This apparatus has a *cis* end that is nearer to the endoplasmic reticulum and a *trans* end that is farther from it. The incoming proteins are received at the *cis* end and the outgoing vesicles bud off from the *trans* end. The lysosomes are the cell's scavengers or garbage cleaners. They help in the disintegration of unwanted or harmful substances in the cell. The mitochondria are the energy-producing plants where the cell's energy currency, ATP, is synthesized. A mitochondrion has a membrane quite similar to that of a prokaryotic bacterium, and inside it there are protruded folds of the membrane called *christae*. The rest of the space inside a mitochondrion is called its *lumen* where ATP synthesis takes place. In a plant cell, a chloroplast also has an outer membrane resembling that of a prokaryotic bacterium, and inside it the lumen containing stacks of flat disc-like structures called *thylakoids*. This is where chlorophyll, a magnesium-containing protein, is found. Chlorophyll plays a central role in photosynthesis—the plant's food production process. Chlorophyll is also the reason why leaves are green. Chromoplasts contain other pigments such as carotene and xanthophylls instead of chlorophyll. These are the reason behind the bright display of yellow, orange and brown in the Fall season. The remaining major organelle in a eukaryotic cell is the nucleus. It is roughly spherical with an outer membrane that has some pores in order to let nuclear products out into the cytoplasm or let cytoplasmic substances in. The central, denser region of a nucleus is called the nucleolus. Most important of all, it is the nucleus that stores an organism's blueprint of life, the DNA, which is the subject of the next section.

2.2 DNA and Genes

DNA (Deoxyribonucleic acid) is present in all living organisms and is of central importance to regulating cellular function and conveying hereditary information, largely by virtue of the *genes* it encodes. As mentioned in the previous section, DNA is a polymer of nucleotides, conceptually arranged as a ladder, as shown in Figure 2.2A. The functionally interesting portion of the nucleotide is its *base*. There are four different bases: *adenine, cytosine, guanine* and *thymine*, often referred to by their single letter abbreviations: A, C, G and T. Pairs of bases, one from each *strand* of the DNA, bind together to form the steps of the ladder. These *base pairs* are held in place by two backbones of sugar (deoxyribose) and phosphate molecules, which form the sides of the ladder. *Complementary base pairing* recognizes that each base binds with only one other kind of base: A and T bind with each other, and G and C bind with each other.

In reality, DNA molecules are not straight and ladder-shaped, as shown in Figure 2.2A, but have a complex three-dimensional (3-D) structure. The two *strands* of the DNA (sides of the ladder) are twisted into the well-known

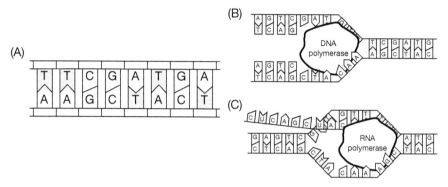

Figure 2.2 *(A) Conceptual structure of DNA as a polymer of nucleotides. Note the complementary base-pairing: A only binds with T, and G only binds with C. (B) Replication of the DNA by DNA polymerase. (C) Transcription of the DNA into RNA, by RNA polymerase.*

double-helix shape, famously discovered in 1953 by James Watson and Francis Crick, aided by X-ray crystallographic studies of Rosalind Franklin. At a larger scale, the DNA double helix is wrapped around barrel-shaped protein molecules called *histones*. The DNA and histones are in turn wrapped into coils and other structures depending on the state of the cell.

In some cells, most or all of the DNA occurs in a single molecule. In prokaryotes, such as bacteria, most of the DNA resides in a single *chromosome*, a long loop of DNA without beginning or end. Bacteria also often contain *plasmids*, much shorter loops of DNA that can be exchanged between bacteria as a means of sharing beneficial genes. In eukaryotes, the DNA is divided into multiple chromosomes, each a linear stretch of DNA with a definite beginning and end. Many eukaryotes are *polyploid*, carrying more than one copy of each chromosome. Humans, for example, are *diploid*, as most human cells carry two copies of each chromosome—one inherited from each parent. Some plants are tetraploid, carrying four copies of each chromosome. The total length of DNA and the number of chromosomes into which it is divided vary by organism. The single chromosome of the bacterium *Escherichia coli* has about 4.6 million base pairs, while humans have over 3 billion base pairs of DNA divided into 23 (pairs of) chromosomes. Fruit flies have less than 200 million base pairs divided into 8 pairs of chromosomes, while rice has roughly 400 million base pairs divided into 24 pairs of chromosomes.

When a cell divides to form two *daughter cells*, as during organism growth or to replace old or damaged tissue, the DNA must be *replicated* in order to provide each daughter cell with its own copy. Complementary base-pairing is key to this process. During replication, a group of proteins including *DNA polymerase* travels along the DNA (see Figure 2.2B). The two strands of DNA are separated, and complementary base pairs are filled in along both strands, resulting in the two needed copies.

Less than 5% of the DNA is believed to encode useful information. The other 95%, sometimes called "junk" DNA, has no apparent function. Of the 5% of functional DNA, the majority specifies genes—regions of the DNA comprising two parts, a *regulatory region* or *promoter* and a *coding region*. The regulatory region is partly responsible for specifying the conditions under which the gene product is produced, or the degree to which it is produced. The coding region specifies the functional molecular product or products—often a protein, but sometimes an RNA (ribonucleic acid). Proteins, discussed in more detail in the next section, are sequences of amino acids. For protein-coding genes, the coding region specifies the amino-acid sequence of the protein. However, the protein is not constructed directly from the DNA. Instead, the DNA is *transcribed* into an RNA intermediate, called *messenger RNA* (mRNA), which is then *translated* into protein. For RNA-coding genes, it is the RNA itself that serves an important function, and the RNA is not translated into protein.

RNA is composed of a single sequence, or strand, of nucleotides. These nucleotides are similar to the DNA nucleotides, except that the backbone uses the sugar ribose, and the base uracil (U) is used instead of thymine (T). Transcription, the construction of an RNA chain from the DNA, is similar to DNA replication (see Figure 2.2C). A group of proteins, collectively called RNA polymerase(s), open up the DNA at the start of the coding region and move along the DNA until reaching the end of the coding region. The *transcript* is produced one nucleotide at a time by complementary base-pairing—an A, C, G or T in the DNA is bonded to a U, G, C or A respectively in the RNA. The RNA molecule does not stay bound to the DNA. As it is constructed, the RNA nucleotide chain separates from the DNA and the two DNA strands then rejoin, leaving them in the same state as before transcription. The RNA strand does not form into any neat geometrical shape as the DNA does. Instead, its nucleotides bind to each other to form a complex 3-D shape unique to each RNA sequence. For RNA-coding genes, features of this 3-D shape determine its functional properties—causing it to bind to specific parts of certain proteins, for example. For protein-coding genes, the creation of the RNA is just the first step in producing a protein.

There are 20 naturally occurring amino acids out of which proteins are composed. The RNA sequence encodes an amino acid sequence in a straightforward manner. Each triplet of nucleotides, called a *codon*, specifies one amino acid. For example, the RNA nucleotide triplet AAA codes for the amino acid lysine, while the triplet AAC codes for asparagine. There are four nucleotides, and thus $4^3 = 64$ distinct possible codons. However, as there are only 20 amino acids, the codon code is redundant. For example, the amino acid leucine is encoded by six different codons: UUA, UUG, CUA, CUC, CUG and CUU. The entire transcript does not code for amino acids. A small amount of RNA at the start and end of the transcript, called untranslated regions or UTRs, are not translated into amino acids. For some genes, especially in eukaryotes, there are untranslated gaps in the middle called *introns*. The portions of the transcript that are translated are *exons*. In eukaryotes, *splicing* removes the introns before the transcript leaves the nucleus for translation. Splicing can result in the elimination of exons as well, a process called *alternative splicing*, resulting in different *splice variants* of a protein. This allows a cell to produce different versions of a protein from the same DNA. Ribosomes translate the RNA transcript into the corresponding amino acid sequence. In prokaryotes, free floating ribosomes can begin translation as soon as the transcript is created, or even while it is being created. In eukaryotes, the transcripts must be transported outside the nucleus and to a ribosome.

Transcription, splicing and translation are regulated in various ways. Characteristic patterns of nucleotides in the DNA indicate where transcription should begin and end. For example, in many bacteria, transcription begins just after a stretch of DNA with the sequence TATAAT. However, some variations on this sequence are allowed, and conversely, not every occurrence of this sequence in the DNA is followed by the coding region of a gene. A different pattern indicates the end of the coding region. In eukaryotes, there is much more variation in these patterns. In either case, even when the DNA of an organism is completely sequenced, there remains uncertainty about the exact locations and number of the genes. Variations in transcription initiation patterns have evolved in order to control the rate, or frequency, with which the RNA polymerase binds to the DNA and begin transcription.

Transcription is primarily regulated, however, by *transcription factors*. These proteins bind the regulatory region of the gene on the DNA, usually within a few hundreds or thousands of base pairs of the transcription initiation site, and influence the frequency or rate of transcription by any number of means: blocking the binding of RNA polymerase, changing the shape of the DNA to expose the transcription initiation site and increase RNA polymerase binding, attracting RNA polymerase to the region of transcription initiation, or acting as a dock or blocker for other proteins which themselves have similar effects. Because transcription factors are proteins and thus generated from genes, conceptually some genes regulate other genes, giving rise to *gene regulatory networks*. Transcription factors bind to the DNA at *transcription factor binding sites* by virtue of their 3-D shape. These are typically between 8 and 20 nucleotides long. An organism may have hundreds of different proteins that act as transcription factors, each of which binds to different characteristic patterns of nucleotides. As with the transcription initiation site, however, there are variations in these patterns, some of which may serve the purpose of influencing the frequency or strength with which the transcription factor binds. Because of the short length of these sites, the variability seen in the patterns, and the virtual sea of DNA in which they can be found, it is difficult to identify transcription factor binding sites in the DNA. Further, binding of a transcription factor to a site does not guarantee any influence on transcription. Thus, identifying *functional* sites is a further complication.

In eukaryotes, and for a relatively few genes in prokaryotes, splicing and alternative splicing follow transcription. The signals that regulate splicing reside partly in the transcribed RNA sequence itself, though clear signals have not been identified. RNAs from RNA-coding genes also play a role, especially in alternative splicing. However, this depends on a complex interplay between the 3-D structures of the regulatory RNAs, the transcript, and the splicing machinery, and is not well understood.

Translation is not as heavily regulated as transcription, and its regulation is better understood. A ribosome assembles the sequence of amino acids specified by the transcript in order they are encountered. Any of three codons, UAA, UAG and UGA indicate the end of the protein and that translation should stop. In eukaryotes, ribosomes bind to the start of the transcript and move along it until encountering the first AUG codon, where translation begins. Sometimes, the translating machinery will skip over the first or even second occurrence of AUG and begin at the next occurrence. This skipping is influenced by the adjacent nucleotides, and is another mechanism by which different variants of a protein are created from the same DNA coding region. In bacteria, the ribosomes do not bind to the start of the transcript. Instead, there is a longer RNA sequence to which they bind (including an AUG codon), and they can bind to that sequence anywhere it occurs in the transcript and begin translation. As a result, it is possible for a single transcript to code for several proteins, each separated by a region that is not translated. Such a gene is termed an *operon*, and is a common means by which proteins that have related functions are co-regulated. For example, the well known *lac* operon in *Escherichia coli* includes three genes, one of which helps to metabolize the sugar lactose and another of which transports lactose into the cell from the extracellular environment. Translation of an RNA can be also regulated by other proteins or RNAs, which can, for example, bind the RNA and block the translation mechanism.

2.3 Proteins

Proteins are involved in virtually every process in cells, including sensing of the cell's environment and communication between cells. Proteins are polymer chains built from amino acids, which have the chemical form shown in Figure 2.3A. A backbone comprising one nitrogen and two carbon atoms is bound to various hydrogen and oxygen molecules. The central carbon is also bound to the unit "R," which can be a single atom or a group of atoms. The R unit distinguishes different amino acids. In glycine, R is simply one hydrogen atom. In methionine, R contains three carbon, seven hydrogen and one sulfur atoms. The R subunits differ in their chemical properties, most relevantly in

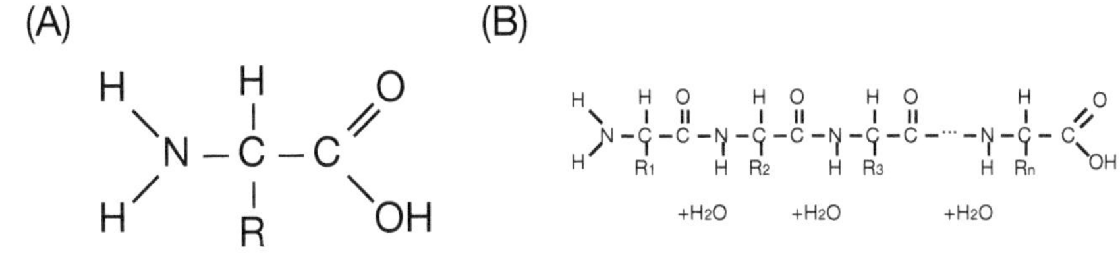

(A)

(B)

Figure 2.3 *(A) An amino acid. (B) Amino acids bind into polymer chains, forming peptides and proteins. The binding of two amino acids releases a water molecule.*

size, acidity, and polarity. When amino acids bind together to form polymers, water molecules are released, and the parts of the amino acids that remain are called *residues* (see Figure 2.3B). Shorter amino acid sequences are *peptides*, while longer ones are *proteins*. Proteins are typically hundreds or thousands of amino acids long.

As a protein is generated by translation from a transcript, it *folds* into a complex 3-D shape, which depends on its amino acid sequence as well as the chemical environment in which it folds. In describing protein structure, four different levels are considered. The *primary structure* of a protein is its amino acid sequence. *Secondary* and *tertiary structure* refer to the 3-D, folded shape of the protein. Tertiary structure is a specification of 3-D coordinates for every atom in the protein, in an arbitrary coordinate system, as well as which atoms are chemically bound to each other. Tertiary structures contain commonly-occurring patterns, such as *α-helices* and *β-sheets* (see Chapter 8 for pictures). α-helices are stretches of the protein that wind into a helical form. A β-sheet is a set of *β-strands* that align lengthwise and to each other, forming a sheet-like shape. The secondary structure of a protein assigns each amino acid to participating in an α-helix, participating in a β-sheet, or neither. Sometimes, other categories are included, such as bends between α-helices. Secondary structure thus abstracts components of the tertiary structure. Many proteins participate in *complexes*, groups of proteins and other molecules that are weakly bound together. A complex may contain proteins of different types, and conversely, one type of protein may participate in different complexes. A specification of how proteins and other molecules organize into complexes is *quaternary structure*.

Proteins play a number of important roles related to DNA and RNA that we have already mentioned: transcription factors regulate transcription rates, RNA polymerase creates the transcripts, DNA polymerase replicates the DNA, histones affect the 3-D structure of the DNA and influence transcription. Proteins are also responsible for repairing damage to DNA, for example caused by radiation, and for untangling DNA.

Proteins also act as *enzymes*, molecules that facilitate the occurrence of chemical reactions—often, very specific reactions. For example, the enzyme lactase facilitates the transformation of the sugar lactose into two other sugars, glucose and galactose. The DNA- and RNA-related roles of proteins mentioned above can also be viewed as enzymatic. The specificity of the enzyme to the reaction or reactions it *catalyzes* (accelerates) depends on its *active sites*, portions of the tertiary structure of the protein that bind to particular *reactants*. Enzymes are discussed at greater length in Section 2.4.

Cell membranes, such as the outer cell membrane or the nuclear membrane, are impermeable to many types of molecules. *Transmembrane* proteins control the flow of molecules across membranes and convey signals from one side to the other. These proteins reside within the membrane and project on either side. *Channels* allow the transport of molecules, often very specific molecules. For example, ion channels in heart muscle or neurons permit the flow of ions such as sodium, potassium, calcium or chlorine. Many of these channels can close, stopping ion flow, or open, allowing flow, depending on triggers such as electrical activity. The dynamical properties of these channels opening and closing, and the resulting changes in ion flows, drive larger-scale electrophysiological phenomenon such as the transmission of action potentials down the axon of a neuron and the beating of the heart. Nuclear pore proteins form complexes that allow molecules such as transcripts and transcription factors across the nuclear membrane. *Signaling* is a related task, but need not involve the transport of a molecule from one side of a membrane to another. Often, the binding of a specific molecule to the transmembrane protein causes a structural change to the part of the protein on the other side of the membrane. This structural change then typically sets off a chain of reactions which convey the signal—about the presence of the molecule on the other side of the membrane—to wherever that signal needs to go.

Structural proteins provide rigidity to cells and perform various mechanical functions. The *cytoskeleton*, for example, is made of long, filamentous protein complexes and maintains the shape of cells and organelles by forming a network beneath the cellular or organellar membrane. The cytoskeleton is also involved in changes to cell shape, as in the formation of *pseudopods*. *Microtubules*, a part of the cytoskeleton, act as conduits or "roads," directing the movement of molecules and organelles within the cell. Structural proteins are also involved in movement at the cellular and whole organism levels. The flagella of bacteria such as *Escherichia coli* and of sperm are composed of microtubules. The movement of myosin protein along actin filaments, part of the cytoskeleton, generates the contractile force in animal muscle cells.

Proteins and peptides are found in many other locations and processes as well—as *antibodies* in the immune system, as hemoglobin in the blood, as hormones, as *neurotransmitters* and as sources of energy. In virtually every cellular and organismal process, the involvement of proteins in some fashion is the rule rather than the exception.

2.4 Metabolism

The totality of all biochemical reactions that occur in a cell is known as metabolism. It involves all four types of fundamental biochemical molecules mentioned in Section 2.1, and includes all the DNA-, RNA- and protein-related processes described above, as well as many others. Most reactions in the cell occur at low spontaneous rates and require the cooperative and coordinated action of various enzymes to catalyze them. In this section, we introduce some broad concepts and terminology for discussing and categorizing the systems of chemical reactions that take place in the cell and discuss enzymes in greater detail.

Sequences of reactions, in which the output of one reaction is the input to the next, can be conceptually organized into *metabolic pathways*. A metabolic pathway is, in some sense, like a roadmap showing the steps in the conversion process of one biochemical compound to another or one form of energy to another. Understanding these pathways and the relationships between them is important for our understanding of the mechanisms behind the effects of drugs, toxins and diseases. Metabolic pathways can be broadly categorized into three groups: *catabolic*, *anabolic* and *central*. Catabolism means disassembly of complex molecules to form simpler products and its main objectives are to produce energy or provide raw materials to synthesize other molecules. The energy produced is temporarily stored in high-energy phosphate molecules (ATP) and high-energy electrons. Anabolism means synthesis of more complex compounds from simpler ingredients and it usually needs the energy derived from catabolic reactions. Central pathways, such as the citric acid or Krebs cycle, are usually involved in interconversions of *substrates* (substances on which enzymes act) and can be regarded as both catabolic and anabolic. Catabolic pathways are *convergent* because through them, a great diversity of complex molecules is converted to a relatively small number of simpler molecules and energy-storing molecules. Anabolic pathways are *divergent* because through them a small number of simple molecules are used to synthesize a variety of complex molecules. Another way of classifying metabolic pathways is to categorize them as *linear*, *branched*, *looped* or *cyclic*. A linear pathway is the simplest of all; it is a single sequence of reactions in which a specific initial input is ultimately converted to a specific end-product, with no possibility of alternative reactions or digressions in the pathway. This kind of pathway is not found very often. A more common type is a branched pathway, in which an intermediate compound can proceed down one branch or another, leading to possibly different end-products. Typically at each branching point there are several enzymes competing for the same substrate and the "winner" determines which branch will be followed. A looped pathway is one that involves many repetitions of a series of similar reactions. A cyclic pathway is one whose end-product is the same as the initial substance it started with. An example is the urea synthesis cycle in humans.

Enzymes are proteins that act as *biological* catalysts, accelerating the rates of various reactions, but they themselves are not irreversibly altered during the reactions. They are usually required in small amounts and have no effect on the thermodynamics of the reactions they catalyze. They differ from inorganic (i.e., non-biological) catalysts in that the latter typically speed up reactions hundreds or thousands of times, whereas enzymes often speed up reactions a billion or a trillion times. There are other important differences, such as the high specificity of an enzyme for its substrate, lack of unwanted products or harmful side-reactions that might possibly interfere with the main reaction, and ability to function in the physiological environment of an organism's interior. To understand how an enzyme accomplishes its task, one needs to know the different types of forces that are at play in a chemical compound. Inside a molecule, there are *ionic* or *covalent* bonds that hold the atoms together and between molecules, there are weaker forces such as *hydrogen bonds* and *Van der Waals forces*. During a chemical reaction, existing covalent bonds are broken and new ones are formed. Breaking covalent bonds requires an energy input in some form, such as heat, light or radiation. This is known as the *activation energy* of the reaction. This energy excites the electrons participating in a stable covalent bond and shifts them temporarily to orbitals further from the atomic nucleus, thereby breaking the bond. These excited electrons then might adopt a different stable configuration by interacting with electrons from other atoms and molecules, thereby forming new covalent bonds and releasing energy. This energy output may be exactly the same as, higher than or lower than the initial activation energy. In the first case, the reaction is called *energetically neutral*, in the second case it is called *exothermic* and in the last case, *endothermic*. It is important to know that a reaction usually does not proceed in one direction only. At least in principle, if two compounds C1 and C2 can react with each other to form two other compounds C3 and C4, the products C3 and C4 can also react to form C1 and C2. In practice, starting with only C1 and C2, first the forward reaction alone will occur producing C3 and C4, but with increasing accumulation of the latter, the reverse reaction will also start taking place, forming C1 and C2. Continuing in this manner, a stage will be reached when the rate of the forward reaction will be identical to that of the reverse reaction. This is known as *equilibrium* and at this stage, the ratio of the total amount of C1 and C2 to the total amount of C3 and C4 will be constant for a given temperature. Any addition or removal of any of the four compounds will temporarily disturb the equilibrium and the reaction will then proceed to restore it. An enzyme lowers the activation energy of a reaction and increases the rate at which the reaction comes to equilibrium, without changing the equilibrium concentrations of the compounds. It does so primarily in the following four ways: (a) by providing a surface on which the molecules participating in a reaction can come together in higher concentrations than in a free solution, so that they are more likely to collide and interact; (b) by providing a microenvironment for the participating molecules that is different from the free-solution environment (e.g., a non-aqueous environment in a watery solution); (c) by taking up electrons from or donating electrons to covalent bonds and (d) by subtly changing the shape of the reactants in a way that encourages the reaction to take place. The interactions between an enzyme and its substrates depend on reactive groups in the side-chains of amino acids that are part of the enzyme's active sites. These reactive side-chains may be far apart in the primary amino acid sequence of the enzyme but come closer together when the enzyme molecule folds and assumes its 3-D shape. This is why the 3-D structure of a protein is important for its function and is what enables the specificity of an enzyme for its substrate(s). The interactions between an enzyme and its substrate(s) are usually non-covalent (i.e., ionic bonds, hydrogen bonds, hydrophobic interactions, etc.), although occasionally transient covalent bonds may be formed. Some key factors affecting the activity of an enzyme are (a) temperature, (b) the pH (acidity) of the solution, (c) concentration of the substrate(s) and (d) presence or absence of inhibitors (e.g., a medical drug or toxic substance that interferes with the enzyme-substrate interaction). Depending on the type of reaction they catalyze, enzymes can be classified into categories such as hydrolases (involved in the hydrolysis of covalent bonds), oxido-reductases (involved in oxidation and reduction), transferases (involved in transferring a reactive group from one substrate to another) and so forth.

2.5 Biological Regulation Systems: When They Go Awry

In order for a highly complex system such as a living organism to survive and function properly, it is crucial that the variety of biochemical pathways that sustain the organism and the countless biomolecules that participate in them be *regulated*. Regulation has evolved because without it the highly coordinated and concerted activities of various groups of biomolecules that is essential for the viability of a living organism would be impossible. There are many different levels and forms of biological regulation. It can happen at the genetic level (e.g., transcription regulation via the binding of transcription factors or translation regulation via the degradation or inactivation of mRNAs) or at the proteomic or metabolomic level through enzymes, hormones and other regulatory agents. Also, there can be many different control mechanisms. For example, control on the quantity of a metabolite can be achieved through a *supply-demand pathway* (where two other metabolites serve as its "source" and "sink" simultaneously) or through *feedback inhibition* (where a sufficient concentration of the end-product of a metabolic pathway inhibits the pathway itself). For a detailed classification of feedback inhibition into *sequential feedback, concerted nested feedback, cumulative nested feedback* and so forth (see [2]). In Section 2.2, we discussed genetic regulation. Here we shed some light on the regulation of cell *proliferation* and describe the consequences of uncontrolled growth.

Cells in almost all parts of our body are constantly proliferating, although most of the time it goes unnoticed because it is a slow process and usually does not result in any visible growth. The primary purpose of this ongoing proliferation is to replenish the cells lost or damaged through daily wear and tear. For example, cells in the outermost layer of our skin (epidermis) and those in the lining (or epithelium) of our intestine are subject to frequent wear and tear and need replenishment. Another purpose is to respond to a trauma or injury, where cell proliferation has to accelerate in order to expedite the healing of a wound. Importantly, as soon as the proliferating cells fill the incision created by an injury, they stop their growth "overdrive" and return to the normal "wear and tear" rate of proliferation. Occasionally, the cells in a wound proliferate a little bit beyond what is needed for complete healing, thereby creating a hypertrophied keloid (or heaped-up scar), but even this is considered normal.

Two processes control cell proliferation. One involves substances called *growth factors* and *growth inhibition factors*. Growth factors stimulate cells to grow and multiply. They are produced all the time for the sake of daily replenishment, but in greater amounts in the case of an injury. It is exactly the opposite for growth inhibition factors whose production is reduced during a trauma and goes back to the everyday level once the healing is complete. The other process is *apoptosis* or programmed cell-death. It allows individual cells within a group to die, thereby leaving the group the same size in spite of new cells produced by proliferation. Some substances enhance apoptosis in certain kinds of tissues.

When cells do not respond to the body's built-in control mechanisms for proliferation, they show uncontrolled growth and produce a *tumor*. Sometimes they cross the normal boundaries of the tissue to which they originally belonged and invade surrounding tissues. Even worse, sometimes they can get into blood vessels or lymph vessels, travel to parts of the body that are distant from their place of origin and spread the phenomenon of uncontrolled growth to those areas. In addition, some of them may produce substances that interfere with the normal functioning of various systems in the body, such as the musculo-skeletal system or the nervous system. When they do all these, we call the resulting condition *cancer* and the mechanism by which cancer spreads from one part of the body to another is called *metastasis*. Upon reaching a distant region in the body, a small lump of metastatic cancer cells break out of the blood-capillary wall, establish themselves there and continue their uncontrolled growth to produce a new tumor. To crown it all, once the new tumor has grown to a certain size, it persuades the body to supply more oxygen and nutrients by growing new blood vessels to it (a process known as *angiogenesis*).

There are different varieties of cancer, but as a whole it remains one of the deadliest diseases in the world. Despite years of intense research, there is no universal cure for cancer. Some types of cancer can be cured, or at least temporarily remedied, depending on the stage at which they were diagnosed.

The traditional methods used to combat the disease include *radiation therapy* (destroying cancerous cells by irradiating them), *chemotherapy* (destroying them by administering a combination of chemicals into the body) and surgery, but all these induce significant collateral damage (i.e., destruction of nearby healthy tissue) and have side effects, not to mention the risk of a relapse (i.e., return of the disease). Recently, more "directed" methods with a higher precision for destroying cancer-cells and fewer side effects have been developed, such as *proton beam therapy* and *drug-induced angiogenesis inhibition*, but these are still not widely available.

In cancer research, the central question is why certain cells in an organism defy the built-in regulatory mechanisms for proliferation and show uncontrolled growth. Although scientists do not have a complete answer yet, they are closer to it today than they were before the genomic era. Certain genes have been identified as *oncogenes* and *tumor suppressor genes* that play a key role in the development of cancer. In many cases, a mutation in one of those genes or some other kind of damage to it will trigger the uncontrolled growth. There are a variety of causes (*mutagens*) for such mutations, including exposure to radiation and certain types of chemicals. Sometimes, if a cell is subjected to oxidative stress (i.e., it is exposed to highly reactive free oxygen-radicals due to an abundance of them in the bloodstream), it will undergo DNA damage. It has also been observed that chronic inflammation increases the risk of cancer.

2.6 Measurement Technologies

Much of the material we have reviewed in this chapter was learned through traditional "low-throughput" laboratory techniques. However, one of the driving forces behind bioinformatics is the advent of "high-throughput" technologies for detecting and quantifying the abundances of various biomolecules and interactions between them. In this section we describe some of the traditional low-throughput—but sometimes more accurate—techniques of molecular biology and along with their more recent, high-throughput brethren.

Perhaps the best-known application area for bioinformatics is in the analysis of DNA sequences. Traditional methods for *sequencing* DNA (i.e., determining the sequence of As, Cs, Gs and Ts comprising a gene, a chromosome or even the entire *genome* of an organism) were laborious, hands-on procedures that could only produce sequences of limited length and at comparatively high cost. DNA sequencing was revolutionized during the 1980's and 90's by the development of machines or robotic systems that could carry out the labwork (semi)automatically, along with computers for storing and analyzing the data. Today, the genomes of many species, including humans, have been completely sequenced. While this trend continues, the emphasis of sequencing has broadened to include the sequencing of individuals' genomes, in order to detect the differences—mostly single-letter changes in the sequence called *single nu-*

cleotide polymorphisms (SNPs)—that are the basis for much of the observable differences between individuals.

Other techniques focus not on static features of an organism, but rather on properties that may change over time or that may differ in different parts of an organism—especially the concentrations of different RNAs or proteins. *Gel electrophoresis* is a traditional method for detecting the presence of DNA, RNA or proteins in tissue samples. In one-dimensional (1-D) electrophoresis (often called *1-D PAGE* after the polyacrylamide gels typically used), a sample of molecules from a tissue is inserted at one end of a rectangular-shaped plate of gel. In one variant, an electric field causes the molecules to migrate towards the other side of the gel. DNA and RNA molecules are naturally negatively charged, and so are accelerated by the field. Proteins are usually bound with substances that make them negatively charged. While the field accelerates the molecules, friction decelerates them. Friction is greater for larger molecules, and so the molecules separate by size, with the smallest molecules traveling farthest through the gel. The separated molecules can be visualized by staining, fluorescence or radiation, and show up as *bands* at different lengths along the gel. If the gel is calibrated with molecules of known sizes, and if the sample contains relatively few kinds of molecules, the presence or absence of bands can be interpreted as the presence or absence of particular molecules. This can be used to detect if a particular gene is being transcribed, for example, by looking for RNAs of the appropriate size, or if the gene's protein is present in the sample. Often, several samples are run side by side in a single gel, to compare the molecules present in each one. If the molecules cannot be identified, due to lack of calibration or other reasons, they can be extracted from the gel again to be identified. Another version of gel electrophoresis for proteins involves a gel with a pH gradient and an electric field. In this case, the proteins move to a position in the gel corresponding to their *isoelectric point*, where the pH balances the protein's charge.

In *2-D gel electrophoresis* (2-D PAGE), usually used with proteins, the sample is inserted at one corner of the gel and sorted in one direction first (often by isoelectric point) and then the electric field is shifted 90 degrees and the proteins are additionally separated by size. After staining, the result is a gel with spots corresponding to proteins of different isoelectric point and size. By comparing gels, one can look for proteins that are present under one condition and absent in another. On the 2-D gels, it is possible to separate hundreds or even thousands of different kinds of proteins, and so 2-D gel electrophoresis is considered a high-throughput measurement technology. 1-D or 2-D gel electrophoresis techniques are usually used to determine the presence or absence of a kind of molecule, rather than to quantify how much is present, although the darkness or spread of a band or spot can sometimes be interpreted as reflecting quantity.

The *Southern blot*, the *Northern blot* and the *Western blot* are extensions of gel electrophoresis that are used to determine the identity of DNA, RNA or protein molecules in the gel respectively. After running the gel, the molecules are transferred onto and bound to a film. A solution of *labeled probes* is washed over the film. For example, if one were interested in testing for the presence of single-stranded DNA for a particular gene, the probes could be DNA strands with basis that are complementary to (that is, binding partners for) some portion of that gene's DNA and not complementary to any portion of the DNA of any other gene. When washed over the film, these probes would bind only to the DNA from the gene of interest. Complementary DNA probes are also used to detect specific RNAs. Antibodies are used as probes to detect specific proteins. After the probes bind to their target molecules, the solution is washed off to remove unbound probes. The locations of the probes are determined based on their fluorescent or radioactive labels, which are easily imaged.

Gene expression *microarrays*, which have revolutionized the monitoring of gene expression, are an adaptation of the blotting idea to a massively parallel scale. Most microarrays measure RNA levels. Microarrays are small glass or silicon slides with many thousands of spots on them. In each spot are probes for different genes, so that a single slide can contain probes for virtually every gene in an organism. There are two main types of microarrays, *one-channel arrays* (also called *one-color* or *oligonucleotide* arrays) and *two-channel arrays* (also called *two-color* or *cDNA* arrays). In the one-channel arrays, RNAs are extracted from a tissue sample, bound with *biotin*, and then washed over the array, where they bind to the probes—ideally, only to probes corresponding to the same gene. Fluorescent molecules are applied, which bind to the biotin on the RNAs. The fluorescent molecules are then excited by a laser and imaged. The more fluorescence coming from a spot, the greater the number of RNAs bound to it, and therefore the greater the expression of the corresponding gene. For two-channel arrays, RNA is extracted from two different samples. The RNA from each sample is converted by *reverse transcription* into *single-stranded complementary DNAs* (cDNAs), and labeled with fluorescent molecules. Each sample is labeled with molecules that fluoresce at different wavelengths. The labeled cDNAs from the two samples are then mixed and washed over the microarray, where they bind to probes. The relative fluorescence in each wavelength of each spot indicates the relative expression of the corresponding gene in the two samples.

Another massively parallel means of measuring gene expression is *Serial Analysis of Gene Expression* (SAGE). In this technique, RNAs are extracted from a sample, and a small stretch from one end of each RNA is converted into cDNA. These cDNAs are then bound together in long chains and sequenced by a DNA sequencing machine. These sequences can then be examined to identify and count the source cDNAs, effectively counting the number of RNA molecules that were so converted.

While microarrays and SAGE are revolutionary in their ability to quantitatively measure the expression of thousands of genes simultaneously, they are not without their drawbacks. The measurements are notorious for being noisy, with significant variability in the data due to inevitable variations in the complex measurement procedure as well as differences in experimental conditions, equipment used and technicians carrying out the experiment. Furthermore, microarray experiments are expensive. While the expression of thousands of genes may be measured in each sample, most studies include only a handful of samples, usually no more than a few tens, or as much as a few 100s for extremely well-funded studies. The machine learning and statistical issues engendered by such data are significant, and have been a major area of study ever since the technologies were developed.

Labeled probes are also used in living or recently-living tissue. In *in situ hybridization*, labeled probes for DNA or RNA sequences are inserted into cells and imaged. This reveals the spatial distribution of the target within the cell or tissue and, if observations are taken over a period of time, the temporal distribution as well. *Immunohistochemistry*, or *immunostaining*, follows the same idea, but the probes are antibodies and the target molecules are proteins. As some proteins are *markers* for—that is, indicative of—specific organelles in a cell or tissues types in a body, immunostaining is often used to localize such structures. While one probe can reveal the spatial/temporal distribution of a single type of molecule, two different probes with different labels are used to study the differences or similarities in the distributions of different molecules. This is often used, for example, to determine whether two different proteins collocate, which could indicate a functional protein-protein interaction, or under which conditions they collocate. It is technically difficult to introduce and image more than a few different kinds of probes, however, so these techniques are strongly limited in the number of different types of molecules that can be studied simultaneously.

A common method for detecting and quantifying proteins in a sample is *mass spectrometry*. Mass spectrometers separate and quantitate ions with different *mass-to-charge ratios*. The principle is similar to that of gel electrophoresis. Electric or magnetic fields accelerate the ions differentially, until they reach a detector. Typically, proteins are enzymatically digested into much smaller fragments, peptides, which are then fed into the mass spectrometer. The result is a measured distribution of ions of different mass-to-charge ratios. In some cases, this *peptide mass fingerprint* is sufficient to identify which protein or proteins are present in the sample. However, identification is not always possible. In tandem mass spectrometry, ions traveling through a first mass analyzer, which separates according to mass-to-charge ratio, can be selectively sent through one or more additional mass analyzers. This allows the successive selection and measurement of specific ranges of peptides, allowing for more definite identification. The most recent techniques allow the enzymatic digestion step to be skipped, instead introducing entire proteins to the first stage of a tandem mass spectrometer. However, ionizing the proteins without breaking them down is a much more delicate process than it is for the peptide fragments, and requires more sophisticated and expensive equipment.

As mentioned in Section 2.3, many proteins act together in complexes. A traditional technique for determining the complexes in which a given protein participates, and under which conditions, is *co-immunoprecipitation*. Immuno-precipitation extracts a specific protein from a solution by binding it with an antibody specific to that protein. The antibodies are in turn bound to insoluble proteins or other constructs, such as agarose beads, which are easily separated from the solution. If the target protein is in a complex with other proteins, then they will also be extracted and can be identified by mass spectrometry.

Another method for determining protein-protein interactions is the *yeast two-hybrid* screen. This procedure relies on an important feature of eukaryotic transcriptional regulation: While many transcription factors contain two do-mains, a DNA binding domain and an activation domain that stimulates tran-scription, these two domains need not be in a specific position or orientation with respect to each other. A yeast two-hybrid screen for the interaction of two proteins, A and B, works by inserting three new genes in a yeast cell. One gene, which we call A′, codes for the protein A fused with a transcription fac-tor binding domain. Another gene, which we call B′, codes for the protein B fused with an activation domain. The third gene, C, is a *reporter gene*, which has two important features: its regulatory region contains a binding site for the A′ protein, and its expression is easily measured (perhaps because the C protein is fluorescent). If proteins A and B interact and form a complex, then the hope is that A′ and B′ will also interact and form a complex. Further, the A′ portion of that complex will bind to C's promoter and the B′ activation domain will stimulate expression of C. The expression or non-expression of C is then an indicator of whether A and B do or do not interact.

A variant of immunoprecipitation, *chromatin immunoprecipitation*, is used to study where transcription factors bind to the DNA. In a given tissue sample, depending on the conditions of the experiment, certain transcription factors will be expressed and will be bound to the DNA at various locations. The transcription factors are first *cross-linked* (bound) to proteins comprising the DNA's chromatin. The cells are then broken open and the DNA broken into fragments by *sonication*, bombardment by sound waves. A chosen transcrip-tion factor can then be immunoprecipitated, bringing with it the chromatin to which it is bound and the DNA. Then the DNA is separated from the chro-matin and the transcription factor, and can be identified. Ideally, this DNA represents only stretches that were bound by the transcription factor, and no others. These DNA segments can be identified by traditional low-throughput means, some of which were discussed in the previous section. Alternatively, they can be identified in a high-throughput way using a special DNA microar-ray with probes designed to cover the whole genome of the organism. This is called *ChIP-on-chip* or *ChIP-chip*, the first "ChIP" referring to the chro-matin immunoprecipitation procedure and the second "chip" referring to the

24 THE BIOLOGY OF A LIVING ORGANISM

microarray. Repeating the procedure for every known transcription factor in the organism yields an estimate of the entire set of transcription factor binding sites on the DNA. However, many sites can be missed, as the transcription factor may not be expressed or may not bind under the conditions of the experiment. Furthermore, transcription factors may bind to many locations on the DNA without affecting transcription, and such non-functional sites are not of interest.

These biotechnologies and techniques, and especially methods such as genome sequencers, 2-D PAGE, gene expression microarrays, SAGE, mass spectrometry, yeast 2-hybrid screens and ChIP-chip, are generating a virtual flood of data, motivating and indeed necessitating the use of machine learning and statistical techniques to extract useful information.

References

[1] B. Alberts, A. Johnson, J. Lewis, M. Raff, K. Roberts, and P. Walter. *Molecular Biology of the Cell, Fourth Edition*. Garland, 2002.
[2] D. Fell. *Understanding the Control of Metabolism*. Portland Press, 1997.

CHAPTER 3

Probabilistic and Model-Based Learning

3.1 Introduction: Probabilistic Learning

We begin this chapter by addressing the question: "Why are probabilistic and model-based learning relevant in the context of biological systems?" This is a pertinent question because, after all, probability theory deals with uncertainty and probabilistic models are a way of quantifying uncertainty. When one thinks about the kind of objects that constitute biological data (e.g., nucleotide sequences in the DNA, amino acid sequences in peptides, the molecular components of carbohydrates and lipids, the metabolites participating in a certain metabolic pathway, etc.), there is a lot of predictability about them in the sense that one will always find the same nucleotide sequence at a specific position on a specific chromosome of a specific organism (except for rarely occurring events called mutations) or the same amino acid sequence in a specific protein molecule. So, where does the uncertainty come from? In fact, it primarily results from the inadequacy of our present state of knowledge compared to what remains to be discovered in the future. Many aspects of a biological system are still partially known and currently known 'facts' often turn out to be wrong in the light of newly discovered knowledge. Also, there is a constant need for extrapolating what we know about a smaller or simpler organism to a larger and more complex one that is still unexplored. In other words, researchers in bioinformatics are constantly faced with the need to use inductive reasoning and to draw inferences. There are three different concepts of knowledge in this world. The philosopher's view is that all knowledge is correct and the distinction between right and wrong depends only on the observer's viewpoint. The scientist's view is that all knowledge is wrong unless it can be experimentally verified by independent observers. To put it in another way, a scientist such as a physicist or a chemist uses deduction to add to existing knowledge (i.e., if A implies B and B can be experimentally shown to be false, then A must be false). On the other hand, the probabilist's or statistician's view of knowledge is based on the principle of induction. It goes like this: if A implies B and B is observed to happen, A is more likely to be true. Probability and statistics enable us to quantify, for example, how much more likely A becomes if B is observed k times (k=1,2,3,...). Often, a classical statistician's approach is to start with a set of competing hypothesis about an unknown quantity or object, choose a probability model describing the relation between the unknown and the observable (i.e., data) and finally, reach a decision regarding the hypotheses based on the evidence from the data, along with an assessment of the uncertainty involved in that decision. This is called *hypothesis testing*. At other times, he/she would try to find out the most likely value(s) of the unknown by maximizing the joint probability model for the data-vector (called the *likelihood function*) with respect to the unknown. This is known as *maximum likelihood estimation*. These two are the central theme of frequentist inference —there are many variations to these themes. There is, however, another parallel approach called the Bayesian approach. A Bayesian would start with some a priori assumptions about the unknown, usually in the form of a probability distribution (called a *prior* distribution) that can be based completely on his/her personal belief or on already existing information or on some kind of 'expert opinion.' The Bayesian would then combine this prior distribution with a probability model relating the unknown and the observed data (i.e., the likelihood function) to get a joint distribution and from it, using the Bayes principle, ultimately derive the conditional distribution of the unknown given the observed data (the *posterior* distribution). To a Bayesian, therefore, acquiring new knowledge basically means updating the prior information about the unknown in the light of the observed data.

Associated with each of these approaches are some important issues that are crucial from the viewpoint of implementation and error assessment. For example, while estimating an unknown parameter, a probabilistic way of quantifying the error in the estimate is to look at its mean squared error (MSE), which is nothing but the average (or expected) squared distance between the unknown parameter and the estimate. Now, it can be easily shown that this MSE is the sum of the estimate's variance and squared bias. Some of these terms will be elaborated on later, but roughly speaking, the *bias* of an estimator (an *estimator* being a computational formula used to compute an estimate) is a measure of how far its probabilistic average value is from the unknown parameter it is meant to estimate. Similarly, the *variance* of an estimator is its average (or expected) squared distance from its own probabilistic average value. But it often turns out that maximum-likelihood estimates have nonzero bias and there is no easy way to compute either the bias or the variance. In such a situation, a technique called *bootstrap* is often used to estimate those two things. The underlying principle of bootstrap is that, if one repeatedly takes random samples from the observed data and computes the value of an estimator from those, the variation in these computed values will be a reasonably good indicator of the estimator's variance under certain conditions. Similarly, the difference between the simple average of these computed values and the original value of the estimator based on the original set of observed data will be a reasonably good indicator of the estimator's bias. Example 3.4 illustrates this bootstrap idea. As far as maximum likelihood estimation is concerned, recall that it involves maximizing the likelihood function with respect to the unknown parameter. Often this maximization turns out to be analytically intractable or becomes difficult because of missing data and some kind of an iterative numerical solution seems to be the only way out. The *expectation-maximization* or EM algorithm is one such method and is described in a later section. Regarding the Bayesian approach, it is so heavily dependent on numerical computation in order to be useful that the phrases 'Bayesian statistics' and 'Bayesian computation' have almost become synonymous. Since the final output of the Bayesian approach is the conditional distribution of the unknown given the observed data (i.e., the posterior distribution), one needs to be able to draw sample observations from this distribution in order to answer more specific questions. Often the posterior distribution is so 'ugly' or 'unmanageable' due to the high dimensionality of the unknown parameter-vector and/or the analytical intractability of the steps involved in its derivation, the standard methods for drawing random deviates from probability distributions are of little use. So, in such cases, a Bayesian resorts to a *Markov Chain Monte Carlo* (MCMC) sampling technique, the idea behind which is basically the following. Starting from an initial set of 'guessed' values for the unknown parameters, one starts a Markov chain (see definition in Section 5) whose states are the updated value-vectors of the parameter-vector and which moves from state to state via some specific mechanism designed to ensure the chain's eventual convergence to the targeted posterior distribution. In other words, the state-to-state propagation mechanism is so chosen that the chain becomes *ergodic* (see definition in Section 5) with its stationary distribution being the targeted posterior distribution. Some widely used iterative propagation mechanisms are Gibbs sampling and Metropolis-Hastings algorithm. Once the Markov chain is set in motion and left alone to run for while (called the *burn-in* period) so that it gets 'close enough' to its stationary distribution, the next several states that the chain visits are used as sample observations from the desired posterior distribution.

The rest of the chapter is organized as follows. Section 2 provides the basic concepts of probability theory that are subsequently needed, including the Bayes theorem. Section 3 introduces discrete and continuous random variables and their probability distributions. Section 4 deals with the basics of information theory. Section 5 summarizes some important definitions and results from stochastic processes. Section 6 is a snapshot of an important class of models called hidden Markov models. Section 7 elaborates on frequentist statistical inference. Section 8 draws attention to some associated computational aspects. Finally, Section 9 talks about Bayesian statistical inference and the associated computational issues.

3.2 Basics of Probability

We live in a world full of uncertainty. We are not sure how the weather will be a week from now, we cannot predict with certainty whether our favorite football team will win its next game and we are equally unsure about how the stock market will behave tomorrow! Since avoiding uncertainty is not an option and surrendering to it helplessly is a bad option, we should try to gain as much control over it as possible. The first step towards controlling uncertainty is quantifying it, that is, measuring it numerically. In order to do that, we must look for any kind of pattern or regularity that uncertainty may exhibit. And fortunately, it often does exhibit discernible patterns in real life. For example, if it is cloudy and raining right now at a certain latitude and longitude, exactly predicting the weather condition at that spot 24 hours from now may not be possible, but we can at least browse through the extensive weather records for the past hundred years and find out how the weather has changed (or not changed) in a 24-hour period at that particular location on rainy days during the same season. If we see a distinct pattern in the history, it will enable us to conclude what kind of weather condition is more likely for tomorrow and what kind is less likely or moderately likely. If we can translate the pattern that we discovered into a mathematical model, we can be even more precise in our prediction. For example, we might be able to make statements such as "Bright, sunny and warm weather is only 30% likely 24 hours from now" or "A cloudy, windy and chilly condition is 75% likely." As soon as we do that, we are entering the realm of probability.

Probability theory, therefore, is a mathematically consistent and coherent way of putting a numerical value to uncertainty. To understand it, we must learn to speak the language first. In the context of probability theory, the word 'experiment' will mean any kind of activity that produces results or outcomes. For example, throwing a ball and tossing a coin are experiments; so is eating dinner or listening to a lecture. There are basically two types of experiments: those with a perfectly predictable outcome and those with many possible outcomes so that one is never sure which one is going to occur. Experiments of the first kind are called *deterministic* and those of the second kind are *random* or *stochastic*. An illustration of a deterministic experiment is to mix together an acid and a base (or alkali), because they will invariably produce a specific salt along with some water. Another example would be to heat a bowl of water to 100 degrees Celsius (or 212 degrees Fahrenheit) under normal atmospheric pressure, at which point it will surely start boiling. However, things are not so predictable when you flip a fair coin. It can either land heads up or tails up. So this is a random experiment. So is throwing a dart onto a dartboard with your eyes closed or rolling a cubic die—it is difficult to predict exactly where the dart will land or exactly what number will turn up on the die. Sometimes whether an experiment should be classified as deterministic or random depends on the extent of details we are willing to allow in the description of it. For example, think of holding a glass bottle several feet above the ground in your hand and then releasing it. If this is how we describe the experiment, it is deterministic since the bottle will invariably fall towards the ground and the law of gravity will exactly tell you its acceleration and velocity the moment it hits the floor (depending on its release-height, of course). However, if we describe the experiment as releasing a glass bottle from several feet above the ground and watching what happens to it as it hits the concrete floor, it becomes a random experiment. Because no matter how much physics you know, it will be impossible to predict with certainty whether the bottle will remain intact after hitting the ground or break into k pieces (k=2, 3, 4,...)

So it should be clear that probability theory deals primarily with a random experiment because there is some uncertainty associated with its outcomes. The first step towards quantifying this uncertainty is a complete enumeration of all the outcomes. Such a complete list of all possible outcomes of a random experiment is known as its *sample space*. Depending on the experiment, the sample space can be finite or countably infinite. For example, when a coin is flipped, the sample space is S = {Heads up, Tails up} or simply $S = \{H, T\}$. When a regular cubic die is rolled, $S = \{1, 2, 3, 4, 5, 6\}$. However, if our experiment is counting how many flips of a coin was needed to see the first 'H,' then S consists of all positive integers and is infinite. Once again, the sample space of a random experiment is determined solely by the description of the experiment. For example, the simple experiment of drawing a card from a (well-shuffled) deck of 52 cards with our eyes closed can have different sample spaces depending on the extent of details we allow. If we draw a card, blindfolded, from a deck of 52 cards and just watch its color, the sample space is simply {red, black}. However, if we do the same experiment and watch only the suit it is coming from, S will be {diamonds, spades, hearts, clubs}. Finally, in the context of the same experiment, if we are interested in all the features of a card (its color, suit as well as the number or picture on it), then S will be a list of 52 distinct cards.

Having written down the sample space of a random experiment, we now must link it to the concept of an *event*. Any subcollection of outcomes listed in the sample space is called an event. In real life, we are more often interested in such collections of outcomes than in individual outcomes. In the language of set theory (the branch of mathematics that deals with the relationships among, and the properties of, various subcollections of a bigger collection), the sample space is the *universal set* and the events are its *subsets*. So, following set-theoretic notations, events are usually denoted by A, B, C, etc. For example, in the context of rolling a cubic die (what was the sample space for it?), A could be the event that an odd number turned up. In other words, $A = \{1, 3, 5\}$. Similarly, B could be the event that a number ≤ 2 turned up, i.e., $B = \{1, 2\}$. In the context of drawing a card from a deck of 52 cards, A could be the event that the card drawn is an ace, B could be the event that it is black, C could be the event of it being a face-card and D, the event of it being a diamond. These are examples of simple events. Given two such simple events, one can construct all sorts of compound events using the set-theoretic operations *union* ($\bigcup$), *intersection* ($\bigcap$) and *complementation*. Given two events (or sets) A and B, their union is defined as the bigger collection containing all the outcomes in A as well as those in B. So, using the notation "$\in$" to denote that an outcome x belongs to an event A, the event $A \bigcup B = \{x : x \in A \text{ or } x \in B\}$. Notice that this "or" is not an exclusive or, i.e., it includes the possibility that an outcome x may belong to both A and B. The event $A \bigcap B$ is defined as the subcollection containing only those outcomes that are common to both A and B. In other words, it is $\{x : x \in A \text{ and } x \in B\}$. For an event A, its complement (A^c or $\bar{A}$) is the collection of all outcomes in the sample space that do not belong to A, i.e., $A^c = \{x : x \notin A\}$. Applying the set-theoretic operations on these, one can create compound events such as $A \bigcap B^c = \{x : x \in A \text{ and } x \notin B\}$ (also called $A - B$), $A^c \bigcap B = \{x : x \notin A \text{ and } x \in B\}$ (also called $B - A$), $A \Delta B = (A - B) \bigcup (B - A)$ (also called the symmetric difference), and so on. Let us point out, for future reference, that the operations of union and intersection are *distributive*, i.e., for any three sets A, B and C, $A \bigcup (B \bigcap C) = (A \bigcup B) \bigcap (A \bigcup C)$ and $A \bigcap (B \bigcup C) = (A \bigcap B) \bigcup (A \bigcap C)$. All these can be nicely depicted by a simple diagram called a *Venn diagram*, which consists of an outer rectangle representing the universal set (or sample space) and a circle representing each simple event $A, B, \ldots$ inside it. If A and B have some elements in common, the two circles representing them should be overlapping. If A and B do not overlap as events (in which case, they are called *mutually exclusive* or *disjoint*), the circles should be drawn as separated too. In the former case, the total area covered by the double circle represents $A \bigcup B$, the overlap between the two circles represents $A \bigcap B$, the crescent-shaped portion of the circle A that is outside the overlap region stands for $A - B$ and the corresponding portion of the other circle stands for $B - A$. If one covers up the circle A, the rest of the rectangular box that is still visible represents A^c. Similarly, for B^c. Notice that the union of any event and its complement is the entire rectangular box. Now the question is: what should we call the 'outside border' region of the rectangular box that is still visible when we cover up the entire double circle (i.e., $A \bigcup B$)? Since, in general, 'outside' means 'complement,' one correctly concludes that it should be called $(A \bigcup B)^c$. But is there another name for it?

Result 3.1. $(A \bigcup B)^c$ is the same as $A^c \bigcap B^c$. Similarly, $(A \bigcap B)^c$ is the same as $A^c \bigcup B^c$.

These are the famous *De Morgan's rules*. It is easy to verify them. For example, if x is an outcome in $(A \bigcup B)^c$, that means $x \notin A \bigcup B$, which in turn means that x is neither in A nor in B. In other words, $x \in A^c$ and $x \in B^c$. But this is equivalent to saying that $x \in A^c \bigcap B^c$. So every outcome in $(A \bigcup B)^c$ also belongs to $A^c \bigcap B^c$. Similarly, by retracing this line of reasoning, one can conclude that the converse is true too. This shows that the two collections are identical. The other De Morgan's rule can be similarly established. But what are these useful for? Here is an example.

Example 3.1. Suppose that, in a two-set Venn diagram, $A - B$ has 15 outcomes (which is often denoted by $n(A - B) = 15$ where "n" means *cardinality*), $(A \bigcap B)^c$ has 43 outcomes, B itself has 20 and $A^c \bigcap B^c$ has 15. Can we find out $n(A)$ and $n(S)$? Notice first of all that $A \bigcup B = (A - B) \bigcup B$, so that $A \bigcup B$ in this case has 15+20=35 outcomes. Now, by De Morgan's first rule, $(A \bigcup B)^c = A^c \bigcap B^c$, so in this case $(A \bigcup B)^c$ has 15. Also, since the union of any event and its complement gives us the whole sample space S, $n(S)$ must be $n(A \bigcup B) + n(A \bigcup B)^c = 35 + 15 = 50$. Finally, since $A \bigcap B$ and $(A \bigcap B)^c$ are complements of each other, their union is the whole sample space. So $n(A \bigcap B) = n(S) - n(A \bigcap B)^c = 50 - 43 = 7$. As a result, $n(A)$ will be 22, the sum of the cardinalities of its two non-overlapping pieces $A - B$ and $A \bigcap B$. In the above example, we have repeatedly used the cardinality formula $n(C) + n(C^c) = n(S)$ for any event C. Another cardinality formula that is often useful is: $n(A \bigcup B) = n(A) + n(B) - n(A \bigcap B)$. It is often called the *union rule* and is easy to verify once the Venn diagram is drawn.

Once the concepts of a sample space and compound events are understood, the next step is to assign a positive fraction to each of the listed outcomes in a mathematically coherent way that is consistent with reality. This positive fraction will be called the *probability* of the associated outcome. In fact, this is the step where believers in the frequentist or classical notion of probability differ philosophically from the believers in subjective probability. The former will define the probability of an outcome as the "long-term" ratio between the number of times that particular outcome has occurred and the number of trials (i.e., the number of times the experiment has been repeated). In more mathematical terms, the probability of an outcome x is $p(x) = \lim_{n \to \infty} \frac{\#x}{n}$, where n is the number of trials and $\#x$ is the frequency of occurrence of that outcome in n trials. The existence of this limit is ensured by the so-called strong law of large numbers (SLLN). On the other hand, a believer in subjective probability may assign a probability to an outcome that reflects his/her degree of confidence in its occurrence. For example, suppose somebody is playing a dice game where each time he/she rolls a "6" with a fair die, he/she wins \$ 100. If a "6" turned up in 5 of the last 10 rolls by chance, he/she may very well think "I am feeling lucky" and assume that the chances are 50% that the next roll will also produce a "6." This is based on perception and has nothing to do with the long-term proportion of occurrence of the outcome "6" for a fair die, which is $\frac{1}{6}$. In this book, we stick to the frequency-based definition of probability, although there are some problems with it in practice. We simply do not know how many times a random experiment has to be repeated in order to be able to determine the "true" probabilities of its outcomes accurately. For example, if one wants to determine the probability of the outcome "6" in a die-rolling experiment by repeated trials, usually the quantity $\frac{\#6}{n}$ (where $n=$ number of trials) fluctuates a lot for small or moderate values of n, thereby producing misleading estimates. This ratio is ultimately guaranteed to stabilize around the 'true' probability, but does 'ultimately' mean after 10000 trials or 1000000? That is why it is more convenient to "reason out" these probabilities using mathematical reasoning, based on the precise description of the experiment. For example, in the experiment of tossing a *fair* or *unbiased* coin, the adjective "fair" or "unbiased" leads us to the logical conclusion that none of the two outcomes H and T should be more likely than the other. As a result, each of them should be assigned a probability of $\frac{1}{2}$, since the total probability of the entire sample space must always be 1 or 100%. Likewise, in the case of rolling a *perfect* or *balanced* die, those adjectives will lead to the logical conclusion that each of the six outcomes should receive the same probability, which must therefore be $\frac{1}{6}$ in order to keep the total 1. However, if a special die is used whose three sides are marked with the number '3,' another two sides are marked with '2' and the remaining side with '1,' our logical reasoning will not lead to assign equal probabilities to the three possible outcomes 1, 2 and 3. Instead we will conclude that '2' should be twice as likely and '3' should be three times as likely as '1,' which automatically determines the probabilities to be $P(1) = \frac{1}{6}$, $P(2) = \frac{1}{3}$ and $P(3) = \frac{1}{2}$. In this case, the outcomes are *unequally likely*, whereas in the previous two examples, the "fairness" adjectives made the outcomes *equally likely*. Probabilities determined in this way are called *theoretical* or *model-based* probabilities. How different these are from the long-term frequency-based probabilities will depend on how realistic our models are and/or how correct our logical reasoning is. For instance, if our random experiment is picking a ball without looking from a box containing 50 balls of identical size and 5 different colors red, black, white, purple and yellow (10 balls of each color), our logical reasoning will lead us to assign a probability of $\frac{1}{5}$ to each of the outcomes $\{R, B, W, P, Y\}$. This will indeed coincide with the long-term frequency of each of the colors because the balls are otherwise identical. However, if our experiment was to catch a fish (without looking) using a line and baits from a tank containing 50 fish of 5 different colors R, B, W, P and S (10 of each color) and releasing it back to the tank, our logical reasoning might once again lead us to assign the same probability (1/5) to each color. That is if we continued to think of the fish as colored balls in a box. But the reality may be different. Some fish may have a bigger appetite than others or be more aggressive in nature, which makes them more likely to bite the bait. In this case, the model-based probabilities may differ from the ones based on long-term frequencies since our model failed to take into account the full reality. Having said this, we should also point out that in many practical applications, modeling the fish as identical balls of different colors will be 'good enough,' in the sense that the resulting discrepancy in the probabilities will hardly matter. That is why we fit probability models to real-life phenomena, knowing that they are semi-realistic approximations at best. As George Box, a renowned statistician, once said: "No model is correct; but some are useful."

Having assigned nonnegative fractions (or probabilities) to the outcomes in a sample space in one way or another, now we go on to define probabilities of events. The probability $P(A)$ of an event A is formally defined as $P(A) = \sum_{x:x \in A} P(x)$. However, for experiments with equally likely outcomes, it reduces to a simple formula: $P(A) = \frac{n(A)}{n(S)}$. For example, in the experiment of rolling a balanced die, if A is the event that an odd number turns

up and B is the event of a number ≤ 2 turning up, $P(A) = \frac{1}{6} + \frac{1}{6} + \frac{1}{6}$
$= \frac{1}{2} = \frac{3}{6} = \frac{n(A)}{n(S)}$. Similarly, $P(B) = \frac{1}{3} = \frac{n(B)}{n(S)}$. But the above formula does
not hold for unequally likely outcomes. Recall the special die with the num-
ber '3' on three sides, '2' on two sides and '1' on one side? For it, if B de-
notes the same event, the correct $P(B)$ is $\frac{1}{6} + \frac{1}{3} = \frac{1}{2}$, which is different from
$\frac{n(B)}{n(S)} = \frac{2}{3}$. Some direct consequences of this computationally convenient formula in the equally likely case are (a) $P(A) = 1 - P(A^c)$ for any event A
and (b) $S(A \bigcup B) == P(A) + P(B) - P(A \bigcap B)$ for any two events A and B,
which come from the corresponding cardinality formulae mentioned earlier.
Probabilities defined in this way satisfy the following:

Probability axioms:
Axiom 1. $0 \leq P(A) \leq 1$ for any event A in S;
Axiom 2. $P(S) = 1$;
Axiom 3. For any countable collection $E_1, E_2, E_3, \ldots$ of mutually exclusive
events in S, $P(E_1 \bigcup E_2 \bigcup E_3 \bigcup \ldots) = P(E_1) + P(E_2) + P(E_3) + \ldots$. In other
words, $P(\bigcup_{i=1}^{\infty} E_i) = \lim_{k \to \infty} \sum_{i=1}^{k} P(E_i)$.
Any way of defining probabilities of events (objective or subjective) should satisfy these axioms in order to be called consistent. The last of the three axioms
is called *countable additivity*. Some prefer replacing it by the *finite additivity*
axiom of De Finetti, but here we stick to the former (actually, countable additivity implies finite additivity but not vice versa). In any case, we are now
in a position to compute probabilities of various compound events.

Example 3.2. For the experiment of drawing a card, without looking, from a
(well-shuffled) deck of 52 cards, recall that we defined events $A = \{$ the card
drawn is an ace $\}$, $B = \{$ the card drawn is black $\}$, $C = \{$ The card drawn
is a face card $\}$ and $D = \{$ the card drawn is a diamond $\}$. Let us compute
the probabilities (a) $P(A \bigcup B)$, (b) $P(A \bigcap D)$, (c) $P(B \bigcap C)$, (d) $P(B \bigcap D)$,
(e) $P(A \bigcup C)$, (f) $P(A \Delta D)$, (g) $P(A \bigcup C \bigcup D)$. $P(A \bigcup B)$ is 28/52 or 7/13
by direct counting, since there are 28 cards altogether in the deck that are
either ace or black (or both). It could also be found using the union rule, as
$P(A) = 1/13$, $P(B) = 1/2$ and $P(A \bigcap B) = 1/26$. $P(A \bigcap D)$ is 1/52, since
there is only one diamond ace. The answers to (c), (d) and (e) are 3/26,
0 and 4/13 respectively (not counting the aces as face cards). $P(A \Delta D)$ is
15/52, since $P(A - B) = 3/52$, $P(B - A) = 12/52 = 3/13$ and $A \Delta D$ is
the disjoint union of these two (notice that for two disjoint events E and
F, the union formula $P(E \bigcup F) = P(E) + P(F) - P(E \bigcap F)$ simply reduces
to $P(E \bigcup F) = P(E) + P(F)$ since the intersection term vanishes). Finally,
$P(P(A \bigcup C \bigcup D)$ is 25/52, by direct counting.

Remark 3.1. The union rule for a three-set Venn diagram is a little more
complicated. Just as a Venn diagram with two overlapping sets A and B has
$2^2 = 4$ disjoint components $A - B$, $B - A$, $A \bigcap B$ and $(A \bigcup B)^c$, a Venn

diagram with three mutually overlapping sets has $2^3 = 8$ disjoint components that can be appropriately named using set-theoretic symbols. In view of this, it is not too difficult to see why $P(E_1 \bigcup E_2 \bigcup E_3) = P(E_1) + P(E_2) + P(E_3) - P(E_1 \bigcap E_2) - P(E_2 \bigcap E_3) - P(E_3 \bigcap E_1) + P(E_1 \bigcap E_2 \bigcap E_3)$.

Let us now focus our attention on event pairs such as A and B, A and D or B and C. Since $P(A) = 1/13$, $P(B) = 1/2$ and $P(A \bigcap B) = 1/26$, clearly an interesting relation holds, namely, $P(A)P(B) = P(A \bigcap B)$. Similar is the case for the other two pairs. One might get the impression that it happens all the time. But what about the event pairs B and D? If we define a new event E as the card drawn being a spade, what about the pair B and E or D and E? For B and D, $P(B \bigcap D) = 0 \neq P(B).P(D)$. Similar is the case for D and E. Regarding B and E, we have $P(B)P(E) = (1/2)(1/4) = 1/8$ which is different from $P(B \bigcap E) = 1/4$. So the equation that the pairs $\{A, B\}$, $\{A, D\}$ and $\{B, C\}$ satisfy is a special one and when it holds, the event-pair involved is called an *independent* pair of events. So the pairs $\{B, E\}$, $\{D, E\}$ and $\{B, D\}$ are *dependent* pairs.

Example 3.3. Suppose our experiment is to choose a whole number randomly from among $1, 2, \ldots, 20$. Now, the phrase "choosing randomly" means choosing in such a way that all possible outcomes are equally likely. How would one carry out the experiment to ensure this? One way could be to write the numbers 1 through 20 on twenty identical-looking paper chips or plastic tokens, put them in a bowl or a hat, mix them well and then draw one without looking. In any case, if A is the event that the chosen number is odd and B, the chosen number is a multiple of 3, then A and B are independent (verify it). If C is the event that the chosen number is ≤ 10 and D is the event that it is a multiple of 4, then C and D are dependent (verify it too).

Now we will probe into it a little bit farther and try to understand *why* some event-pairs are independent and others are not. If we examine the independent event-pairs in the above examples carefully, we will see that within each pair, the events do *not* carry any information regarding one another. For instance, if your instructor is standing at the lecture podium and drawing a card, without looking, from a (well-shuffled) deck of 52 cards, and you are sitting in the audience, trying to predict whether the card drawn will be an ace. The instructor looks at the card drawn but does not tell you what it is. To you, the chances of it being an ace are 1/13. If, at this point, the instructor changes his/her mind and reveals that the card drawn is actually black, does it change the probability of it being an ace in your mind? Earlier, you were thinking that it would be 1 in 13 since there are 4 aces in the deck of 52 cards. Now, in view of this additional information about the color, your brain will automatically stop thinking about the 26 red cards that are irrelevant and focus only on the black cards. But the proportion of aces is still the same—2 out of 26 or 1/13. In other words, the color of a card has *no* information about the "ace or non-ace" classification, and vice versa. When you are trying to predict the color of a card drawn by somebody else, any information about whether it is an ace does not affect your chances of making a correct prediction either. This *lack of mutual information* is the real explanation behind probabilistic independence. The color of a card does not carry any information about whether it is a face card or a number card, nor does the name of its suit about whether it is an ace.

If independence is another name for lack of information, a pair of dependent events must contain some information about each other. The question is: *how much?* One way of quantifying the information that an event carries about another is to measure the amount of change in the latter's probability, given the knowledge of the former's occurrence. In Example 3.3, the ordinary probability of D is $5/20$ or $1/4$. But as soon as it is revealed that the chosen number has turned out to be ≤ 10, our focus shifts to only the outcomes $1, \ldots, 10$ and since there are only two multiples of 4 in this range, the new probability of D is 2 out of 10 or $1/5$. So the additional piece of information that C has occurred has a "shrinking effect" on the denominator of the familiar formula $P(D) = \frac{n(D)}{n(S)}$ and changes it from the total count of the entire sample space to the count of the "relevant" part of the sample space (i.e., $n(C \bigcap S)$ or $n(C)$). At the same time, the numerator changes from $n(D)$ to the count of the "relevant" part of D (i.e., $n(D \bigcap C)$). The resulting formula $\frac{n(D \bigcap C)}{n(C)}$ is called the *conditional probability* of D given C and is denoted by $P(D \mid C)$. It is easy to see that $P(D \mid C) = \frac{P(D \bigcap C)}{P(C)}$, which yields the following important equation:

$$P(C \bigcap D) = P(C)P(D \mid C). \tag{3.1}$$

This is the so-called *long multiplication rule*, which in fact generalizes to any finite number of events. If $E_1, E_2, \ldots, E_n$ are n events, then

$$P(\bigcap_{i=1}^{n} E_i) = P(E_1)P(E_2 \mid E_1)P(E_3 \mid E_1 \cap E_2) \ldots P(E_n \mid E_1 \cap \ldots \cap E_{n-1}).$$

$$\tag{3.2}$$

Notice that if two events A and B are independent, $P(A \mid B)$ reduces to $P(A)$ and, similarly, $P(B \mid A)$ reduces to $P(B)$. In fact, this can be used as the definition of independence. Let us now see some more examples of conditional probabilities.

Example 3.4. In the context of tossing a fair coin twice, so that $S = \{HH, HT, TH, TT\}$ and all four outcomes are equally likely, what is the conditional probability that both tosses show 'heads up,' given that none of them produces a tail? Since the intersection of the two events has probability $= 1/4$ and that for the latter event is $3/4$, the desired conditional probability will be $1/3$.

Example 3.5. Suppose our random experiment is rolling a balanced die twice. The sample space S will have 36 outcomes, each being an ordered pair of whole numbers (x, y) with $1 \leq x, y \leq 6$. And they are all equally likely, each having probability $1/36$. What will be the conditional probability of exactly one of x and y being odd (call this event A), given that x and y are different (call it B)? Once again, since the probability of their intersection is $18/36$ or $1/2$ (can be verified by direct counting or reasoned out easily) and the denominator probability is $30/36$ or $5/6$ (all but the outcomes $(1, 1), (2, 2), \ldots, (6, 6)$), the answer is $3/5$. Do you think that A and B are independent?

Example 3.5. Suppose you are drawing two cards one by one, without looking, from a (well-shuffled) deck of 52 cards *with* replacement. The sample space will have $52^2 = 2704$ ordered pairs of cards, all of which are equally likely. Now, what is the conditional probability of both being face cards, given that both are spades? Clearly, the intersection of these events has $3^2 = 9$ outcomes in it whereas the second event itself has $13^2 = 169$. So the answer is $9/169$.

Example 3.6. The file cabinet in your office is locked and you have a bunch of n keys in a key-ring, exactly one of which will open it. You start trying the keys one by one and once a key has been tried, it is not tried again. What are the chances that you succeed at the kth attempt $(k \leq n)$? If we define events $E_1, E_2, \ldots, E_{k-1}$ as $E_i = \{$failing at the ith attempt$\}$ and define E_k as "succeeding at the kth attempt," then the desired event $E = \bigcap_{i=1}^{k} E_i$. But by (3.2),

$$P(E) = P(E_1)P(E_2 \mid E_1) \ldots P(E_{k-1} \mid \bigcap_{i=1}^{k-2} E_i)P(E_k \mid \bigcap_{i=1}^{k-1} E_i),$$

which in this case is $(\frac{n-1}{n})(\frac{n-2}{n-1}) \ldots (\frac{n-k+1}{n-k+2})(\frac{1}{n-k+1})$. But this is simply $1/n$. In other words, somewhat counter-intuitively, the chances of succeeding at any particular attempt are the same.

Example 3.7. Let us assume that in a certain population, 5% of the people carry a certain virus (i.e., the virus infection has a *prevalence* rate of 5%). A clinical diagnostic test that is used to detect the infection will pick up the infection correctly 99% of the times (i.e., has a *false negative* rate of 1%). The test raises a false alarm (i.e., gives a positive result even when the infection is not there) 2% of the times. If a person walks into the clinic, takes the test and gets a positive result, what are the chances that he/she actually has the infection? It is a legitimate question since the test appears to be slightly imperfect. Now, let us translate this story to symbols. Let $A = \{$a randomly chosen person from that population actually has the infection$\}$ and $B = \{$a randomly chosen person from that population tests positive$\}$. Then, in terms of these, what do we know and what is the question? We know that $P(A) = 0.05$, $P(B \mid A) = 0.99$ and $P(B \mid A^c) = 0.02$. We are supposed to find $P(A \mid B)$. As we know, it is simply $\frac{P(A \cap B)}{P(B)}$. But, according to (3.1), the numerator is $P(A)P(B \mid A) = (0.05)(0.99) = 0.0495$. What about the denominator? Notice that $B = B \bigcap S = B \bigcap (A \bigcup A^c) = (B \bigcap A) \bigcup (B \bigcap A^c)$ due to the distributive property. So $P(B) = P(B \bigcap A) + P(B \bigcap A^c)$. We already know the first term. The second term can be found similarly as $P(A^c)P(B \mid A^c) = (1 - 0.05)(0.02) = 0.019$. So $P(B) = 0.0495 + 0.019 = 0.0685$. So the answer is 0.7226 or 72.26%.

There is something interesting in this last example, namely, the way we computed $P(B)$. Realizing that the sample space $S = A \bigcup A^c$, we decomposed B into two disjoint pieces $(B \bigcap A$ and $B \bigcap A^c)$ and computed $P(B)$ as the sum of the probabilities of those pieces. This can be generalized to the following scenario. Suppose $A_1, A_2, \ldots, A_n$ are mutually disjoint events such that $\bigcup_{i=1}^{n} A_i = S$. Such a collection of events is called a *partition* of the sample space. So, in Example 3.7, A and A^c formed a partition of the sample space. Any event B can be written as $B = \bigcup_{i=1}^{n} B \cap A_i$, which is a disjoint union. So

$$P(B) = \sum_{i=1}^{n} P(B \cap A_i) = \sum_{i=1}^{n} P(B \mid A_i)P(A_i), \text{ due to (3.1)}. \qquad (3.3)$$

This is known as the *law of total probability*. Using this, any of the "reverse" conditional probabilities $P(A_j \mid B)$ can be computed as

$$P(A_j \mid B) = \frac{P(A_j \cap B)}{P(B)} = \frac{P(B \mid A_j).P(A_j)}{\sum_{i=1}^{n} P(B \mid A_i)P(A_i)}. \qquad (3.4)$$

This is known as the *Bayes theorem*. In this context, the $P(A_i)$'s are called the *prior* probabilities of the partition-cells, the "forward" conditional probabilities $P(B \mid A_i)$'s are called the *likelihoods* and the "reverse" conditional probabilities $P(A_i \mid B)$'s are called the *posterior* probabilities. This simple result is so powerful that an entire subfield of statistics is based on this idea. Bayesian inference now dominates the cutting-edge applications of statistics in the highly challenging data-analysis problems posed by modern science. We will see more of it later. For the time being, let us see some examples of the Bayes theorem in action.

Example 3.8. In the experiment of drawing a card, without looking, from a (well-shuffled) deck of 52 cards, what are the chances that the card is a diamond, given that it is a face card? Let A_1, A_2, A_3 and A_4 be the events that the card drawn belongs to the diamond suit, the spade suit, the heart suit and the club suit respectively. Together, they form a partition of the sample space. Let B be the event that the card drawn is a face card. Then the prior probabilities are $P(A_i) = \frac{1}{4}$ for $i = 1, \ldots, 4$, the likelihoods are $P(B \mid A_i) = \frac{3}{13}$ and, therefore, the posterior probabilities are $\frac{(3/13)(1/4)}{(4)(3/13)(1/4)}$ or $\frac{1}{4}$. In other words, the prior and the posterior probabilities are identical. What does it tell you? That the event B is independent of the partition cells (we already knew that, right?).

Example 3.9. If you are trying to buy an air ticket the day before the 4th

of July weekend to fly from Green Bay, Wisconsin, to Chicago, Illinois, there are several choices: United Airlines (UA), Northwest Airlines (NA), Midwest Airlines (MA), American Airlines (AA) and Delta ComAir (DCA). They respectively operate 15%, 20%, 25%, 30% and 10% of the flights that depart from the Green Bay airport daily. The chances of getting a seat on a day like that are small: 3% on UA, 5% on NA, 3% on MA, 2% on AA and 5% on DCA. Given that you ultimately managed to fly to Chicago, what are the chances that you chose American Airlines? Well, the five airlines form a five-cell partition with prior probabilities given by the percentages of the daily flights out of Green Bay that they operate. Using the likelihoods of getting a seat on various airlines, The Bayes theorem yields the following posterior probability of having chosen AA, given that you managed to fly:

$$\frac{(0.3)(0.02)}{(0.15)(0.03) + (0.2)(0.05) + (0.25)(0.03) + (0.3)(0.02) + (0.1)(0.05)} = \frac{0.006}{0.033},$$

which is about 0.182. Notice that in an example like this, the prior probabilities of the partition cells are more likely to be subjective, because often one's choice of a carrier is motivated by lower fares or the desire to earn frequent flier miles or the quality of in-flight service. The percentages of daily flights operated by the airlines would indeed be the correct priors if one chose a carrier randomly, which is seldom the case. In real-life applications of the Bayes theorem, there is often a lot of controversy regarding the appropriate choice of prior probabilities, but the theorem works irrespective of how they are chosen.

Since we compute the probability of an event A as $n(A)/n(S)$ in the 'equally likely' scenario, efficient computation of probabilities depends on our ability to quickly enumerate sample spaces and events. In the examples discussed so far, counting the number of outcomes in a sample space or an event has been a piece of cake, but the degree of difficulty quickly increases with the size of the set. For example, any counting for experiments such as 5 rolls of a die or 10 tosses of a coin is not so trivial. So we must develop some clever counting tricks. Here is a motivating example.

Example 3.10. Suppose you are synthesizing proteins *in vitro* in a laboratory. Recall from Chapter 2 that the building blocks of proteins are amino acids (AA) and there are 20 of them. If you decide to synthesize a 10AA oligopeptide chain consisting of distinct AA's by randomly choosing from the full collection of 20, with the restriction that the first AA in the chain must be a methionine, what are the chances of ending up with the chain methionine-valine-threonine-glycine-alanine-arginine-leucine-ceistine-proline-isoleucine? To answer this question, the first order of business is to find out $n(S)$. What is a typical outcome here? It is an ordered arrangement of 10 distinct AA's chosen out of 20 so that the first one is a methionine. So the best way to find $n(S)$ is to first find the number of different subcollections of 9 AA's chosen out of 19 (i.e., all but methionine) and then multiply it with the number of different ways in which 9 distinct AA's can be rearranged. In what follows, we first address the 'subcollections' question and then the 'rearrangement' question.

If you have a bag containing just one item (let it be an apple or A), how many subcollections of 1 item can you form from it? Just one (i.e., $\{A\}$). If you have two items in your bag instead of one (say, an apple and a banana, or simply A and B), how many single-item subcollections can you form from it? Two (i.e., $\{A\}$ and $\{B\}$). How many double-item subcollections can be formed? Only one (namely, $\{A, B\}$). Next, if you have three items in your bag (say, an apple, a banana and a cucumber, or simply A, B and C), the numbers of single-item, double-item and triple-item subcollections that can be formed are 3, 3 and 1 respectively. List them all and verify. If you have one additional item in the bag so that it now contains four items (say, A, B, C and D), once again the different single-item, double-item, triple-item and quadruple-item subcollections that can be obtained from it are easy to list. Their numbers are 4, 6, 4 and 1 respectively. Now, in order to facilitate the recognition of a pattern in these counts, let us ask the 'silly' question: "How many *empty* subcollections can be formed?" in each of the above cases. The answer is always 1, since an empty subcollection is an empty subcollection—there is no variety of it. With this, let us write those counts in the following way:

$$
\begin{array}{ccccccc}
 & & & 1 & & 1 & & \\
 & & 1 & & 2 & & 1 & \\
 & 1 & & 3 & & 3 & & 1 \\
1 & & 4 & & 6 & & 4 & & 1
\end{array}
$$

The pattern that emerges in this triangular arrangement of numbers (widely known as a *Pascal's triangle*) is that each non-terminal number in it is the sum of its two nearest neighbors from the preceding row (called its 'parents'). The terminal numbers in each row have just one 'parent' apiece. So, how does it help answer our 'subcollection' question in general? In order to find the number of different k-item subcollections that can be formed from a bag containing n distinct items, we just need to look up the $(k+1)^{th}$ entry in the nth row of this table. That entry, usually denoted by $_nC_k$, is nothing but $\frac{n!}{k!(n-k)!}$.

Now, to rearrangements. If you just have an apple (or A), the number of different rearrangements of it is just 1. If, instead, you have two items (A and B), there are two possible rearrangements (AB and BA). For three items A, B and C, the number is 6 (ABC, ACB, BAC, BCA, CAB and CBA). For 4 items A,B,C and D, it is 24. If we still try to list all the 24 rearrangements (we will quit this habit shortly, once we know the formula), the following is the most efficient way of doing it. For the moment, ignore D and focus on the remaining three items A, B and C. From the previous step, we know that they have 6 distinct rearrangements. List them all and add the missing letter D at the end of each of them. You have generated 6 distinct rearrangements of A,B,C and

D. Next, repeat this process with A,B and D only, ignoring C. You generate 6 more. Ultimately you will generate another 6+6=12 by ignoring B and A one at a time. That is why the total number is 24. In view of this "ignoring one at a time" algorithm, the number of different rearrangements of 5 distinct items will be 24+24+24+24+24=120. In any case, is there a pattern emerging here too? The numbers of distinct rearrangements of n items were 1 for $n = 1$, 2=(2)(1) for $n = 2$, 6=(3)(2)(1) for $n = 3$, 24=(4)(3)(2)(1) for $n = 4$, and so forth. So in general, the number of different rearrangements (also called *permutations*) of n distinct items is $n(n-1)(n-2)\ldots(3)(2)(1)$, usually denoted by $n!$. This simple formula, however, fails if some of the items concerned are indistinguishable. For example, two identical A's can be arranged in only one way, not two. Similarly, two identical A's and a B can be arranged in just three ways (AAB, ABA, BAA), not six. If you have n objects, k_1 of which are identical of type 1, ..., k_L of which are identical of type L ($\sum_{i=1}^{L} k_i = n$), then the number of different rearrangements drastically reduces from $n!$ to $\frac{n!}{(k_1!)\ldots(k_L!)}$.

Having said all these, let us go back to our original question of oligopeptide synthesis. 9 distinct amino acids can be chosen from a "bag" of 19 in $_{19}C_9 = \frac{19!}{(9!)(10!)}$ different ways. A set of 9 distinct amino acids can be permuted in 9! different ways. So, the total number of possible 10AA oligopeptide chains consisting of distinct AA's and starting with methionine is $(\frac{19!}{(9!)(10!)})(9!) = \frac{19!}{10!}$. In general, the number of different *ordered* permutations of k distinct objects chosen out of n distinct objects is $\frac{n!}{(n-k)!}$, which is denoted by $_nP_k$.

Example 3.11. In the experiment of drawing two cards, without looking, from a (well-shuffled) deck of 52 cards one by one without replacement, what are the chances that both are hearts? The answer is $n(A)/n(S)$ where $n(S) =_{52} C_2 = \frac{52!}{(2!)(50!)} = 1326$ and $n(A) =_{13} C_2 = \frac{13!}{(2!)(11!)} = 78$. So the chances that both of them are hearts are about 5.88%.

Example 3.12. If you are randomly permuting the four letters of our genetic alphabet (i.e., the four nucleotides adenine, guanine, thymine and cytosine or A,G,T and C), what are the chances that the purines and the pyrimidines are together? For this problem, $n(S)$ is of course $4! = 24$. To find $n(A)$, let us look at it this way. The 'purine block' and the 'pyrimidine block' can be permuted in $2! = 2$ ways and within each block, the two bases can be permuted in $2! = 2$ ways. So $n(A) = (2)(2)(2) = 8$ and the answer is $1/3$.

3.3 Random Variables and Probability Distributions

Now let us play a different 'game' with sample spaces, events and probabilities. For many real-life random experiments, the sample space is very large and the individual outcomes are not of our interest; instead we are interested in clusters of outcomes sharing a common feature. So it is enough to summarize the sample space as a list of those clusters and the corresponding probabilities. Usually, each such cluster corresponds to a unique value of a 'quantity of interest' (called a *random variable*), so listing the clusters is equivalent to listing the distinct values of the associated random variable. The resulting table, with values of the random variable in one row and the corresponding probabilities in the other, is called a *probability mass function* table or p.m.f. table. Here are some examples.

Example 3.13. In the experiment of tossing a fair coin three times, which has $S = \{$ HHH, HHT, HTH, THH, HTT, THT, TTH, TTT$\}$ with equally likely outcomes, if the 'quantity of interest' or random variable (call it X) is simply the number of heads among the three tosses, the p.m.f. table will be

Values	0	1	2	3
Probs	$\frac{1}{8}$	$\frac{3}{8}$	$\frac{3}{8}$	$\frac{1}{8}$

Example 3.14. In the experiment of rolling a balanced die twice, with S having 36 equally likely ordered pairs of whole numbers (x, y), $1 \le x, y \le 6$, if the random variable of interest (call it Y) is the sum of the two numbers that turned up, the p.m.f. table will be

Values	2	3	4	5	6	7	8	9	10	11	12
Probs	$\frac{1}{36}$	$\frac{2}{36}$	$\frac{3}{36}$	$\frac{4}{36}$	$\frac{5}{36}$	$\frac{6}{36}$	$\frac{5}{36}$	$\frac{4}{36}$	$\frac{3}{36}$	$\frac{2}{36}$	$\frac{1}{36}$

Example 3.15. Suppose ten million tickets of a state lottery have been sold for a dollar a piece. Only one of them carries a super bumper prize of $ 5000000, 5 of them carry mega-prizes of $ 500000 each, 25 of them carry second prizes of $ 50000 each, 50 of them carry third prizes of $ 10000 each and another 100 of them carry consolation prizes of $ 1000 each. If you go to a store and randomly buy a ticket for this lottery, what is the p.m.f. table of your earning (call it W)? Notice that irrespective of whether you win a prize or not, you always pay the ticket price. So the p.m.f. table of your earning will be as follows:

Amounts	$ 4999999	$ 499999	$ 49999	$ 9999	$ 999	$ -1
Probs	$\frac{1}{10000000}$	$\frac{1}{2000000}$	$\frac{1}{400000}$	$\frac{1}{200000}$	$\frac{1}{100000}$	$\frac{9999819}{10000000}$

In this last example, the important question is what your *average* (or *expected*) earning will be if you repeatedly buy tickets for this lottery. Will it be as high as the grand prizes this lottery advertises? The *expected value* or *mean* of a random variable W (denoted by $E(W)$ or μ_W) is defined as $\sum_{i=1}^{m} w_i p_i$ if W

takes the values $w_1, \ldots, w_m$ with probabilities $p_1, \ldots, p_m$ respectively. It is simply a weighted average of its values, with the weights being the probabilities. For this lottery, $E(W) = -0.065$. That is, on an average you *lose* six and a half cents! This is why the lottery business is profitable to its host.

While the expected value is a useful device for comparing two random variables X and Y (after all, we cannot compare them value-by-value or probability-by-probability since their p.m.f. tables may not even be of the same length), it is by no means the whole story. Two completely different random variables may end up having the same expected value, so we need some other summary-measures to capture the other differences. Any p.m.f. table can be pictorially represented by a *stick plot* or a *probability histogram*. In a stick plot, the distinct values are marked along the horizontal axis and a stick is erected over each value whose height is the corresponding probability. In a probability histogram, the sticks are replaced by rectangular bars of equal width centered at the values. Various features of this plot such as how 'fat' or 'thin' its tails are, how asymmetric it is around its expected value and how peaked or flat-topped it is, are connected to its *moments*. The rth *raw moment* of X for any real number r is defined as $E(X^r)$, i.e., as the expected value of the random variable X^r which takes the values $x_1^r, x_2^r, \ldots$ with probabilities $p_1, p_2, \ldots$ ($x_1, x_2, \ldots$ being the values of X). The rth *central moment* of X is defined as $E(X - E(X))^r$, i.e., as the expected value of the random variable $(X - E(X))^r$ which takes the values $(x_1 - \mu_X)^r, (x_2 - \mu_X)^r, \ldots$ with probabilities $p_1, p_2, \ldots$. The 2nd central moment of a random variable is called its *variance* (denoted by σ_X^2) and it is related to the 'fatness' or 'thinness' of the histogram tail. Usually the heavier the tail, the greater the variance. The square root of the variance is known as the *standard deviation* (denoted by σ_X). The 3rd central moment is related to the degree of *skewness* (i.e., lack of symmetry) of the histogram and the 4th central moment is related to its degree of peakedness. Two important properties of expected values are (i) $E(X + Y) = E(X) + E(Y)$ for any two random variables X and Y and (ii) $E(cX) = cE(X)$ for any random variable X and any constant c. Simple algebraic expansions and repeated use of the above two properties will show that the raw moments and the central moments are related. For example, $\sigma_X^2 = E(X^2) - (E(X))^2$. Also, $E(X - E(X))^3 = E(X^3) - 3E(X)E(X^2) + 2(E(X))^3$. All the raw moments of a random variable can be conveniently obtained from its *moment generating function* (MGF). Just as a thin popcorn-bag from the supermarket shelf, when microwaved, swells up and produces lots of delicious popcorns, the MGF is like a handy storage device that produces raw moments of many different orders when repeatedly differentiated. Formally, the MGF of X is defined as $M_X(t) = E(e^{tX})$ for t in some interval around 0 on the real line. So it is the expected value of the random variable e^{tX} that takes the values $e^{tx_1}, e^{tx_2}, \ldots$ with probabilities $p_1, p_2, \ldots$. Its rth derivative, evaluated at $t = 0$, is the rth raw moment of X. An important and useful fact about MGF's is that there is a one-to-one correspondence between the form of the MGF and the form of the associated p.m.f., so it is possible to identify the p.m.f. by looking at the MGF (call it "MGF fingerprinting" if you will!).

Example 3.16. For the random variable Y in Example 3.14, $\mu_Y = E(Y) = \frac{252}{36} = 7$, $\sigma_X^2 = E(Y^2) - (E(Y))^2 = \frac{1974}{36} - 7^2 = 5.833$, the 3rd raw moment $E(Y^3) = \frac{16758}{36} = 465.5$ and the 3rd central moment $E(Y - E(Y))^3 = 465.5 - 3(7)(\frac{1974}{36}) + 2(7^3) = 0$. The 3rd central moment is 0 for any p.m.f. that is *symmetric* with respect to its mean (the converse is not necessarily true, though). Finally, the MGF of Y is

$$M_Y(t) = \frac{e^{2t} + e^{12t} + 2(e^{3t} + e^{11t}) + 3(e^{4t} + e^{10t}) + 4(e^{5t} + e^{9t}) + 5(e^{6t} + e^{8t}) + 6e^{7t}}{36}.$$

Its first derivative of it w.r.t. t, evaluated at $t = 0$, is $E(Y) = 7$, its second derivative evaluated at $t = 0$ is $E(Y^2) = 54.833$ and its third derivative evaluated at $t = 0$ is $E(Y^3) = 465.5$ (verify these).

Now consider the random experiment of closing your eyes and touching a 6-inch ruler with a needle-tip. What if we define our random variable X to be the distance between the point where the needle touched the ruler and the left end-point (or zero-point) of the ruler? A needle-tip being so sharp and pointed, X can actually take any value between 0 and 6. That is, the set of possible values of X is the entire interval $[0, 6]$ and is, therefore, uncountably infinite. This immediately implies that if we wanted to write down a p.m.f. table for X or draw its probability histogram, we would fail. Also, despite the fact that your eyes being closed gives this experiment an 'equally likely' flavor, in the sense that intuitively each real number in the interval $[0, 6]$ should have the same probability, what *is* that probability? Just as each of the 6 outcomes in the roll of a balanced die has probability $\frac{1}{6}$, will it be $\frac{1}{\infty}$ or 0 in this case? Then we are in the strange situation that each single outcome has probability 0, yet one of those countless outcomes will definitely occur! This kind of random variables, which lands us in this bizarre situation, is called the class of *continuous* random variables. In contrast, the random variables with p.m.f. tables that we saw earlier will be called *discrete*. By the way, from our examples above, one should not get the wrong impression that *all* discrete random variables have finite value-sets (or *supports*). Examples of discrete random variables with countably infinite supports are forthcoming.

In any case, getting back to continuous random variables, how does one compute probabilities for them? For example, for the X in the 'needle and ruler' experiment, what is $P(X \leq 2)$ or $P(3 \leq X \leq 5)$? Well, think about the natural discrete counterpart of this experiment. Since this experiment basically says: "Pick any real number randomly from within $[0, 6]$," its natural discrete counterpart would be: "Pick any whole number randomly from among 1,2,3,4,5 and 6". It should look familiar since it is nothing but the balanced die-rolling experiment. If for this experiment, we defined X to be the whole number that turned up, how would we compute $P(X \leq 2)$ and $P(3 \leq X \leq 5)$? One way would be to compute them directly from the p.m.f. table of X or by adding the heights of appropriate sticks in its stick plot, but these are not relevant to our present situation where there is no p.m.f. table or stick plot. Another way of computing those probabilities would be by the *area method* from the probability histogram of X. The probability histogram of an 'equally likely' experiment such as rolling a balanced die looks like a flat or rectangular box. To find $P(X \leq 2)$, we just need to find the total area of the bars in this histogram that correspond to this event (i.e., the two leftmost bars). Since each rectangular bar has a base-width of 1 unit and height $= \frac{1}{6}$, the total area of those two bars is $\frac{1}{3}$. Similarly, $P(3 \leq X \leq 5) = \frac{1}{2}$, the sum of the areas of three such bars. If we now imagine following the same area method to compute those probabilities in the continuous case, except at a much, much finer scale with 'needle-thin' bars (because there are countless values crammed together in a small space), we begin to understand how one deals with continuous random variables. We are actually taking the limit of a histogram with n bars as $n \to \infty$, keeping the base-width of the histogram fixed. As we do this, the upper contour of the histogram (which would have a rugged broken-line structure in a discrete histogram) starts to appear like a smooth curve. This curve, which still encloses a total area of 1 underneath it like the discrete histograms, will be called the *probability density curve* of X. The function $f(x) : D \to \mathcal{R}$, whose graph is the probability density curve, will be called the *probability density function* (p.d.f.) of X (D being the value-range or support of X). In general, any nonnegative-valued integrable function defined on the real line, which encloses a total area of 1, could potentially be the p.d.f. of some continuous random variable coming from some underlying experiment.

So, in our 'needle and ruler' experiment, the density curve is a flat line on the interval $[0, 6]$ and the p.d.f. is $f(x) = \frac{1}{6}I(x \in [0, 6])$, where $I(.)$ is the indicator function that takes the value 1 only if the condition in it is satisfied (otherwise 0). This is known as the *Uniform* p.d.f. $P(X \leq 2)$ is the area underneath this flat line between 0 and 2. Similarly, $P(3 \leq X \leq 5)$ is the area between 3 and 5. Formally speaking, for any two real numbers a and b,

$$P(a \leq X \leq b) = \int_a^b f(x)dx. \tag{3.5}$$

This gives us the formal reason why, for a continuous random variable X, $P(X = x) = 0$ for any particular value x, because it is simply the integral in (3.5) with $a = b$. Now, if we define a function $F(t) : \mathcal{R} \to [0, 1]$ as $F(t) = \int_{-\infty}^t f(x)dx$, then $F(t)$ is nothing but the area underneath the p.d.f. $f(x)$ up to the point t or, equivalently, $P(X \leq t)$. This F is called the *cumulative distribution function* (c.d.f.) corresponding to the p.d.f. f. In this case, the Uniform[0,6] c.d.f. is

$$F(t) = 0 \text{ if } t \leq 0; \quad \int_0^t \frac{1}{6}dx \text{ or } \frac{t}{6} \text{ if } t \in (0, 6); \ 1 \text{ if } t \geq 6. \tag{3.6}$$

By the fundamental theorem of integral calculus, such an $F(t)$ is a continuous (if fact differentiable) function with $F'(x) = f(x)$. Incidentally, we could also have defined a c.d.f. for a discrete random variable, but those c.d.f.'s would be *step functions*, i.e., their graphs could consist of flat line-segments with jumps in-between (explain to yourself why). But irrespective of discrete or continuous, all c.d.f.'s have the following properties:

Result 3.2. The c.d.f. $F(.)$ of any random variable X satisfies
(i) $\lim_{t\to-\infty} F(t) = 0$ and $\lim_{t\to\infty} F(t) = 1$;
(ii) $F(t)$ is *right-continuous*, i.e., $\lim_{t\to s+} F(t) = F(s)$, where "$t \to s+$" means that t approaches s from its right side;
(iii) $F(a) \le F(b)$ for any $a, b \in \mathcal{R}$ with $a \le b$.

The density curves of different continuous random variables can have a great variety of geometric shapes. Here is another example.

Example 3.17. Suppose you touch a 6-inch ruler with a needle with your eyes closed, and your friend does the same thing independently of you. Let Y_1 be the distance between the touching point and the zero-point on the ruler in your case, and Y_2 be that in your friend's case. If we define $Y = Y_1 + Y_2$, this Y will have a *triangular* density on the support $[0,12]$ with p.d.f. $f(y)$ given by

$$f(y) = \frac{y}{36}I(0 \le y \le 6) + \frac{12 - y}{36}I(6 < y \le 12). \tag{3.7}$$

How do we know? One way would be to go through the same limiting process that led us to the Uniform[0,6] density earlier. For this experiment, the natural discrete counterpart is two persons rolling a balanced die each, independently of one another. The discrete histogram corresponding to the sum of the two numbers is essentially triangle-shaped (ignoring the rugged upper contours of the bars).

Now that we know how continuous random variables behave, all the numerical summaries we defined for a discrete random variable (i.e., raw and central moments) can be easily generalized to them. For a continuous X with p.d.f. $f(x)$ and support D, the rth raw and central moments are defined as

$$E(X^r) = \int_D x^r f(x)dx, \; E(X - \mu_X)^r = \int_D (x - \mu_X)^r f(x)dx, \tag{3.8}$$

provided that the integrals exist. Like before, various features of the density curve (e.g., lightness or heaviness of tails, skewness, flatness, etc.) are connected to its moments and the central moments are expressible in terms of the raw moments. The MGF of X is also defined analogously as $M_X(t) = E(e^{tX}) = \int_D e^{tx} f(x)dx$ if this integral exists on some interval around 0, and the mechanism by which the raw moments are generated from this MGF is the same as before.

Example 3.18. The MGF of the Uniform[0,6] random variable X mentioned earlier is

$$M_X(t) = \int_0^6 e^{tx}\frac{1}{6}dx = \frac{1}{6}\frac{e^{6t} - 1}{t} \text{ for } t \ne 0.$$

Remember the definition of two events being independent? It can be easily generalized to any finite number of events. The events $E_1, \ldots, E_n$ are independent if and only if $P(\bigcap_{i=1}^n E_i) = \prod_{i=1}^n P(E_i)$. The definition of the independence of a bunch of random variables is analogous. The random variables $X_1, \ldots, X_n$ with supports $D_1, \ldots, D_n$ respectively are said to be independent if for any subsets A_i of D_i $(i = 1, \ldots, n)$, $P(X_1 \in A_1, \ldots, X_n \in A_n) = \prod_{i=1}^{n} P(X_i \in A_i)$. An immediate consequence of this definition is that the expected value of the product of a bunch of independent random variables turns out to be the product of the individual expected values. We will now focus on some frequently encountered and useful random variables— both discrete and continuous.

A vast majority of real-life random experiments producing discrete random variables can be modeled by the following:
(i) the number of heads in n independent tosses of a (possibly biased) coin whose $P(\text{head}) = p$ in a single toss;
(ii) the number of independent tosses of a (possibly biased) coin with $P(\text{head}) = p$ that are needed to get the kth head;
(iii) the number of black sheep in a random sample of n sheep drawn without replacement from a population of $N(\geq n)$ sheep containing r black sheep;
(iv) the (limiting) number of heads in a very large number (n) of independent tosses of a highly biased coin with a very small $P(\text{head}) = p$ so that np is moderate.
We will briefly describe the resulting random variable and its properties in each of these cases.

If X is the number of heads in n independent tosses of a coin whose $P(\text{head})$ in each toss is $p \in (0, 1)$, the p.m.f. of X is given by

$$P(X = x) =_n C_x p^x (1 - p)^{n-x}, \text{ for } x = 0, 1, \ldots, n. \tag{3.9}$$

This can be easily derived using the independence between tosses and the fact that there are $_n C_x$ different permutations of x identical H's and $(n - x)$ identical T's that give rise to the same value of X. This is known as the Binomial(n, p) p.m.f., which is also called a Bernoulli trial for $n = 1$. For this X, $\mu_X = E(X) = np$, $\sigma_X^2 = np(1 - p)$ and $M_X(t) = (1 - p + pe^t)^n$.

If X is the number of independent tosses of a coin with $P(\text{head}) = p$ that are needed to see the first head, the p.m.f. of X is given by

$$P(X = x) = p(1 - p)^{x-1}, \text{ for } x = 1, 2, \ldots . \tag{3.10}$$

This is once again easy to derive using the independence between tosses. This is known as the Geometric(p) p.m.f., which has mean $\mu_X = \frac{1}{p}$, variance $\sigma_X^2 = \frac{1-p}{p^2}$ and MGF $M_X(t) = \frac{pe^t}{1-(1-p)e^t}$. An interesting and useful feature of this distribution is its memoryless property. Given that no head has appeared until the kth toss, the conditional probability that there will be no head until the $(k + m)^{th}$ toss is the same as the (unconditional) probability of seeing no head in the first m tosses, for any two positive integers k and m. That is, $P(X > k + m \mid X > k) = P(X > m) = (1 - p)^m$.

RANDOM VARIABLES AND PROBABILITY DISTRIBUTIONS

If X is the number of independent tosses of a coin with $P(\text{head}) = p$ that are needed to see the rth head ($r \geq 1$), the p.m.f. of X is given by

$$P(X = x) =_{x-1} C_{r-1} p^r (1-p)^{x-r}, \text{ for } x = r, r+1, \ldots . \quad (3.11)$$

This is equally easy to derive using the independence between tosses and the fact that the first $r - 1$ heads could occur anywhere among the first $x - 1$ tosses. This is known as the *Negative Binomial*(r, p) p.m.f., with mean $\mu_X = \frac{r}{p}$, variance $\sigma_X^2 = \frac{r(1-p)}{p^2}$ and MGF $M_X(t) = (\frac{pe^t}{1-(1-p)e^t})^r$. It is related to the Geometric(p) p.m.f. in the sense that if $Y_1, \ldots, Y_r$ are independent and identically distributed (i.i.d.) random variables each having a Geometric(p) p.m.f., $\sum_{i=1}^r Y_i$ will have a Negative Binomial(r, p) distribution.

Suppose a population of N sheep consists of r black sheep and $N - r$ white sheep. If a random sample of n ($\leq r$) is drawn without replacement from this population and X is the number of black sheep in this sample, then X has the p.m.f.

$$P(X = x) = (_r C_x)(_{N-r} C_{n-x})/_N C_n, \text{ for } x = 0, 1, \ldots, n. \quad (3.12)$$

This is not difficult to see, in view of the Pascal's triangle and the associated formula for finding the number of different subcollections. This is known as the *Hypergeometric* (N, r, n) p.m.f., with mean $\mu_X = \frac{nr}{N}$ and variance $\sigma_X^2 = \frac{nr(N-r)(N-n)}{N^2(N-1)}$. The MGF does not simplify to a convenient expression, but can still be written down as a summation.

If X counts how many heads there are in a very large number n of independent tosses of a highly biased coin with $P(\text{head}) = p$ very small such that np is moderate (call it λ), then the binomial p.m.f. of X can be well-approximated by the following p.m.f.:

$$P(X = x) = \frac{e^{-\lambda}\lambda^x}{x!}, \text{ for } x = 0, 1, \ldots . \quad (3.13)$$

This can be shown formally by taking the double limit of a Binomial(n, p) p.m.f. as $n \to \infty$ and $p \to 0$ in such a way that np remains a constant λ. This is called the *Poisson*(λ) p.m.f., with mean $\mu_X = \lambda$, variance $\sigma_X^2 = \lambda$ and MGF $M_X(t) = e^{\lambda(e^t - 1)}$. This is a widely used probability model for random experiments producing count data, except that it is only suitable for datasets with the mean approximately equal to the variance. However, there are ways to modify this p.m.f. so that it can accommodate datasets with the variance higher or lower than the mean (see, for example, Shmueli et al., *Applied Statistics* (2005)). Also, the Poisson(λ) p.m.f. has the *reproductive* property that if X_1 and X_2 are independent random variables with $X_i \sim$ Poisson(λ_i), then $X_1 + X_2 \sim$ Poisson$(\lambda_1 + \lambda_2)$.

Now, the continuous case. A vast majority of real-life random experiments that produce continuous random variables can be modeled by the following probability density functions, in addition to the Uniform$[a, b]$ p.d.f. discussed a short while ago (with mean $\frac{a+b}{2}$ and variance $\frac{(b-a)^2}{12}$).

A random variable X is said to have a *Normal* (or *Gaussian*, after Carl Friedrich Gauss) p.d.f. with mean $\mu_X \in \mathcal{R}$ and variance σ_X^2 ($\sigma_X > 0$) if its p.d.f. is given by

$$f(x) = \frac{1}{(2\pi\sigma^2)^{0.5}} e^{-(x-\mu)^2/2\sigma^2}, \text{ for } x \in (-\infty, \infty), \tag{3.14}$$

where we have omitted the subscript X to reduce clutter. The MGF of X is $M_X(t) = e^{t\mu_X + t^2\sigma_X^2/2}$. The special case with mean 0 and variance 1 is called a *standard normal* p.d.f. and the associated random variable is usually denoted by Z. The $N(\mu_X, \sigma_X^2)$ p.d.f. has the *linearity* property, that is, $X \sim N(\mu_X, \sigma_X^2) \implies aX + b \sim N(a\mu_X + b, a^2\sigma_X^2)$. This can be shown by first deriving the MGF of $aX + b$ from that of X and then using "MGF fingerprinting." As a result, $X \sim N(\mu_X, \sigma_X^2) \implies \frac{X - \mu_X}{\sigma_X} \sim N(0, 1)$ or standard normal. The ratio $\frac{X - \mu_X}{\sigma_X}$ is usually called the *z-score* of X. Another celebrated result that plays a fundamental role in probability and statistics is the *central limit theorem* (CLT). In its most simplified form, it basically says that if $\{X_1, \ldots, X_n\}$ is a random sample from some p.m.f. or p.d.f. having mean μ_X and standard deviation σ_X (i.e., if $X_1, \ldots, X_n$ are i.i.d. random variables from the same p.m.f. or p.d.f.), then the p.d.f. of $(\frac{1}{n}\sum_{i=1}^{n} X_i - \mu_X)/(\sigma_X/n^{\frac{1}{2}})$ approaches that of $N(0, 1)$ in the limit as $n \to \infty$. In other words, for n sufficiently large, the p.d.f. of the z-score of $\bar{X} = \frac{1}{n}\sum_{i=1}^{n} X_i$ can be well-approximated by a standard normal p.d.f. This has far-reaching consequences, one of which is the large-sample normal approximation to discrete p.m.f.'s. A normal p.d.f. also has the *reproductive* property that if $X_i \sim N(\mu_i, \sigma_i^2)$ for $i = 1, \ldots, n$ and they are independent, the sum $\sum_{i=1}^{n} X_i$ has a $N(\sum_{i=1}^{n} \mu_i, \sum_{i=1}^{n} \sigma_i^2)$. Since probability computation using a normal p.d.f. is not possible analytically, extensive tables are available for ready reference.

A random variable X is said to have an *Exponential*(β) p.d.f. if the p.d.f. is given by

$$f(x) = \frac{1}{\beta} e^{-\frac{x}{\beta}}, \text{ for } x \in (0, \infty) \text{ and } \beta > 0. \tag{3.15}$$

This p.d.f. has mean β, variance β^2 and MGF $(1 - \beta t)^{-1}$ for $t < \frac{1}{\beta}$. The c.d.f. is $F(t) = 1 - e^{-\frac{t}{\beta}}$. Two interesting and useful facts about this p.m.f. are (a) its *memoryless* property and (b) its relation to the geometric p.m.f. mentioned earlier. The fact that $P(X > t + s \mid X > t) = P(X > s) = e^{-\frac{s}{\beta}}$ for any positive t and s is known as the memoryless property (verify it). If Y is defined as the largest integer $\leq X$ (also called the "floor" of X), then $Y + 1$ has a Geometric(p) p.m.f. with $p = 1 - e^{-\frac{1}{\beta}}$. This is because $P(Y + 1 = y + 1) = P(Y = y) = P(y \leq X < y + 1) = (1 - e^{-\frac{1}{\beta}})e^{-\frac{1}{\beta}y}$. Sometimes the exponential p.m.f. is reparametrized by setting $\lambda = \beta^{-1}$, so the p.d.f. looks like $f(x) = \lambda e^{-\lambda x} I(x > 0)$ with mean λ^{-1} and variance λ^{-2}. If $X_1, \ldots, X_n$ are i.i.d. random variables having the p.d.f. in (3.15), then the minimum of $\{X_i\}_{i=1}^{n}$ also has an exponential p.d.f. and the sum $\sum_{i=1}^{n} X_i$ has a *Gamma*$((n, \beta))$ p.d.f.

A random variable X is said to have a *Gamma*(α, β) p.d.f. if the p.d.f. is given by

$$\frac{1}{\Gamma(\alpha)\beta^\alpha}e^{-\frac{x}{\beta}}x^{\alpha-1} \text{ for } x > 0, \, \alpha > 0 \text{ and } \beta > 0, \tag{3.16}$$

where $\Gamma(\alpha) = \int_0^\infty e^{-y}y^{\alpha-1}dy$ is the gamma function with the property that $\Gamma(\nu + 1) = \nu\Gamma(\nu)$. This, in particular, implies that $\Gamma(n + 1) = n!$ for any nonnegative integer n. The parameter α is called the *shape parameter* as the density curve has different shapes for its different values. Its mean is $\alpha\beta$, variance is $\alpha\beta^2$ and MGF is $(1 - \beta t)^{-\alpha}$ for $t < \frac{1}{\beta}$. Clearly, an exponential p.d.f. is a special case of (3.16) with $\beta = 1$. Also, for $\alpha = \frac{n}{2}$ and $\beta = 2$, it is called a *chi-squared* (χ^2) p.d.f. with n *degrees of freedom*, which is widely useful because of the connection that $Z \sim N(0, 1) \implies z^2 \sim \chi^2$ with 1 degree of freedom. The Gamma(α, β) p.d.f. has the *reproductive* property in the sense that if $X_i \sim$ Gamma(α_i, β) for $i = 1, \ldots, n$ and they are independent, then $\sum_{i=1}^n X_i \sim$ Gamma$(\sum_{i=1}^n \alpha_i, \beta)$.

A random variable X is said to have a *Beta*(a, b) p.d.f. if the p.d.f. looks like

$$f(x) = \frac{\Gamma(a + b)}{\Gamma(a)\Gamma(b)}x^{a-1}(1 - x)^{b-1}, \text{ for } x \in (0, 1), a > 0 \text{ and } b > 0. \tag{3.17}$$

This p.d.f. has mean $\frac{a}{a+b}$ and variance $\frac{ab}{(a+b)^2(a+b+1)}$. Notice that for $a = b = 1$, this is nothing but the Uniform[0,1] p.d.f. It also has a connection with the Gamma(α, β) p.d.f. To be precise, if $X \sim$ Gamma(α_1, β), $Y \sim$ Gamma(α_2, β) and X and Y are independent, the ratio $\frac{X}{X+Y}$ will have a Beta(a, b) p.d.f. with $a = \alpha_1$ and $b = \alpha_2$. More generally, if $X_1, \ldots, X_k, X_{k+1}, \ldots, X_m$ are independent random variables with $X_i \sim$ Gamma(α_i, β) $(\beta > 0$ and $\alpha_i > 0$ for $i = 1, \ldots, n)$, then the p.d.f. of $\sum_{i=1}^k X_i / \sum_{i=1}^m X_i$ is Beta(a, b) with $a = \sum_{i=1}^k \alpha_i$ and $b = \sum_{i=k+1}^m \alpha_i$.

So far we have dealt with individual random variables. We now move on to the case where we have a bunch of them at once. If X and Y are two discrete random variables, then the *joint p.m.f.* of the *random vector* (X, Y) is given by $P(\{X = x\} \cap \{Y = y\})$ for all value-pairs (x, y). It can be imagined as a two-dimensional table with the values of X and Y being listed along the left margin and top margin respectively and the joint probabilities being displayed in various cells. So each row of this table corresponds to a particular value of X and each column corresponds to a particular value of Y. If we *collapse* this table column-wise, i.e., add all the probabilities displayed in each row, thereby ending up with a single column of probabilities for the values of X, this single column of probabilities gives us the *marginal* p.m.f. of X. Similarly, by collapsing the table row-wise (i.e., summing all the probabilities in each column), we get the *marginal* p.m.f. of Y. Now, for each fixed value x_i of X, the *conditional* p.m.f. of Y given that $X = x_i$ is nothing but $\{\frac{P(Y=y_j \text{ and } X=x_i)}{P(X=x_i)}\}_{j=1}^m$, assuming that Y takes the values $y_1, \ldots, y_m$. Similarly, for each fixed value y_i of Y, the conditional p.m.f. of X given that $Y = y_i$ can be defined. The raw and central moments and the MGF of X computed using such a conditional p.m.f. will be called its conditional moments and conditional MGF. Similar is the terminology for Y.

All these concepts can be easily generalized to the case of an $n \times 1$ random vector $(X_1, \ldots, X_n)$. The joint p.m.f. of these can be imagined as an n-dimensional table, whose cells contain $P\{\bigcap_{i=1}^{n}(X_i = x_i)\}$ for various values x_i of X_i $(i = 1, \ldots, n)$. We can now talk about several types of marginal and conditional p.m.f.'s, such as the *joint marginal* p.m.f. of $X_{i_1}, \ldots, X_{i_k}$ which is obtained by collapsing the n-dimensional table w.r.t. all other indices except $i_1, \ldots, i_k$, or the *joint conditional* p.m.f. of $X_{i_1}, \ldots, X_{i_k}$ given $X_{j_1}, \ldots, X_{j_l}$ for two disjoint subsets of indices $\{i_1, \ldots, i_k\}$ and $\{j_1, \ldots, j_l\}$ which is obtained by dividing the joint marginal p.m.f. of $X_{i_1}, \ldots, X_{i_k}, X_{j_1}, \ldots, X_{j_l}$ with that of $X_{j_1}, \ldots, X_{j_l}$. We can also talk about joint moments (marginal or conditional). For example, the $(j_1, \ldots, j_k)^{th}$ order joint raw moment of $X_{i_1}, \ldots, X_{i_k}$ is defined as $E(\prod_{s=1}^{k} X_{i_s}^{j_s})$, where the expected value is based on the joint marginal p.m.f. of $X_{i_1}, \ldots, X_{i_k}$. In the case of a conditional joint moment, this expected value would be based on the conditional joint p.m.f. of $X_{i_1}, \ldots, X_{i_k}$ given another disjoint set of variables.

In a situation where we have a bivariate random vector (X, Y), the numerical summary that describes the nature (i.e., direction) of the *joint variability* of X and Y is called the *covariance* (denoted by $COV(X, Y)$ or σ_{XY}). It is defined as

$$COV(X, Y) = E\{(X - \mu_X)(Y - \mu_Y)\} = E(XY) - \mu_X \mu_Y. \qquad (3.18)$$

A positive value of it indicates a *synergistic* or *direct* relation between X and Y, i.e., in general X increases if Y does so. A negative covariance, on the other hand, shows an *antagonistic* or *inverse* relation whereby an increase in X will be associated with a decrease in Y in general. In order to measure the strength of the linear association between X and Y (which is difficult to assess from the covariance since it is an absolute measure, not a relative one), we convert the covariance to *correlation* (denoted by ρ_{XY}). It is defined as

$$\rho_{XY} = \frac{\sigma_{XY}}{\sigma_X \sigma_Y}, \qquad (3.19)$$

where σ_X and σ_Y are the marginal standard deviations of X and Y. It is bounded below and above by -1 and 1 respectively (which is easy to see by the *Cauchy-Schwarz inequality* that says: $E \mid UV \mid \leq (E(U^2))^{\frac{1}{2}}(E(V^2))^{\frac{1}{2}}$ for any random variables U and V), so that any value close to its boundaries is considered evidence of a strong linear association between X and Y. Values close to the center of this range testify to a relatively weak linear association. In the case where we have an $n \times 1$ random vector, there are $_nC_2 = \frac{n(n-1)}{2}$ pairwise covariances to talk about. If we construct an $n \times n$ matrix Σ whose $(i, j)^{th}$ entry is $COV(X_i, X_j) = \sigma_{ij}$ (say), then it will be a symmetric matrix with the ith diagonal entry being the variance of X_i or σ_i^2 $(i = 1, \ldots, n)$. This matrix is referred to as the *variance-covariance matrix* or the *dispersion* *matrix* of the random vector. Sometimes it is more convenient to work with the matrix obtained by dividing the $(i,j)^{th}$ entry of Σ with $\sigma_i\sigma_j$ (i.e., the product of the standard deviations of X_i and X_j). This symmetric $n \times n$ matrix with all diagonal entries $= 1$ is called the *correlation matrix* of $(X_1, \ldots, X_n)$ due to obvious reasons. In case the dispersion matrix Σ has an inverse (an $n \times n$ matrix W is said to be the *inverse* of another $n \times n$ matrix W^* if $WW^* = W^*W = I$, the $n \times n$ identity matrix with all diagonal entries $= 1$ and all off-diagonal entries $= 0$), the inverse is called the *precision matrix*. Here are some examples.

Example 3.19. Suppose the joint p.m.f. of (X, Y) is given by the table

	5	10	15	20
0	$\frac{1}{48}$	$\frac{1}{48}$	$\frac{1}{12}$	$\frac{1}{8}$
1	$\frac{1}{12}$	$\frac{1}{12}$	$\frac{1}{6}$	$\frac{1}{6}$
2	$\frac{1}{8}$	$\frac{1}{12}$	$\frac{1}{48}$	$\frac{1}{48}$

Here, the marginal p.m.f. of X is actually Binomial$(2, \frac{1}{2})$ and that of Y is

value	5	10	15	20
prob	$\frac{11}{48}$	$\frac{9}{48}$	$\frac{13}{48}$	$\frac{15}{48}$

So X has mean$=(2)(1/2)=1$ and variance$=(2)(1/2)(1-1/2)=1/2$, and those for Y are 13.33 and respectively. The conditional p.m.f. of X given that $Y = 15$ is $P(X = 0 \mid Y = 15) = \frac{4}{13}$, $P(X = 1 \mid Y = 15) = \frac{8}{13}$ and $P(X = 2 \mid Y = 15) = \frac{1}{13}$.

Example 3.20. An agency that conducts nationwide opinion polls has decided to take a random sample (with replacement) of 1000 people from the entire U.S. population and ask each sampled individual about the effectiveness of the Kyoto protocol to control global warming. Each individual can say "effective" or "not effective" or "no opinion" (let us classify answers such as "Don't even know what the Kyoto protocol is!" as "no opinion" for the sake of simplicity). If the true (but unknown) percentage of people in the U.S. who think that the Kyoto protocol would be effective is 25%, that of those who consider it ineffective is 35% and the rest are in the third category, what are the chances that in a sample of size 1000, half will have no opinion (including those who are unaware of the whole issue) and an equal number of people will consider the protocol effective or ineffective? Let X and Y be the number of sampled individuals in the "effective" and "ineffective" groups respectively. Then the answer will come from the joint p.m.f. of (X, Y). What is it? Since we are dealing with randomly sampled individuals from a huge population, it is reasonable to assume that each person's opinion is independent of everybody else's. Since there are three possible answers, asking each individual about the Kyoto protocol is like tossing an unbalanced "three-sided coin" for which, the probabilities associated with the three sides are 0.25, 0.35 and 0.4. Also, since it is sampling without replacement, these probabilities remain unchanged from "toss" to "toss." So, just as the probability of k heads in n tosses of a regular (i.e., two-sided) coin with $P(\text{head}) = p$ is $_nC_k p^k (1 - p)^{n-k}$ (see 3.9), where the coefficient $_nC_k$ is nothing but the number of different rearrangements of k identical H's and $n - k$ identical T's, the answer in the present case will be

$$P(X = 250, Y = 250) = \frac{1000!}{(250!)(250!)(500!)}(0.25)^{250}(0.35)^{250}(1-0.25-0.35)^{500},$$

where the coefficient in front is the number of different rearrangements of 250 identical "yes" answers, 250 identical "no" answers and 500 identical "no opinion" answers. In general, when the sample-size is n, $P(X = k_1$ and $Y = k_2)$ has the same expression with 250 and 250 replaced by k_1 and k_2 for any k_1 and k_2 that add up to something $\leq n$. This is known as the *Trinomial*(n, p_1, p_2) p.m.f. (here $n = 1000, p_1 = 0.25$ and $p_2 = 0.35$). It can be easily extended to the situation where the question has $m(> 3)$ possible answers and the true (but unknown) population-proportions associated with those answers are $p_1, \ldots, p_m$ respectively ($\sum_{i=1}^{m} p_i = 1$). If X_i counts the number of sampled individuals that gave the ith answer ($i = 1, \ldots, m - 1$), then the joint p.m.f. of the X_i's looks like

$$P(x_1, \ldots, x_{m-1}) = \frac{n!}{(x_1!) \ldots (x_{m-1}!)((n - \sum_{i=1}^{m-1} x_i)!)} \left(\prod_{i=1}^{m-1} p_i^{x_i} \right) p_m^{n - \sum_{i=1}^{m-1} x_i},$$

for any nonnegative integers $x_1, \ldots, x_{m-1}$ with $\sum_{i=1}^{m-1} x_i \leq n$. This is called the *Multinomial*$(n, p_1, \ldots, p_{m-1})$ p.m.f. Marginally, each X_i is Binomial(n, p_i), so that its mean is np_i and variance is $np_i(1 - p_i)$. The covariance between X_i and X_j ($i \neq j$) is $-np_i p_j$, which intuitively makes sense since a large value of X_i will force X_j to be small in order to keep the sum constant ($= n$).

For continuous random variables, the analogous concepts are defined as follows. The joint p.d.f. $f(x_1, \ldots, x_n)$ of the continuous random variables $X_1, \ldots, X_n$ is a nonnegative-valued function defined on a subset D of $\mathcal{R}^n$ such that its n-fold integral over D is 1. The subset D, which is the Cartesian product of the value-ranges of the individual X_i's, is called the support of this joint p.d.f. The marginal joint p.d.f. of the random variables $X_{i_1}, \ldots, X_{i_k}$ is obtained by integrating the joint p.d.f. with respect to all the other variables over their full ranges. The conditional joint p.d.f. of a subcollection of random variables $\{X_{i_1}, \ldots, X_{i_k}\}$ given another disjoint subcollection $\{X_{j_1}, \ldots, X_{j_l}\}$ is obtained by dividing the marginal joint p.d.f. of the combined collection $X_{i_1}, \ldots, X_{i_k}, X_{j_1}, \ldots, X_{j_l}$ by that of $\{X_{j_1}, \ldots, X_{j_l}\}$ only. The marginal and conditional joint moments are the expected values of products of random variables computed using their marginal and conditional joint p.m.f.'s respectively. Pairwise covariances and correlations are defined in the same way as in (3.18) and (3.19). Here is an example.

Example 3.21. A random vector $\mathbf{X} = (X_1, X_2)$ is said to have a *bivariate normal* joint p.d.f. with mean vector $\boldsymbol{\mu} = (\mu_1, \mu_2)$ and dispersion matrix $\Sigma = \begin{bmatrix} \sigma_1^2 & \rho\sigma_1\sigma_2 \\ \rho\sigma_1\sigma_2 & \sigma_2^2 \end{bmatrix}$ for some $\sigma_1 > 0$, $\sigma_2 > 0$ and $\rho \in (-1, 1)$ if its joint p.d.f. looks like

$$f(x_1, x_2) = \frac{1}{2\pi\sigma_1\sigma_2(1 - \rho^2)^{1/2}} e^{-\frac{1}{2}(\mathbf{X}-\boldsymbol{\mu})\Sigma^{-1}(\mathbf{X}-\boldsymbol{\mu})^t},$$

the superscript "t" meaning "transpose." The exponent $-\frac{1}{2}(\mathbf{X} - \boldsymbol{\mu})\Sigma^{-1}(\mathbf{X} - \boldsymbol{\mu})^t$

$$= -\frac{1}{2(1-\rho^2)}\left[\left(\frac{x_1 - \mu_1}{\sigma_1}\right)^2 - 2\rho\frac{(x_1 - \mu_1)(x_2 - \mu_2)}{\sigma_1\sigma_2} + \left(\frac{x_2 - \mu_2}{\sigma_2}\right)^2\right]$$

is a *quadratic form* in x_1 and x_2. It can be shown by integrating this joint p.d.f. with respect to x_2 over its full range $(-\infty, \infty)$ that the marginal p.d.f. of X_1 is Normal(μ_1, σ_1^2). Similarly, integrating out the variable x_1 will yield the marginal p.d.f. of X_2, which is Normal(μ_2, σ_2^2). From the structure of the dispersion matrix, it should be clear that the correlation between X_1 and X_2 is ρ. Also, the conditional p.d.f. of X_2 given $X_1 = x_1$ is normal with mean and variance

$$E(X_2 \mid X_1 = x_1) = \mu_2 + \rho\frac{\sigma_2}{\sigma_1}(x_1 - \mu_1); \ \mathrm{Var}(X_2 \mid X_1 = x_1) = \sigma_2^2(1 - \rho^2).$$

The above formula for $E(X_2 \mid X_1 = x_1)$ is called the *regression* of X_2 on X_1. Likewise, the conditional p.d.f. of X_1 given $X_2 = x_2$ will be normal with mean $E(X_1 \mid X_2 = x_2) = \mu_1 + \rho\frac{\sigma_1}{\sigma_2}(x_2 - \mu_2)$ and variance $\mathrm{Var}(X_1 \mid X_2 = x_2) = \sigma_1^2(1-\rho^2)$. All these can be generalized to an $n\times 1$ random vector $(X_1, \ldots, X_n)$ whose joint p.d.f. will be called *multivariate normal* (MVN) with mean vector $\boldsymbol{\mu} = (\mu_1, \ldots, \mu_n)$ and dispersion matrix $\Sigma_{n\times n}$. The $(i, j)^{th}$ and $(j, i)^{th}$ entries of Σ would both be $\rho_{ij}\sigma_i\sigma_j$, where ρ_{ij} is the correlation between X_i and X_j. Once again, the marginal p.d.f.'s and the conditional p.d.f.'s will all be normal. The MVN$(\boldsymbol{\mu}, \Sigma)$ joint p.d.f. looks like

$$f(x_1, \ldots, x_n) = \frac{1}{(2\pi)^{n/2}(\mid \Sigma \mid)^{1/2}} e^{-\frac{1}{2}(\mathbf{X}-\boldsymbol{\mu})\Sigma^{-1}(\mathbf{X}-\boldsymbol{\mu})^t},$$

where $\mid \Sigma \mid$ denotes the *determinant* of Σ (see, for example, Leon (1998) for the general definition of the determinant of a $n \times n$ matrix).

This MVN$(\boldsymbol{\mu}, \Sigma)$ p.d.f. and its univariate version (3.14) are members of a much more general class that includes many other familiar p.d.f.'s and p.m.f.'s such as (3.13), (3.15) and (3.16). It is called the *exponential family* of distributions. An $n \times 1$ random vector $\mathbf{X}$ is said to have a distribution in the exponential family if its p.d.f. (or p.m.f.) has the form

$$f(\mathbf{x}) = c(\boldsymbol{\theta})d(x)e^{\sum_{i=1}^{k} \pi_i(\boldsymbol{\theta})t_i(\mathbf{x})} \tag{3.20}$$

for some (possibly vector) parameter $\boldsymbol{\theta}$ and some functions $\pi_1, \ldots, \pi_k, t_1, \ldots, t_k,$ c and d. The vector $(\pi_1(\boldsymbol{\theta}), \ldots, \pi_k(\boldsymbol{\theta}))$ is called the *natural parameters* for (3.20). Often (3.20) is written in a slightly different way as:

$$f(\mathbf{x}) = \exp\left\{\sum_{i=1}^{k} \pi_i(\boldsymbol{\theta})t_i(\mathbf{x}) + C(\boldsymbol{\theta}) + D(\mathbf{x})\right\}, \qquad (3.21)$$

where $C(\boldsymbol{\theta}) = \ln c(\boldsymbol{\theta})$ and $D(\mathbf{x}) = \ln d(\mathbf{x})$. This is known as the *canonical form* of the exponential family. Can you identify the natural parameters $\boldsymbol{\theta}$ and the functions $C(.), D(.), \pi_i(.)$ and $t_i(.)$ $(i = 1, \ldots, n)$ for the p.m.f.'s and p.d.f.'s in (3.13)-(3.16)?

Two quick notes before we close our discussion of multivariate random vectors. From the definition of covariance, it should be clear that two independent random variables have zero covariance (and hence, zero correlation). So, if the components of a random vector $\{X_1, \ldots, X_n\}$ are independent random variables, their dispersion matrix Σ is diagonal (i.e., has zeroes in all off-diagonal positions). However, the converse is not necessarily true. There exist random vectors with dependent components that have diagonal dispersion matrices (can you construct such an example?). Also, for a random vector with independent components, the joint p.m.f. (or p.d.f.) turns out to be the *product* of the individual p.m.f.'s (or p.d.f.'s). And in general, the joint p.d.f. $f(x_1, \ldots, x_n)$ of a continuous random vector $(X_1, \ldots, X_n)$ can be written as

$$f(x_1)f(x_2 \mid X_1 = x_1)f(x_3 \mid X_1 = x_1 \& X_2 = x_2) \ldots f(x_n \mid X_1 = x_1, \ldots, X_{n-1}$$
$$(3.22)$$

$= x_{n-1})$, which is analogous to (3.2).

Next we turn to the relation between the p.m.f.'s (or p.d.f.'s) of two random variables that are functionally related. We will explore some methods of deriving the p.m.f. (or p.d.f.) of a function of a random variable that has a familiar p.m.f. or p.d.f. If we were only interested in computing expected values, this would not be essential, since for a discrete random variable X with values $x_1, x_2, \ldots$ and corresponding probabilities $p_1, p_2, \ldots$, the expected value of $Y = g(X)$ is simply $E(g(X)) = \sum_i g(x_i)p_i$. Likewise, for a continuous random variable X with support D and p.d.f. $f(x)$, we have $E(g(X)) = \int_D g(x)f(x)dx$. But sometimes we do need to know the exact p.m.f. or p.d.f. of $g(X)$ and there are three main methods to derive it from that of X: (a) the *c.d.f.* method, (b) the *Jacobian* method and (c) the *MGF* method. We illustrate each of these by means of an example.

Example 3.22. Let X have an Exponential(β) p.d.f. Suppose we want to derive the p.d.f. of $Y = aX + b$ for some $a > 0$ and $b \in \mathcal{R}$. Let us start with its c.d.f. $G(t)$ (say). $G(t) = P(Y \le t) = P(aX + b \le t) = P(X \le \frac{t-b}{a}) = 1 - e^{-\frac{(t-b)/a}{\beta}}$. So the p.d.f. of Y will be $G'(t) = \frac{1}{a\beta}e^{-\frac{(t-b)}{a\beta}}$. This is sometimes called the *negative exponential* p.d.f. with location parameter b and scale parameter $a\beta$.

Example 3.23. We will once again derive the p.d.f. of $Y = aX + b$ where $X \sim$ Exponential(β), but by a different method this time. If we call the mapping $X \longrightarrow aX + b = Y$ the *forward* transformation, the *inverse* transformation will be $Y \longrightarrow \frac{Y-b}{a} = X$. The *Jacobian* of this inverse transformation is the quantity $\mid \frac{d}{dy} \frac{y-b}{a} \mid$, which is simply $\frac{1}{a}$ here. Now, if we write the p.d.f. of X having replaced all the x's in it by $\frac{y-b}{a}$ and then multiply it with the Jacobian obtained above, we get $(\frac{1}{\beta} e^{-\frac{(y-b)/a}{\beta}})(\frac{1}{a})$. This is precisely the p.d.f. we got in Example 3.22.

Example 3.24. It was mentioned earlier that moment generating functions have a one-to-one correspondence with p.m.f.'s (or p.d.f.'s), so that "MGF fingerprinting" is possible. For instance, an MGF of the form $M(t) = e^{t\mu + t^2 \sigma^2/2}$ corresponds only to a Normal(μ, σ^2) p.d.f. So, if $X \sim$ Normal(μ, σ^2), what is the p.d.f. of $Y = aX + b$ for some $a > 0$ and $b \in \mathcal{R}$? Let us first find its MGF. $M_Y(t) = E(e^{tY}) = E(e^{t(aX+b)}) = E(e^{taX} e^{tb}) = e^{tb} E(e^{taX})$, the last equality following from the linearity property of expected values mentioned earlier in this section. But $E(e^{taX})$ is nothing but the MGF of X evaluated at "ta," i.e., it is $M_X(ta) = e^{ta\mu + t^2 a^2 \sigma^2/2}$. So $M_Y(t) = e^{tb} e^{ta\mu + t^2 a^2 \sigma^2/2} = e^{t(a\mu+b) + t^2 a^2 \sigma^2/2}$. This corresponds only to a Normal$(a\mu+b, a^2\sigma^2)$ p.d.f., which must be the p.d.f. of Y.

Example 3.25. Let X be a continuous random variable with a p.d.f. $f(x)$ and a (strictly monotonically increasing) c.d.f. $F(t)$. What will be the p.d.f. of $Y = F(X)$? Let us derive it via the c.d.f. method. The c.d.f. of Y is $G(t) = P(Y \leq t) = P(F(X) \leq t)$. Take some $t \in (0, 1)$. If F^{-1} is the inverse function of F (two functions $h(x)$ and $k(x)$ are called inverses of one another if $h(k(x)) = x$ for all x in the domain of k and $k(h(x)) = x$ for all x in the domain of h), then $P(F(X) \leq t) = P(F^{-1}(F(X)) \leq F^{-1}(t)) = P(X \leq F^{-1}(t))$. But this is, by definition, $F(F^{-1}(t))$, which is nothing but t. In other words, $G(t) = t$ for all $t \in (0, 1)$. Also, obviously, $G(t) = 0$ for $t \leq 0$ and $G(t) = 1$ for $t \geq 1$. Can you identify this c.d.f.? It should be clear from (3.6) that $G(t)$ is the Uniform(0,1) c.d.f. (verify it through differentiation). In other words, $F(X)$ has a Uniform(0,1) p.d.f. The transformation $X \longrightarrow F(X)$ is known as the *probability integral transform*. Turning it around, let us now start with an $X \sim$ Uniform(0,1) and let $Y = F^{-1}(X)$. What distribution will it have? Its c.d.f. is $P(Y \leq t) = P(F^{-1}(X) \leq t) = P(F(F^{-1}(X)) \leq F(t)) = P(X \leq F(t)) = F(t)$, the last equality following from the nature of the Uniform(0,1) c.d.f. Hence we conclude that Y has the c.d.f. $F(t)$.

This last example is particularly important in the context of statistical simulation studies, since it enables one to generate random samples from various probability distributions that have closed-form expressions for their c.d.f.'s. In order to generate n i.i.d. random variables $X_1, \ldots, X_n$ from the Exponential(β) p.d.f., simply generate n i.i.d. Uniform(0,1) random variables $U_1, \ldots, U_n$ (which computers can readily do) and set $X_i = F^{-1}(U_i)$ $(i = 1, \ldots, n)$, where

$F(t) = 1 - e^{-x/\beta}$. Even if the c.d.f. is not strictly monotonically increasing (so that its inverse does not exist), as is the case for discrete random variables, the above technique can be slightly modified to serve the purpose.

We conclude this section by listing a number of useful probability inequalities that are relevant to our book. The second one is a generalization of the Cauchy-Schwarz inequality mentioned earlier.

Result 3.3. Let X and Y be random variables such that $E(| X |^r)$ and $E(| Y |^r)$ are both finite for some $r > 0$. Then $E(| X + Y |^r)$ is also finite and $E(| X + Y |^r) \leq c_r(E(| X |^r) + E(| Y |^r))$, where $c_r = 1$ if $0 < r \leq 1$ or $c_r = 2^{r-1}$ if $r > 1$. This is often called the c_r inequality.

Result 3.4. Let $p > 1$ and $q > 1$ be such that $\frac{1}{p} + \frac{1}{q} = 1$. Then $E(| XY |)$ $\leq (E(| X |^p))^{1/p}(E(| Y |^q))^{1/q}$. This is well known as the Holder inequality. Clearly, taking $p = q = 2$, we get the Cauchy-Schwarz inequality.

Result 3.5. For any number $p \geq 1$, $(E(| X + Y |^p))^{1/p} \leq (E(| X |^p))^{1/p} + (E(| Y |^p))^{1/p}$. This is popularly known as the Minkowski inequality.

Result 3.6. $P(| X | \geq \varepsilon) \leq \frac{E(|X|^r)}{\varepsilon^r}$ for any $\varepsilon > 0$ and $r > 0$ such that $E(| X |^r)$ is finite. This is known as Markov's inequality. If Y is a random variable with mean μ and standard deviation σ, then letting $Y - \mu$ and $c\sigma$ play the roles of X and ε respectively in the above inequality with $r = 2$, we get: $P(| Y - \mu | \geq c\sigma) \leq \frac{1}{c^2}$. This is called the Chebyshev-Bienayme inequality (or simply Chebyshev's inequality).

3.4 Basics of Information Theory

Let X be a discrete random variable with values $x_1, x_2, \dots$ and corresponding probabilities $p_1, p_2, \dots$ Then the *entropy* H of this p.m.f. (call it P) is defined as

$$H(P) = -\sum_i p_i \log(p_i), \qquad (3.23)$$

where "log" denotes natural logarithm (although logarithm with base 2 is used in some applications). It is easy to see that $H(P)$ is 0 when the p.m.f. is degenerate (i.e., takes a single value with probability 1). It can also be verified that the entropy is the maximum for an 'equally likely' p.m.f. where all the p_i's are the same. This indicates that $H(P)$ measures the "extent of information" in a p.m.f. by assessing the "unevenness" in the probabilities. The more "uneven" they are, the more "specialized information" we get about the uncertainty in the associated random variable X. Notice that $H(P)$ is not measuring the *variability* in the values of X as it does not involve the values at all.

If we have two discrete random variables X and Y with the same support or value-set $\{z_1, z_2, \ldots\}$ but different probabilities ($\{p_1, p_2, \ldots\}$ and $\{p_1^*, p_2^*, \ldots\}$ respectively), we can define the *relative entropy* of one w.r.t. the other. Denoting these two p.m.f.'s by P_X and P_Y respectively, the two possible relative entropies are defined as

$$H_{X\|Y}(P_X, P_Y) = \sum_i p_i \log \frac{p_i}{p_i^*}, \quad H_{Y\|X}(P_X, P_Y) = \sum_i p_i^* \log \frac{p_i^*}{p_i}. \qquad (3.24)$$

These are also known as the *Kullback-Leibler divergences* between P_X and P_Y. $H_{X\|Y}$ and $H_{Y\|X}$ are *not* equal in general, and that is one of the reasons why it is not a proper *metric* (or distance measure) between two p.m.f.'s, the other reason being its failure to satisfy the *triangle inequality*. In any case, it is good enough as a measure of the 'dissimilarity' between two p.m.f.'s, since it is 0 if and only if the two p.m.f.'s are identical. The fact that it is > 0 otherwise can be verified using the power-series expansion $\log \frac{1}{1-x} = x + \frac{x^2}{2} + \frac{x^3}{3} + \ldots$ for $\mid x \mid < 1$, which in particular implies that $-\log(1-x) \geq x$.

The continuous counterparts of the entropy and the relative entropy are defined analogously. For a continuous random variable X with support D and p.d.f. $f(x)$, the entropy is $H(f) = -\int_D f(x) \log f(x) dx$. For two continuous random variables X and Y with common support D and p.d.f.'s $f(x)$ and $g(x)$ respectively, the two relative entropies are $H_{X\|Y}(f, g) = \int_D f(x) \log \frac{f(x)}{g(x)} dx$ and $H_{Y\|X}(g, f) = \int_D g(x) \log \frac{g(x)}{f(x)} dx$.

Example 3.26. (Schervish (1996)). Let $f(x)$ be the standard normal p.d.f. and $g(y)$ be the *bilateral exponential* or *Laplace* p.d.f. with parameters 0 and 1 (i.e., $g(y) = \frac{1}{2} e^{-|y|}$) on $\mathcal{R}$. Suppose $X \sim f(x)$ and $Y \sim g(y)$. Then it can be shown that $E(X^2) = 1$, $E(\mid X \mid) = (\frac{2}{\pi})^{1/2}$, $E(Y^2) = 2$ and $E(\mid Y \mid) = 1$. Since $\log \frac{f(x)}{g(x)} = \frac{1}{2} \log \frac{2}{\pi} - \frac{1}{2} x^2 + \mid x \mid$, it turns out that $H_{X\|Y}(f, g) = \frac{1}{2} \log \frac{2}{\pi} - \frac{1}{2} + (\frac{2}{\pi})^{1/2} = 0.07209$ and $H_{Y\|X}(g, f) = -\frac{1}{2} \log \frac{2}{\pi} = 0.22579$. In other words, if our data actually come from a bilateral exponential p.d.f., it is easier to figure out that they did not come from a standard normal p.d.f. than the other way around.

Suppose, now, that $X \sim f(x)$ where $f(x)$ involves an m-dimensional parameter $\boldsymbol{\theta}$ in some parameter space $\boldsymbol{\Theta}$. To emphasize this, we will write it as $f(x; \boldsymbol{\theta})$. Suppose also that for each i ($i = 1, \ldots, m$), the partial derivative of $f(x; \boldsymbol{\theta})$ w.r.t. θ_i exists and the conditions necessary for interchanging the operations of differentiation and integration are satisfied while taking the derivative of $\int f(x; \boldsymbol{\theta}) dx$ w.r.t. each θ_i. Then $m \times m$ matrix whose $(i, j)^{th}$ entry is the covariance between the partial derivatives of $\log f(x; \boldsymbol{\theta})$ w.r.t. θ_i and θ_j is called the *Fisher information matrix* (FIM) for $\boldsymbol{\theta}$ based on X. The $m \times 1$ random vector whose i^{th} coordinate is the partial derivative of $\log f(x; \boldsymbol{\theta})$ w.r.t. θ_i is often called the *score function*. So the FIM is simply the variance-covariance matrix of the score function.

Example 3.27. Let $X \sim \text{Binomial}(n, p)$. Then its p.m.f. looks like $P(X = x) = {}_nC_x p^x (1-p)^{n-x}$ and the log of it is $\log {}_nC_x + x\log p + (n-x)\log(1-p)$. So $\frac{d}{dp}\log(p.m.f.) = \frac{x}{p} - \frac{n-x}{1-p} = \frac{x-np}{p(1-p)}$. So the Fisher information in this case is $\frac{n}{p(1-p)}$. Since $p \in (0, 1)$, the largest value of $p(1-p)$ is 0.25, which corresponds to the minimum Fisher information. The closer p is to 0 or 1, the higher is the information.

3.5 Basics of Stochastic Processes

Suppose you are a meteorologist at a local radio station and you read your weather bulletin every half hour around the clock everyday, which includes a report on current weather conditions. The weather condition that you report at a particular time of the day (say, 7:00 A.M.) is, in some sense, a 'random variable' because it can be different on different days (sunny, foggy, rainy, muggy, clear but cool, etc.). Perhaps today it is sunny at that early morning hour, but 24 hours from now, it will be raining at 7:00 A.M. In other words, there may be a *transition* from one state of weather to another in the 24-hour period. A similar transition may occur between tomorrow and the day after tomorrow. So your life-long collection of weather reports at 7:00 A.M. for that radio station (assuming that you did not get a better job!) is nothing but a sequence of random variables, each of which takes 'values' in the same 'value-set' and successive random variables in this sequence may have different 'values.' Also, the weather condition at 7:00 A.M. today may be influenced by how the weather was 24 hours ago or 48 hours ago, but it is unlikely to be affected by the morning weather condition a fortnight ago or a month ago. In other words, it is reasonable to assume that the 'value' of any random variable in your sequence is dependent on a limited number of adjacent random variables in the recent past, but not on the entire "history."

More formally, a sequence of random variables $\{X_1, X_2, X_3, \ldots\}$ with the same support or value-set (say, $V = \{v_1, v_2, v_3, \ldots\}$) is called a *stochastic process*. V is called its *state space*. The index $i = 1, 2, 3, \ldots$ usually refers to time. Many stochastic processes have the property that for any indices i and j, the conditional probability of $X_i = v_j$ given the entire past history (i.e., $X_1, \ldots, X_{i-1}$) is only a function of $X_{i-1}, \ldots, X_{i-m}$ for some fixed m. If $m = 1$, the stochastic process is called a *Markov chain* (after the famous Russian mathematician A.A. Markov). For a Markov chain, the conditional probabilities of state-to-state transitions, i.e., $P(X_i = v_j \mid X_{i-1} = v_k) = p_{kj}^{(i)}$ (say) are often assumed to be independent of the time-index i so we can drop the superscript "(i)." In that case, it is called a *time-homogeneous* Markov chain. For such a chain, these p_{kj}'s can be arranged in a matrix A whose $(k, j)^{th}$ entry is p_{kj}. This A is called the *one-step transition matrix*. Each row of A sums up to 1, since there will definitely be a transition to *some* state from the current state. Such a matrix is called a *row stochastic* matrix. For some Markov chains, the columns of A also add up to 1 (i.e., a transition is guaranteed to each state from some other state). In that case, it is called a *doubly stochastic* matrix. Notice that, although the current state of a Markov chain only has a direct influence on where the chain will be at the very next time-point, it is possible to compute probabilities of future transitions (e.g., 2-step or 3-step transitions) by multiplying the matrix A with itself repeatedly. For example, the $(k, j)^{th}$ entry of A^2 is $\sum_i p_{ki} p_{ij}$ and this is nothing but $P(X_{n+2} = v_j \mid X_n = v_k)$. To understand why, write it as

$$\frac{P(X_{n+2} = v_j \ \& \ X_n = v_k)}{P(X_n = v_k)} = \frac{\sum_i P\{(X_{n+2} = v_j) \cap (X_{n+1} = v_i) \cap (X_n = v_k)\}}{P(X_n = v_k)}$$

and notice that the latter quantity is nothing but

$$\sum_i P\{X_{n+2} = v_j \mid (X_{n+1} = v_i) \cap (X_n = v_k)\} P(X_{n+1} = v_i \mid X_n = v_k),$$

which, by the Markov property, is $\sum_i P(X_{n+2} = v_j \mid X_{n+1} = v_i) P(X_{n+1} = v_i \mid X_n = v_k) = \sum_i p_{ij} p_{ki}$. Similarly, it can be shown that $P(X_{n+m} = v_j \mid X_n = v_k)$ is the $(j, k)^{th}$ entry of A^m. Now we will see some examples.

Example 3.28. In a laboratory, a certain kind of bacteria are grown in a nutrient-rich medium and monitored every half hour around the clock. The bacteria can either be in a dormant state (call it state 1) or a vegetative growth state (state 2) or a rapid cell division state (state 3), or it can be dead (state 4). Suppose it is reasonable to assume that the observed state of the bacteria colony at any particular inspection time is determined solely by its state at the previous inspection time. If the colony is in state 1 now, there is a 60% chance that it will move to state 2 in the next half hour, a 30% chance that it will move to state 3 and a 10% chance that it will move to state 4. If it is in state 2 at the moment, the chances of it moving to state 1, state 3 and state 4 in the next half hour are 70%, 25% and 5% respectively. If it is in state 3 at present, the chances of it moving to state 1, state 2 and state 4 are 20%, 78% and 2% respectively. However, once it gets to state 4, there is no moving back to any other state (such a state is known as an *absorbing barrier*). So the one-step transition probability matrix for this Markov chain is

$$A = \begin{bmatrix} 0 & 0.6 & 0.3 & 0.1 \\ 0.7 & 0 & 0.25 & 0.05 \\ 0.2 & 0.78 & 0 & 0.02 \\ 0 & 0 & 0 & 1 \end{bmatrix}$$

Notice that the above transition scheme does not allow any "self-loop" for the states 1, 2 and 3, but it would if the first three diagonal entries were positive. In any case, the two-step transition probabilities will be given by A^2 which is

$$\begin{bmatrix} 0 & 0.6 & 0.3 & 0.1 \\ 0.7 & 0 & 0.25 & 0.05 \\ 0.2 & 0.78 & 0 & 0.02 \\ 0 & 0 & 0 & 1 \end{bmatrix} \begin{bmatrix} 0 & 0.6 & 0.3 & 0.1 \\ 0.7 & 0 & 0.25 & 0.05 \\ 0.2 & 0.78 & 0 & 0.02 \\ 0 & 0 & 0 & 1 \end{bmatrix} = \begin{bmatrix} 0.48 & 0.234 & 0.15 & 0.136 \\ 0.05 & 0.615 & 0.21 & 0.125 \\ 0.546 & 0.12 & 0.255 & 0.079 \\ 0 & 0 & 0 & 1 \end{bmatrix}$$

Example 3.29. Here is a Markov chain with a countably infinite state space. Suppose an ant is crawling on an infinite sheet of graphing paper that has an one-inch grid (i.e., has horizontal and vertical lines printed on it with each two consecutive horizontal or vertical lines being 1 inch apart). Starting from the origin at time 0, the ant crawls at a speed of an inch a minute and at each grid-point (i.e., the intersection of a vertical and a horizontal line), it randomly chooses a direction from the set { left, right, vertically up, vertically down }. If we observe the ant at one-minute intervals and denote its position after i minutes by X_i, then $\{X_1, X_2, \ldots\}$ is a Markov chain with state space $\{(j, k) : j \in \mathbb{Z}, k \in \mathbb{Z}\}$, where $\mathbb{Z}$ is the set of all integers. Given that it is currently at the grid-point (j^*, k^*), the one-step transition probability to any of its four nearest neighbor grid-points is $\frac{1}{4}$ and the rest are all zeroes. This is known as a *two-dimensional random walk*.

The k-step transition probability matrix $(k \geq 1)$ of a Markov chain is filled with the *conditional* probabilities of moving from one state to another in k steps, but what about the *unconditional* probabilities of being in various states after k steps? To get those, we need an *initial distribution* for the chain which specifies the probabilities of it being in various states at time 0. In our bacteria colony example, suppose the probability that the colony is in a dormant state at time 0 is $\pi_0(1) = 0.7$, that it is in a vegetative growth state is $\pi_0(2) = 0.2$ and that it is in a rapid cell-division state is $\pi_0(3) = 0.1$ $(\pi_0(4)$ being zero). Let us write it as $\pi = (0.7, 0.2, 0.1, 0)$. Then the (unconditional) probability that the chain will be in a dormant state (state 1) at time 1 is $\sum_{i=1}^4 P(\text{state 1 at time 1 \& state i at time 0})$

$$= \sum_{i=1}^4 P(\text{state 1 at time 1} \mid \text{state i at time 0})P(\text{state i at time 0}), \quad (3.25)$$

which is nothing but $\sum_{i=1}^4 a_{i1}\pi_0(i)$ or the first entry of the 4×1 vector πA. In this case, it is 0.16. Similarly, the (unconditional) probability that the chain will be in state j $(j = 2, 3, 4)$ is the jth entry of πA. It should be clear from (3.25) that the (unconditional) probabilities of being in various states at time k are given by πA^k.

A Markov chain with a finite state-space is said to be *regular* if the matrix A^k has all nonzero entries for some $k \geq 1$. One can prove the following important result about regular Markov chains (see, for example, Bhat (1984)):

Result 3.7. Let $\{X_1, X_2, \ldots\}$ be a regular Markov chain with an $m \times m$ one-step transition probability matrix A. Then there exists an $m \times m$ matrix A^* with identical rows and nonzero entries such that $\lim_{k\to\infty} A^k = A^*$.

The entries in any row of this limiting matrix (with identical rows) are called the *steady-state* transition probabilities of the chain. Let us denote the $(i, j)^{th}$ entry a_{ij}^* of this matrix simply by a_j^*, since it does not vary with the row-index i. The vector $(a_1^*, \ldots, a_m^*)$ represents a p.m.f. on the state space, since

$\sum_{j=1}^{m} a_j^* = 1$. It is called the *steady state distribution* (let us denote it by $\boldsymbol{\Pi}$). If k is large enough so that A^k is equal to or very close to A^*, multiplying A^k with A will yield A^* again because $A^k A = A^{k+1}$ will also be exactly or approximately equal to A^*. Since the rows of A^* are nothing but $\boldsymbol{\Pi}$, this explains the following result:

Result 3.8. Let $\{X_1, X_2, \ldots\}$ be a regular Markov chain with one-step transition matrix A. Its steady-state distribution $\boldsymbol{\Pi}$ can be obtained by solving the system of equations: $\boldsymbol{\Pi} A = \boldsymbol{\Pi}$, $\boldsymbol{\Pi} \mathbf{1}^t = 1$ (where $\mathbf{1}^t$ is the $m \times 1$ column vector of all ones).

Example 3.29. Consider a two-state Markov chain with $A = \begin{bmatrix} 0.6 & 0.4 \\ 0.2 & 0.8 \end{bmatrix}$.
In order to find its steady-state distribution, we need to solve the equations $0.6a_1^* + 0.2a_2^* = a_1^*$, $0.4a_1^* + 0.8a_2^* = a_2^*$ and $a_1^* + a_2^* = 1$. The solutions are $a_1^* = \frac{1}{3}$ and $a_2^* = \frac{2}{3}$.

We conclude our discussion of Markov chains by stating a few more results and introducing a few more concepts.

Result 3.9. Let $\{X_1, X_2, \ldots\}$ be a regular Markov chain with steady-state distribution $\boldsymbol{\Pi} = (a_1^*, \ldots, a_m^*)$. If τ_i denotes the time taken by the chain to return to state i, given that it is in state i right now, then $E(\tau_i) = \frac{1}{a_i^*}$ for $i = 1, \ldots, m$.

Result 3.10. Recall from Example 3.28 that a state i is called an *absorbing barrier* if the $(i, i)^{th}$ entry in the one-step transition probability matrix A is 1. Let B be the set of all absorbing barriers in the (finite) state space of a Markov chain. Assume that there is a *path* from every state outside B to at least one state in B (a path being a sequence of states $i_1 i_2 \ldots i_L$ such that $p_{i_j i_{j+1}} > 0$ for $1 \le j \le L-1$). Then, letting ν_s denote the time needed by the chain to go from a state s outside B to some state in B, we have: $P(\nu_s \le k) = \sum_{r \in B} p_{ir}^{(k)}$, where $p_{ir}^{(k)}$ is the $(i, r)^{th}$ element of A^k. This ν_s is often called the *time to absorption* of a non-absorbing state s.

For any state s in the state space of a Markov chain $\{X_1, X_2, \ldots\}$, define $T_s^{(0)} = 0$ and $T_s^{(k)} = \inf\{n > T_s^{(k-1)} : X_n = s\}$ for $k \ge 1$. This $T_s^{(k)}$ is usually known as the time of the kth return to s. Let $\eta_{rs} = P(T_s^{(1)} < \infty \mid X_0 = r)$. Then intuitively it should be clear that $P(T_s^{(k)} < \infty \mid X_0 = r) = \eta_{rs} \eta_{ss}^{(k-1)}$. In other words, if we start at r and want to make k visits to s, we first need to go to s from r and then return $k - 1$ times to s. A state s is said to be *recurrent* if $\eta_{ss} = 1$. If $\eta_{ss} < 1$, it is called *transient*. A subset Δ of the state space is called *irreducible* if $r \in \Delta$ and $s \in \Delta \implies \eta_{rs} > 0$. A subset Δ^* of the state space is called *closed* if $r \in \Delta^*$ and $\eta_{rs} > 0 \implies s \in \Delta^*$.

Result 3.11. (Contagious nature of recurrence). If r is a recurrent state and $\eta_{rs} > 0$, then s is recurrent as well and $\eta_{sr} = 1$.

Result 3.12. (Decomposition of the recurrent states). If $R = \{r : \eta_{rr} = 1\}$ is the collection of recurrent states in a state space, then R can be written as a disjoint union $\cup_i R_i$ where each R_i is irreducible and closed.

A stochastic process $\{X_0, X_1, X_2, \ldots\}$ is called *stationary* if for any two non-negative integers k and n, the joint distributions of the random vectors $(X_0, \ldots, X_n)$ and $(X_k, \ldots, X_{k+n})$ are the same. Let $\{X_0, X_1, X_2, \ldots\}$ be a Markov chain with steady-state distribution $\Pi = (a_1^*, a_2^*, \ldots)$. If the initial distribution (i.e., the distribution of X_0) is also Π, then the chain will be stationary. This, incidentally, is the reason why the steady-state distribution is also often called the *stationary distribution*.

Suppose a Markov chain $\{X_1, X_2, \ldots\}$ has a stationary distribution $\Pi = (a_1^*, a_2^*, \ldots)$ such that $a_i^* > 0$ for all i. It can be shown that all its states will be recurrent and we will have a decomposition of the state space as suggested in Result 3.11. If, in that decomposition, there is only one component (i.e., if the entire sample space is irreducible), the chain is called *ergodic*. Actually, the definition of ergodicity is much more general and under the above conditions, the irreducibility and the ergodicity of a Markov chain are equivalent. Since the full generality is beyond the scope of this book, we use this equivalence to define ergodicity.

3.6 Hidden Markov Models

Let us now stretch our imagination a little bit more and think of a scenario where $\{X_1, X_2, \ldots\}$ is a Markov chain but is no longer directly observable. Instead, when the event $\{X_i = s\}$ occurs for any state s, we only get to see a 'manifestation' of it. The question that immediately comes to mind is how accurately one can guess (or 'estimate') the true state of the underlying Markov chain. Going back to your daily job as a meteorologist at the local radio station, suppose a person is not directly listening to your 7:00 A.M. weather report (perhaps he/she does not have a radio in his/her room or is too lazy to get off the bed and turn it on). Instead, he/she is trying to guess the content of your report by watching his/her family members' reactions. For example, if the person hears his/her spouse calling her/his office to announce a late arrival plan this morning, it may be because a torrential downpour is going on and the rush-hour traffic will be a mess. Or it may be because a wonderful, sunny morning has inspired his/her spouse to spend a few hours at the local golf course before reporting to work today. More formally, when the underlying chain visits the state s, an observable manifestation M_i is chosen according to a p.m.f. on the set of possible manifestations $\mathbf{M} = \{M_1, \ldots, M_L\}$. This is actually a conditional p.m.f. given s. So there are five items to keep in mind here: the state space $\{s_1, s_2, \ldots, s_m\}$ of the underlying Markov chain, its one-step transition probability matrix A, the set of manifestations $\mathbf{M}$ (also sometimes called an *alphabet* of *emitted letters*), the conditional p.m.f. $\mathbf{P}_i = (p_i(M_1), \ldots, p_i(M_L))$ on $\mathbf{M}$ given the true underlying state s_i and the initial distribution $\boldsymbol{\pi} = (\pi_1, \ldots, \pi_m)$.

Example 3.30. Suppose you are playing a dice game with a partner who occasionally cheats. The rule of the game is that you earn $ 100 from your partner if you roll a 4, 5 or 6 and you pay him/her $ 100 if any of the other three numbers turn up. If played with a 'perfect' or 'balanced' die, it is a fair game, which is easy to see once you write down the p.m.f. table for your earning and compute the expected value. But your partner, who supplies the die, is occasionally dishonest and secretly switches to an unbalanced die from time to time. The unbalanced die has $P(1) = P(2) = P(3) = \frac{1}{4}$ and $P(4) = P(5) = P(6) = \frac{1}{12}$. If the latest roll has been with the balanced die, he/she will switch to the unbalanced one for the next roll with probability 0.05 (i.e., stay with the balanced one with probability 0.95). On the other hand, the chances of his/her switching back from the unbalanced one to the balanced one is 0.9. Here, if Y_i denotes the outcome of your i^{th} roll, then $\{Y_1, Y_2, \ldots\}$ follows a hidden Markov model (HMM) with an underlying Markov chain $\{X_1, X_2, \ldots\}$ whose state space is { balanced, unbalanced } and one-step transition probability matrix is $A = \begin{bmatrix} 0.95 & 0.05 \\ 0.9 & 0.1 \end{bmatrix}$. The alphabet of emitted letters here is $\{1, 2, 3, 4, 5, 6\}$ and the conditional p.m.f. on it switches between $\{\frac{1}{6}, \frac{1}{6}, \frac{1}{6}, \frac{1}{6}, \frac{1}{6}, \frac{1}{6}\}$ and $\{\frac{1}{4}, \frac{1}{4}, \frac{1}{4}, \frac{1}{12}, \frac{1}{12}, \frac{1}{12}\}$, depending on the true nature of the die used. If the chances are very high (say, 99%) that your partner will not begin the game by cheating, the initial distribution for the underlying chain will be $(\pi_1 = 0.99, \pi_2 = 0.01)$.

For an HMM like this, there are three main questions: (a) Given A, $\mathbf{P}$ and $\boldsymbol{\pi} = (\pi_1, \ldots, \pi_m)$, how to efficiently compute the likelihood (or joint probability) of the observed manifestations $\{Y_1, Y_2, \ldots, Y_K\}$? (b) Given $\{Y_1, Y_2, \ldots\}$, how to efficiently estimate the true state-sequence $\{x_1, x_2, \ldots\}$ of the underlying Markov chain with reasonable accuracy? (c) Given the 'connectivity' or 'network structure' of the state space of the underlying chain (i.e., which entries of A will be positive and which ones will be 0), how to efficiently find the values of A, $\mathbf{P}$ and $\boldsymbol{\pi}$ that maximize the observed likelihood mentioned in (a)? Here we will briefly outline the algorithms designed to do all these. More details can be found, for example, in Ewens and Grant (2001).

The forward and backward algorithms: The goal is to efficiently compute the likelihood of the observed Y_i's given A, $\mathbf{P}$ and $\boldsymbol{\pi}$. We begin by computing $\gamma_i^{(k)} = P(Y_1 = y_1, \ldots, Y_k = y_k$ and $X_k = s_i)$ for $1 \leq i \leq m$ and $1 \leq k \leq K$, because the desired likelihood is nothing but $\sum_{i=1}^{m} \gamma_i^{(K)}$. The first one, $\gamma_i^{(1)}$, is clearly $\pi_i p_i(y_1)$. Then, realizing that

$$\gamma_i^{(k+1)} = \sum_{j=1}^m P(Y_1 = y_1, \ldots, Y_{k+1} = y_{k+1}, X_k = s_j \text{ and } X_{k+1} = s_i),$$

we immediately get the *induction* relation: $\gamma_i^{(k+1)} = \sum_{j=1}^m \gamma_j^{(k)} p_{ji} p_i(y_{k+1})$ where p_{ji} is the $(j, i)^{th}$ entry of A. This is an expression for $\gamma_i^{(k+1)}$ in terms of $\{\gamma_j^{(k)}; j = 1, \ldots, m\}$. So, first we compute $\gamma_i^{(1)}$ for $i = 1, \ldots, m$ and then, using the induction formula, compute $\gamma_i^{(2)}$ for all i's which will subsequently produce the values of $\gamma_i^{(3)}$'s through the induction formula again. Continuing in this manner, we will get the $\gamma_i^{(K)}$'s for all i's and then we are done. This is called the *forward* algorithm because of the forward induction it uses.

For the *backward* algorithm, our aim is to compute $\delta_i^{(k)} = P(Y_k = y_k, \ldots Y_K = y_K \mid X_k = s_i)$ for $i = 1, \ldots, m$ and $1 \le k \le K - 1$. We initialize the process by setting $\delta_j^{(K)} = 1$ for all j's. Then we notice that $\delta_i^{(k-1)} = \sum_{j=1}^m p_{ij} p_j(y_{k-1}) \delta_j^{(k)}$, due to the conditional independence of the Y_j's given the state of the underlying chain. Using this backward induction formula, we gradually compute $\delta_i^{(K-1)}$ for $1 \le i \le m$, $\delta_i^{(K-2)}$ for $1 \le i \le m$, and so forth. Once we get $\delta_i^{(0)}$ for all i's, we multiply each of them by the corresponding initial probability π_i to obtain $P(Y_1 = y_1, \ldots, Y_K = y_K \text{ and } X_1 = s_i)$. The desired likelihood is nothing but the sum of this quantity over i from 1 to m.

The Viterbi algorithm. Given an observed sequence of manifestations $y_1, \ldots,$ y_K, our goal is to efficiently estimate the state-sequence $x_1, \ldots, x_K$ of the underlying chain that has the highest conditional probability of occurring. The Viterbi algorithm first computes

$$\max_{x_1, \ldots, x_K} P(X_1 = x_1, \ldots, X_K = x_k \mid Y_1 = y_1, \ldots, Y_K = y_K) \qquad (3.26)$$

and then traces down a vector $(x_1, \ldots, x_K)$ that gives rise to this maximum. We begin by defining

$$\xi_i^{(k)} = \max_{x_j: 1 \le j \le k-1} P(X_1 = x_1, \ldots, X_{k-1} = x_{k-1},$$
$$X_k = s_i, Y_1 = y_1, \ldots, Y_k = y_k)$$

for some k $(1 \le k \le K)$ and some i $(1 \le i \le m)$. For $k = 1$, it is simply $P(X_1 = s_i \text{ and } Y_1 = y_1)$. Then it is not difficult to see that (3.26) is simply $\max_i \xi_i^{(K)}$ divided by $P(Y_1 = y_1, \ldots, Y_K = y_K)$. So the sequence of states $(x_1, \ldots, x_K)$ that maximizes the conditional probability in (3.26) will also correspond to $\max_i \xi_i^{(K)}$. So we focus on the $\xi_i^{(k)}$'s and compute them using forward induction. Clearly, $\xi_i^{(1)} = \pi_i p_i(y_1)$ for all i. Then, $\xi_i^{(k)} = \max_{1 \le j \le m} \xi_j^{(k-1)} p_{ji} p_i(y_k)$ for $k \in \{2, \ldots, K\}$ and $i \in \{1, \ldots, m\}$. Now suppose that $\xi_{i_1}^{(K)} = \max_{1 \le i \le m} \xi_i^{(K)}$. Then we put $X_K = s_{i_1}$. Next, if i_2 is the index for which $\xi_i^{(K-1)} p_{i i_1}$ is the largest, we put $X_{K-1} = s_{i_2}$. Proceeding in this manner, we get all the remaining states. This algorithm does not produce *all* the state sequences of the underlying chain that give rise to the maximum in (3.26)—just one of them.

The Baum-Welch algorithm. Here the goal is to efficiently estimate the unknown parameters π_i's, p_{ij}'s and $p_i(M_j)$'s using the observed manifestations $y_1, \ldots, y_K$. Actually, the data we need will be a collection of observed sequences $\{y_1^{(t)}, \ldots, y_K^{(t)}\}_{t=1,2,\ldots}$. We will denote the corresponding true-state sequences for the underlying chain by $\{x_1^{(t)}, \ldots, x_K^{(t)}\}_{t=1,2,\ldots}$. The first step will be to initialize all the parameters at some values chosen arbitrarily from the respective parameter spaces or chosen to reflect any *a priori* knowledge about them. Using these initial values, we compute $\hat{\pi}_i$ = the expected proportion of times in the state s_i at the first time-point, given the observed data,

$$\hat{p}_{ij} = \frac{E(U_{ij}|y_1^{(t)}, \ldots, y_K^{(t)})}{E(U_i|y_1^{(t)}, \ldots, y_K^{(t)})} \text{ and } \hat{p}_i(M_j) = \frac{E(U_i(M_j)|y_1^{(t)}, \ldots, y_K^{(t)})}{E(U_i|y_1^{(t)}, \ldots, y_K^{(t)})}, \text{ where } U_{ij} \text{ is the ran-}$$

dom variable counting how often $X_r^{(t)} = s_i$ and $X_{r+1}^{(t)} = s_j$ for some t and r, U_i is the random variable counting how often $X_r^{(t)} = s_i$ for some t and r, and $U_i(M_j)$ is the random variable counting how often the events $\{X_r^{(t)} = s_i\}$ and $\{Y_r^{(t)} = M_j\}$ jointly occur for some t and r. Once these are computed, we set these to be the updated values of the parameters. It can be shown that by doing so, we increase the likelihood function of the observed data $\{Y_1, \ldots, Y_K\}$. As a result, if we continue this process until the likelihood function reaches a local maximum or the increment between successive iterations is very small, we will get the maximum likelihood estimates of the parameters. So we now focus on the computations of the above quantities. Let us define $\zeta_r^{(t)}(i,j)$ as

$$P(X_r^{(t)} = s_i, X_{r+1}^{(t)} = s_j \mid y_1^{(t)}, \ldots, y_K^{(t)}) = \frac{P(X_r^{(t)} = s_i, X_{r+1}^{(t)} = s_j, y_1^{(t)}, \ldots, y_K^{(t)})}{P(y_1^{(t)}, \ldots, y_K^{(t)})}.$$
(3.27)

Notice that both the numerator and the denominator of (3.27) can be efficiently computed using the forward and backward algorithms. For example, one can write the numerator as

$$P(X_r^{(t)} = s_i, X_{r+1}^{(t)} = s_j \& y_1^{(t)}, \ldots, y_K^{(t)}) = \gamma_i^{(r)} p_{ij} p_j(y_{r+1}^{(t)}) \delta_j^{(r+1)}.$$

We then define indicators $I_r^{(t)}(i)$ as $I_r^{(t)}(i) = 1$ only if $X_r^{(t)} = s_i$ (and zero otherwise). It is easy to see that $\sum_t \sum_r I_r^{(t)}(i)$ is the number of times the underlying Markov chain visits the state s_i, whereas $\sum_t \sum_r E(I_r^{(t)}(i) \mid y_1^{(t)}, \ldots, y_K^{(t)})$ is the conditional expectation of the number of times the state s_i is visited by the underlying chain given $\{y_1^{(t)}, \ldots, y_K^{(t)}\}$. This conditional expectation is actually $= \sum_t \sum_r \sum_{j=1}^{m} \zeta_r^{(t)}(i,j)$, since

$$E(I_r^{(t)}(i) \mid y_1^{(t)}, \ldots, y_K^{(t)}) = P(X_r^{(t)} = s_i \mid y_1^{(t)}, \ldots, y_K^{(t)}) = \sum_{j=1}^{m} \zeta_r^{(t)}(i,j).$$

Likewise, $\sum_t \sum_r \varsigma_r^{(t)}(i,j)$ can be shown to be the expected number of transitions from s_i to s_j given the observed manifestations. Finally, defining indicators $J_r^{(t)}(i, M_s)$ as $J_r^{(t)}(i, M_s) = 1$ only if the events $\{X_r^{(t)} = s_i\}$ and $\{Y_r^{(t)} = M_s\}$ jointly occur (and zero otherwise), we realize that $E(J_r^{(t)}(i, M_s) \mid y_1^{(t)}, \dots, y_K^{(t)})$ is the (conditional) expected number of times the tth underlying process $\{X_1^{(t)}, X_2^{(t)}, \dots\}$ is in the state s_i at time r and the corresponding manifestation is M_s, given the observed data $\{y_1^{(t)}, \dots, y_K^{(t)}\}$. So, the $E(U_i(M_s) \mid y_1^{(t)}, \dots, y_K^{(t)})$ from the previous page is nothing but

$$\sum_t \sum_r E(J_r^{(t)}(i, M_s) \mid y_1^{(t)}, \dots, y_K^{(t)}) = \sum_t \sum_r \sum_{(t,r):y_r^{(t)}=M_s} \sum_{j=1}^{m} \varsigma_r^{(t)}(i,j).$$

This concludes our discussion of hidden Markov models.

3.7 Frequentist Statistical Inference

We now turn to various methods of learning about, or drawing inferences regarding, unknown parameters from data *without* assuming any prior knowledge about them. Usually this is done by either *estimating* the parameters or *testing* the plausibility of statements/assertions made about them (called *hypotheses*). Sometimes our goal is just to identify the dataset (among a bunch of datasets) associated with the largest or smallest parameter-value, which is known as a *selection* problem. But here we focus primarily on estimation and hypothesis-testing. As indicated in Section 1, the link between the unknown quantities (i.e., the parameters) and the observed data is a probability model which is assumed to have generated the data. Under the simplest setup, the data $\{X_1, X_2, \dots, X_n\}$ are considered to be an i.i.d. sample from the assumed probability model. In other words, the X_i's are considered independent and identically distributed, each following that probability model. In case the data are discrete (i.e., the values are on a countable grid such as whole numbers), the X_i's will have a common p.m.f.. In the continuous case, they will have a common p.d.f. This p.m.f. or p.d.f. will depend on the unknown parameter(s). We will denote both a p.m.f. and a p.d.f. by $f(x; \boldsymbol{\theta})$, where $\boldsymbol{\theta}$ is a vector of parameters. As defined in an earlier section, the *likelihood function* of the observed data is the joint p.m.f. or p.d.f. of $(X_1, \dots, X_n)$, which boils down to $\prod_{i=1}^{n} f(x_i; \boldsymbol{\theta})$ due to independence. The $n \times n$ dispersion matrix of $(X_1, \dots, X_n)$ is $I_{n \times n}$. Under this setup, we first talk about estimation.

The raw and central moments of $f(x; \boldsymbol{\theta})$ will involve one or more of the parameters. If we have k parameters, sometimes equating the formula for the rth raw moment of $f(x; \boldsymbol{\theta})$ to the corresponding raw moment of the sample $(r = 1, \dots, k)$ gives us a system of k equations and solving this system, we get the *method of moments* (MM) estimates of the parameters. Notice that, for an i.i.d. sample from a univariate p.m.f. or p.d.f. $f(x; \boldsymbol{\theta})$, if we define a function

$\hat{F}_n(t) : \mathcal{R} \to [0,1]$ as $\hat{F}_n(t) = \frac{\#X_i \text{ that are } \leq t}{n}$, this function satisfies all the properties of a c.d.f. in Result 3.2. This is known as the *empirical c.d.f.* of the data and the moments corresponding to it are called *sample moments*. For instance, the first raw sample moment is the sample mean $\bar{X} = \frac{1}{n}\sum_{i=1}^n X_i$. In general, the rth raw sample moment is $\frac{1}{n}\sum_{i=1}^n X_i^r$. The second central sample moment is the sample variance $s^2 = \frac{1}{n}\sum_{i=1}^n(X_i - \bar{X})^2$. For reasons that will be clear shortly, some people use $n-1$ instead of n in the sample variance denominator, which then becomes algebraically the same as $\frac{n\sum_i^n X_i^2 - (\sum_{i=1}^n X_i)^2}{n(n-1)}$. Here are some examples of MM estimates.

Example 3.31. Let $X_1, \ldots, X_n$ be i.i.d. from a Poisson(λ) p.m.f. Then $E(X_i) = \lambda$ and equating it to the first raw sample moment, we get $\hat{\lambda}_{MM} = \frac{1}{n}\sum_{i=1}^n X_i$.

Example 3.32. Let $X_1, \ldots, X_n$ be i.i.d. from a Gamma(α, β) p.d.f. Then the mean $E(X_i) = \alpha\beta$ and the variance $E(X_i - E(X_i))^2 = \alpha\beta^2$. So, equating these two to the corresponding sample moments, we get:

$$\alpha\beta = \frac{1}{n}\sum_{i=1}^n X_i = \bar{X}; \; \alpha\beta^2 = \frac{1}{n-1}\sum_{i=1}^n(X_i - E(X_i))^2 = s^2.$$

Solving these, we get $\hat{\beta}_{MM} = s^2/\bar{X}$ and $\hat{\alpha}_{MM} = \bar{X}^2/s^2$.

In our discussion of the Viterbi algorithm for hidden Markov models, we saw how to find the true underlying state-sequence that maximizes the probability of joint occurrence of the observed manifestations. If we apply the same principle here and try to find those values of the parameter(s) $\boldsymbol{\theta}$ that maximize the joint probability of occurrence (or the likelihood function) of the observed data $x_1, \ldots, x_n$, they will be called the *maximum likelihood* (ML) estimate(s) of the parameter(s). Usually, once the likelihood is written down as a function of the parameter(s) (treating the observed data as fixed numbers), setting its partial derivative w.r.t. each parameter equal to zero gives us a system of equations. Solving it, we get the critical points of the likelihood surface. Some of these will correspond to local maxima, others to local minima and still others will be saddle points. Using standard detection techniques for local maxima and finding the one for which, the likelihood surface is the highest, we can get ML estimates of the parameter(s). Often the above procedure is applied to the natural logarithm of the likelihood function, which still produces the correct answers because the logarithmic transformation is monotone (i.e., preserves "increasingness" or "decreasingness"). Clearly, the ML estimate of a parameter is not necessarily unique. It will be so if the likelihood surface is, for example, unimodal or log-concave. Here are some examples of ML estimation.

Example 3.33. Let $X_1, \ldots, X_n$ be i.i.d. Poisson(λ). Then the likelihood function is $L(\lambda; x_1, \ldots, x_n) = e^{-n\lambda} \lambda^{\sum_1^n x_i} / \prod_1^n x_i!$. So,

$$\frac{d}{d\lambda} \log L(\lambda; x_1, \ldots, x_n) = -n + \frac{\sum_{i=1}^n x_i}{\lambda} = 0 \Rightarrow \lambda = \frac{1}{n} \sum_{i=1}^n X_i,$$

which indeed corresponds to the global maximum of the log-likelihood function (and hence, of the likelihood function) since $\frac{d^2}{d\lambda^2} \log L(\lambda; x_1, \ldots, x_n) = -\lambda^{-2} \sum_{i=1}^n X_i$ is negative for all $\lambda \in (0, \infty)$. So, at least in this case, the MM estimator of λ coincided with its ML estimator. But this is not true in general.

Example 3.34. Let X_i's be i.i.d. Gamma$(1, \beta)$, that is, Exponential(β). Then the log-likelihood function is $\log L(\beta; x_1, \ldots, x_n) = -n\log\beta - \frac{\sum_{i=1}^n x_i}{\beta}$, so that $\frac{d}{d\beta} \log L(\beta; x_1, \ldots, x_n) = 0 \Rightarrow \hat{\beta} = \frac{1}{n} \sum_{i=1}^n x_i$.

Example 3.35. Let X_i's be i.i.d. Normal(μ, σ^2). Then the log-likelihood function is $\log L(\mu, \sigma^2; x_1, \ldots, x_n) = -\frac{n}{2}\log(2\pi) - \frac{n}{2}\log\sigma^2 - \frac{\sum_{i=1}^n (x_i - \mu)^2}{2\sigma^2}$. We can certainly take its partial derivatives w.r.t. μ and σ^2 and equate them to zero, thereby getting a system of two equations. But in this case, we can play a different trick. For each fixed value of σ^2, we can maximize the log-likelihood w.r.t. μ and then, having plugged in whatever value of μ maximizes it, we can treat it as a function of σ^2 alone and find the maximizing value of σ^2. The first step is easy, once we observe that

$$\sum_{i=1}^n (x_i - \mu)^2 = \sum_{i=1}^n (x_i - \bar{x} + \bar{x} - \mu)^2 = \sum_{i=1}^n (x_i - \bar{x})^2 + \sum_{i=1}^n (\bar{x} - \mu)^2,$$

the last equality following from the fact that $\sum_{i=1}^n (x_i - \bar{x})(\bar{x} - \mu) = 0$. So it is clear that $\hat{\mu}_{ML} = \bar{x}$. Now, having replaced μ with $\bar{x}$ in the original log-likelihood function, we maximize it via differentiation w.r.t. σ^2 and get $\sigma^2_{ML} = \frac{1}{n} \sum_{i=1}^n (x_i - \bar{x})^2$.

Two quick points about ML estimators before we move on. The ML estimator of a one-to-one function of a parameter θ is the same one-to-one function of $\hat{\theta}_{ML}$. This is an advantage over method-of-moments estimators, since the latter do not have this property. Also, as we see in the case of $\hat{\sigma}^2_{ML}$ in Example 3.35, the ML estimator of a parameter is not necessarily *unbiased* (an unbiased estimator being defined as one whose expected value equals the parameter it is estimating). It can be shown that $E(\sigma^2_{ML}) = \frac{n-1}{n}\sigma^2$, so that an unbiased estimator of σ^2 is $\hat{\sigma}^2_U = \frac{n}{n-1}\hat{\sigma}^2_{ML} = \frac{1}{n-1} \sum_{i=1}^n (x_i - \bar{x})^2$. This is the reason why the commonly used sample variance formula has $n-1$ instead of n in the denominator.

But is 'unbiasedness' a desirable criterion for deciding whether an estimator is 'good' or 'bad'? Are there other criteria? Before we get to these questions, let us talk about another frequently used estimation technique called the *least squares* method. In real life, often we have multivariate data where each observation is a $k \times 1$ vector of measurements on k variables. Some of these variables are our primary focus, in the sense that we want to study how they vary (or whether they significantly vary at all) in response to changes in the underlying experimental conditions and want to identify all the sources contributing to their variability. These are usually called *response* variables. The other variables in the observation-vector may represent the experimental conditions themselves or the factors contributing to the variability in the responses. These are usually called the *explanatory* variables or *design* variables. Sometimes there is no underlying experiment and all the variables are measurements on different characteristics of an individual or an object. In that case, our primary concern is to determine the nature of association (if any) between the responses and the explanatory variables. For example, we may have bivariate measurements (X_i, Y_i) where the response Y_i is the yield per acre of a certain crop and X_i is the amount of rainfall or the amount of fertilizer used. Or X_i could simply be a *categorical* variable having values 1,2 and 3 (1=heavily irrigated, 2=moderately irrigated, 3=no irrigation). In all these cases, we are primarily interested in the "conditional" behavior of the Y's given the X's. In particular, we want to find out what kind of a functional relationship exists between X and $E(Y \mid X)$. In case X is a continuous measurement (i.e., not an indicator of categories), we may want to investigate whether a linear function such as $E(Y \mid X) = \gamma + \beta X$ is adequate to describe the relation between X and Y (if not, we will think of nonlinear equations such as polynomial or exponential). Such an equation is called the *regression* of Y on X and the adequacy of a linear equation is measured by the *correlation* between them. In case X is a categorical variable with values $1, \ldots, m$, we may investigate if a linear equation such as $E(Y \mid X = i) = \mu + \alpha_i$ is adequate to explain most of the variability in Y (if not, we will bring in the indicator(s) of some other relevant factor(s) or perhaps some additional continuous variables). Such an equation is called a *general linear model*. In case all the X's are categorical, we carry out an *analysis of variance* (ANOVA) for this model. If some X's are categorical and others are continuous (called *covariates*), we perform an *analysis of covariance* (ANCOVA) on this model. If such a linear model seems inadequate even after including all the relevant factors and covariates, the conditional expectation of Y may be related to them in a nonlinear fashion. Perhaps a suitable nonlinear transformation (such as log, arcsine or square-root) of $E(Y \mid \mathbf{X})$ will be related to $\mathbf{X}$ in a linear fashion. This latter modeling approach is known as a *generalized linear model* and the nonlinear transformation is called a *link function*.

In all these cases, the coefficients β, γ, μ and α_i are to be estimated from the data. In the regression scenario with response Y and explanatory variables $X_1, \ldots, X_k$, we write the conditional model for Y given $\mathbf{X}$ as

$$Y_i = \gamma + \beta_1 X_1^{(i)} + \ldots + \beta_k X_k^{(i)} + \epsilon_i \text{ for } i = 1, \ldots, n, \qquad (3.28)$$

where the ϵ_i's are zero-mean random variables representing random fluctuations of the individual Y_i's around $E(Y \mid X_1, \ldots, X_k)$. In the ANOVA scenario with one categorical factor (having k levels or categories), we write the conditional model

$$Y_{ij} = \mu + \alpha_i + \epsilon_{ij} \text{ for } 1 \le i \le k \text{ and } 1 \le j \le n_i \left(\sum_{i=1}^{k} n_i = n \right) \quad (3.29)$$

and in the ANCOVA scenario with a categorical factor X and a continuous covariate Z, we write the conditional model

$$Y_{ij} = \mu + \alpha_i + \beta Z_{ij} + \epsilon_{ij} \text{ for } 1 \le i \le k \text{ and } 1 \le j \le n_i \left(\sum_{i=1}^{k} n_i = n \right). \quad (3.30)$$

In the simplest case, the ϵ_i's are assumed to be i.i.d. having a Normal$(0, \sigma^2)$ p.d.f. But more complicated models may assume correlated ϵ_i's with unequal variances and/or a non-normal p.d.f. (which may even be unknown). In these latter cases, maximum likelihood estimation of the coefficients is analytically intractable and numerically a daunting task as well. So we resort to the following approach. In the regression case, we look at the *sum of squared errors*

$$\sum_{i=1}^{n} (Y_i - \gamma - \beta_1 X_1^{(i)} - \dots - \beta_k X_k^{(i)})^2 = \sum_{i=1}^{n} \epsilon_i^2 \quad (3.31)$$

and try to minimize it w.r.t. γ and the β_i's. It can be done in the usual way, by equating the partial derivatives of (3.31) w.r.t. each of those parameters to zero and solving the resulting system of equations. The solution turns out to be $(\mathbf{X}^t \mathbf{V}^{-1} \mathbf{X})^{-1} \mathbf{X}^t \mathbf{V}^{-1} \mathbf{Y}$, where $\sigma^2 V$ is the variance-covariance matrix of the ϵ_i's. This is known as the *least-squares* estimator. In case the ϵ_i's are assumed to be normal, the MLE's of the parameters $\gamma, \beta_1, \dots, \beta_k$ actually coincide with their least-squares estimators. In almost all real-life scenarios, the matrix V will be unknown and will, therefore, have to be estimated. In the special case where V is simply $I_{n \times n}$, the least-squares estimator of $(\gamma, \beta_1, \dots, \beta_k)$ takes the more familiar form $(\mathbf{X}^t \mathbf{X})^{-1} \mathbf{X}^t \mathbf{Y}$.

Example 3.36. In order to verify if a heavily advertised brand of oatmeal indeed affects the blood cholesterol levels of its consumers, it was given to a randomly selected group of 30 volunteers twice a day for a few days. Different volunteers tried it for different numbers of days. Once each volunteer stopped eating the oatmeal, the change in his/her blood cholesterol level (compared to when he/she started this routine) was recorded. If X_i denotes the number of days for which the ith volunteer ate the oatmeal and Y_i denotes the change in his/her blood cholesterol level, the following are the (X_i, Y_i) pairs for the 20 volunteers: (7,1.6), (15,3.1), (5,0.4), (18,3.1), (6,1.2), (10,3.9), (12,3.0), (21,3.9), (8,2.0), (25, 5.2), (14,4.1), (4,0.0), (20,4.5), (11,3.5), (16,5.0), (3,0.6), (21,4.4), (2,0.0), (23,5.5), (13,2.8). First of all, the correlation between X and Y is 0.908, which is an indication of the strong linear association between them. If we write the regression model of Y on X in the matrix notation $\mathbf{Y} = \mathbf{X}\boldsymbol{\beta} + \boldsymbol{\epsilon}$, the response vector $\mathbf{Y}$ will consist of the 20 cholesterol levels, the design matrix $\mathbf{X}$ will look like

$$\begin{bmatrix} 1 & 1 & 1 & 1 & 1 & 1 & 1 & 1 & 1 & 1 & 1 & 1 & 1 & 1 & 1 & 1 & 1 & 1 & 1 & 1 \\ 7 & 15 & 5 & 18 & 6 & 10 & 12 & 21 & 8 & 25 & 14 & 4 & 20 & 11 & 16 & 3 & 21 & 2 & 23 & 13 \end{bmatrix}$$

and the least-squares estimate of β will be $(\mathbf{X}^t\mathbf{X})^{-1}\mathbf{X}^t\mathbf{Y} = (0.0165, 0.22626)$. So the regression equation of Y on X is $Y = 0.0165 + 0.22626X$, which indicates that for a one-unit change in X, our best estimate for the change in Y (in the 'least squares' sense) is 0.22626. This equation also gives us our best prediction for Y corresponding to a hitherto-unseen value of X. Notice that all of the variability in Y cannot be explained by X, since Y will even vary a little bit for individuals with the same value of X. For example, look at the 8th and the 17th volunteers in the dataset above. The proportion of the overall variability in Y that can be explained by X alone is called the *coefficient of determination* and, in this simple model, it equals the square of the correlation (i.e., 82.45%).

There are other methods of finding estimators, such as the *minimum chi-squared* method and the *minimum l_∞ distance* method. But they have their own drawbacks, such as failing to use the full information in a continuous dataset (due to ad hoc discretization of the value-range of a continuous random variable) or being analytically intractable. For more details on them, see Cramer (1946), Rao (1965) or Mood et al. (1974). But we now move on to another important aspect of these estimators—their own random behavior. After all, an estimator is nothing but a formula involving the random sample $\{X_1, \ldots, X_n\}$ and so it is a random variable itself. Its value will vary from one random sample to another. What p.d.f. will it have? One way of getting an idea about this p.d.f. is to look at a histogram of many, many different values of it (based on many, many different random samples) and try to capture the "limiting" shape of this histogram as the number of bars in it goes to infinity. The p.d.f. of an estimator obtained in this way will be called its *sampling distribution*. For instance, the sampling distribution of the MM estimator in Example 3.31 approaches the shape of a normal (or Gaussian) distribution. Similar is the case for the MLE's in Examples 3.33, 3.34 and 3.35 (for the parameter μ). Is it a mere coincidence that the sampling distributions of so many estimators look Gaussian in the limit? Also, if the limiting distribution is Gaussian, what are its mean and variance? The answer to this latter question is easier, since in all the examples mentioned above, the estimator concerned is the sample mean $\bar{X} = \frac{1}{n}\sum_{i=1}^{n} X_i$. Due to the linearity property of expected values, $E(\bar{X}) = \frac{1}{n}E(\sum_{i=1}^{n} X_i) = \frac{1}{n}\sum_{i=1}^{n} E(X_i)$, so that the mean of the sampling distribution of $\bar{X}$ is the *same* as the mean of each X_i. If Σ denotes the variance-covariance matrix of the random vector $\mathbf{X} = (X_1, \ldots, X_n)$, it can be shown that the variance of any linear combination $\mathbf{a}^t\mathbf{X} = a_1X_1 + \ldots + a_nX_n$ is nothing but $\mathbf{a}^t\Sigma\mathbf{a}$. In particular, the variance of $\sum_{i=1}^{n} X_i$ is $\sum_{i=1}^{n} \mathrm{VAR}(X_i) + \sum_{i=1}^{n}\sum_{j\neq i} \mathrm{COV}(X_i, X_j)$, which reduces to just $\sum_{i=1}^{n} \mathrm{VAR}(X_i)$ if the X_i's are i.i.d. (i.e., the covariances are zero). Since $\mathrm{VAR}(cY) = c^2\mathrm{VAR}(Y)$ for any random variable Y and any constant c, it easily follows that $\mathrm{VAR}(\frac{1}{n}\sum_{i=1}^{n} X_i) = \frac{1}{n}\mathrm{VAR}(X_i)$ for i.i.d. X_i's. Now, to the question of the sampling distribution being Gaussian.

It turns out that the sampling distributions of the estimators in Examples

3.31, 3.33, 3.34 and 3.35 (for μ) were not Gaussian by coincidence. If X_i's are i.i.d. from a p.m.f. or p.d.f. with finite mean μ and finite variance σ^2, the sampling distribution of $n^{1/2}(\bar{X} - \mu)/\sigma$ will always be standard normal (i.e., $N(0,1)$) in the limit as $n \to \infty$. This is the gist of the so-called *central limit theorem* (CLT) proved by Abraham DeMoivre and Pierre-Simon Laplace. It is one of the most celebrated results in probability theory, with far-reaching consequences in classical statistical inference. Along with its multivariate counterpart for i.i.d. random vectors, this result opens up the possibility of another kind of estimation, called *interval estimation* in the case of a scalar parameter or *confidence-set estimation* in the case of a parameter-vector. The idea is the following. If $\boldsymbol{\theta}$ is the mean of a p.d.f. $f(x; \boldsymbol{\theta}, \sigma^2)$ having variance σ^2, and $X_1, \ldots, X_n$ are i.i.d. observations from this p.d.f., then according to the CLT, $P(-z_{\alpha/2} \leq n^{1/2}(\bar{X} - \boldsymbol{\theta})/\sigma \leq z_{\alpha/2})$ will be approximately $1 - \alpha$ for a 'sufficiently large' sample size n, where $z_{\alpha/2}$ is the $100(1 - \alpha/2)^{th}$ percentile of the $N(0,1)$ p.d.f. (i.e., the point beyond which the tail-area of a $N(0,1)$ p.d.f. is $\alpha/2$). In other words,

$$P(\bar{X} - \frac{z_{\alpha/2}\,\sigma}{n^{1/2}} \leq \boldsymbol{\theta} \leq \bar{X} + \frac{z_{\alpha/2}\,\sigma}{n^{1/2}}) = 1 - \alpha,$$

and the interval $(\bar{X} - \frac{z_{\alpha/2}\,\sigma}{n^{1/2}}, \bar{X} + \frac{z_{\alpha/2}\,\sigma}{n^{1/2}})$ is called a $100(1 - \alpha)$ % *confidence interval* for $\boldsymbol{\theta}$. However, the variance σ^2 will typically be unknown, rendering the computation of the two end-points of this confidence interval impossible. Fortunately, it can be shown that the distribution of $n^{1/2}(\bar{X} - \boldsymbol{\theta})/s$ is also approximately $N(0,1)$ for large values of n, where $s^2 = \frac{1}{n-1}\sum_{i=1}^{n}(X_i - \bar{X})^2$ is the sample variance. As a result, we get the approximate $100(1 - \alpha)$ % confidence interval

$$(\bar{X} - \frac{z_{\alpha/2}\,s}{n^{1/2}}, \bar{X} + \frac{z_{\alpha/2}\,s}{n^{1/2}}) \qquad (3.32)$$

for $\boldsymbol{\theta}$.

Actually, it can be shown using the multivariate version of the 'Jacobian of inverse transformation' technique (see Example 3.23) that the p.d.f. of $\frac{U}{(V/\nu)^{1/2}}$, where U and V are two independent random variables following $N(0,1)$ and $\chi^2(\nu)$ respectively, is *Student's t* with ν degrees of freedom. It is a p.d.f. for which extensive tables of percentiles are available. So, in case the sample size n is not large, we should replace $z_{\alpha/2}$ in the above confidence interval formula by $t_{\alpha/2}(\nu)$, provided that the data came from a normal distribution. This is because if $\{X_1, \ldots, X_n\}$ is a random sample from a $N(\boldsymbol{\theta}, \sigma^2)$ distribution, it can be shown using a technique called Helmert's orthogonal transformation (see Rohatgi and Saleh (2001), p 342) that $\bar{X}$ and $(n-1)S^2/\sigma^2$ are independent and, of course, $n^{1/2}(\bar{X} - \boldsymbol{\theta})/\sigma \sim N(0,1)$ and $(n-1)S^2/\sigma^2 \sim \chi^2(n-1)$.

Next we look at some criteria for picking the 'best' estimator (if there is one) from a bunch of competing estimators. One criterion (unbiasedness) has already been mentioned. But there are others that may be more desirable or justifiable than unbiasedness under different circumstances. One example is *concentration*. Let $X_1, \ldots, X_n$ be a random sample from a p.d.f. or p.m.f.

$f(x; \boldsymbol{\theta})$ and suppose that we are estimating $\eta = \eta(\boldsymbol{\theta})$, a certain function of $\boldsymbol{\theta}$. An estimator $T = T(X_1, \ldots, X_n)$ is said to be a *more concentrated* estimator of η than another estimator T^* if

$$P_{\boldsymbol{\theta}}(\eta(\boldsymbol{\theta}) - \delta < T \leq \eta(\boldsymbol{\theta}) + \delta \geq P_{\boldsymbol{\theta}}(\eta(\boldsymbol{\theta}) - \delta < T^* \leq \eta(\boldsymbol{\theta}) + \delta)$$

for all $\delta > 0$ and all values of $\boldsymbol{\theta}$. If there is an estimator T^{**} that is more concentrated than any other, it will be called the *most concentrated* estimator. A slightly different but related criterion is *Pitman closeness* (after E.G.J. Pitman). An estimator T of $\eta(\boldsymbol{\theta})$ is called *Pitman-closer* to η than another estimator T^* if

$$P_{\boldsymbol{\theta}}(| T - \eta | < | T^* - \eta |) \geq \frac{1}{2}$$

for all values of $\boldsymbol{\theta}$. Once again, if there exists a T^{**} that is Pitman-closer to η than all its competitors, it will be called the *Pitman-closest*. However, while they are both intuitively desirable, the trouble is that they do not define a *total order* on the set of all possible estimators of η, in the sense that given any two estimators T_1 and T_2, one is not necessarily more concentrated (or Pitman-closer) than the other. As a result, a 'champion' (i.e., a most concentrated or Pitman-closest estimator) seldom exists.

Another popular criterion is *minimum mean-squared-error*, which tries to pick a 'winner' on the basis of how close an estimator is to its 'target' on the average. In other words, the mean squared error (MSE) of an estimator $T = T(X_1, \ldots, X_n)$ for $\eta(\boldsymbol{\theta})$ is $E_{\boldsymbol{\theta}}(T - \eta)^2$, and an estimator is preferable to another if it has a smaller MSE than its competitor. This criterion also fails to induce a total order on the set of all possible estimators, since typically a plot of $\text{MSE}(\boldsymbol{\theta})$ versus $\boldsymbol{\theta}$ for two estimators T_1 and T_2 will show a crisscrossing pattern (i.e., T_1 is preferable for some values of $\boldsymbol{\theta}$ and T_2, for others). Also, according to this criterion, the 'champion' will be the 'impossible' estimator $T(X_1, \ldots, X_n) \equiv \eta(\boldsymbol{\theta})$ that always correctly estimates $\eta(\boldsymbol{\theta})$ regardless of the sample observations. A generalization of the concept of MSE is to assess the 'goodness' of an estimator T by computing its average distance from its 'target' η with respect to any valid distance measure—not necessarily $(T - \eta)^2$. Such a general distance measure is called a *loss function* and its average value (computed using the p.d.f. or p.m.f. that is indexed by $\boldsymbol{\theta}$) is called the corresponding *risk function*. The word 'function' emphasizes the fact that these are indeed functions of the underlying parameter $\boldsymbol{\theta}$. Some examples of loss functions are the *absolute error* loss $| T - \eta |$, the *polynomial loss* $\zeta | T - \eta |^r$ with $\zeta(\boldsymbol{\theta})$ a nonnegative function of $\boldsymbol{\theta}$ and r a positive constant, and the *exponential loss* $e^{|T-\eta|}$. Based on this generalization, another criterion is introduced for comparing estimators—that of *admissibility*. An estimator T_1 is *inadmissible* if there is another estimator T_2 such that $\mathcal{R}_{T_2}(\boldsymbol{\theta}) \leq \mathcal{R}_{T_1}(\boldsymbol{\theta})$ for all $\boldsymbol{\theta}$, with a strict inequality for at least one value of $\boldsymbol{\theta}$ ($\mathcal{R}_{T_i}$ being the risk associated with T_i, $i = 1, 2$). An estimator T will, therefore, be called *admissible* if there is no other competitor to render it inadmissible. Once again, admissibility fails to induce a total order on the set of all estimators, due to their crisscrossing risk functions. This lack of a total order, common to all the criteria discussed so far, is the motivation behind exploring new avenues for comparing estimators. One avenue is to summarize the risk function in some way and just compare two such summaries corresponding to two estimators. Another avenue is to try to apply the comparison criteria described above to a restricted subset of the set of all estimators. These two ideas are briefly discussed below. A third avenue would be to somehow 'average out' $\boldsymbol{\theta}$, but further discussion on it is postponed until the next section. Incidentally, the branch of statistics that deals with the choice of an optimal *decision rule* (a general name for estimators and other functions of the sample observations) by minimizing the risk associated with various loss functions is formally known as *statistical decision theory*.

The first idea mentioned above leads to the concept of a *minimax* estimator. An estimator T is said to be a minimax estimator if

$$\sup_{\boldsymbol{\theta}} \mathcal{R}_T(\boldsymbol{\theta}) \leq \sup_{\boldsymbol{\theta}} \mathcal{R}_{T'}(\boldsymbol{\theta})$$

for all other estimators T'. In other words, loosely speaking, the highest point or 'peak' of the risk function is being used as its summary. The second idea leads to a *uniformly minimum variance unbiased estimator* (UMVUE). It is the 'winner' (according to the MSE criterion) in the restricted class of all unbiased estimators. As has been mentioned earlier, the MSE of an estimator is the sum of its variance and squared bias; hence it reduces to just the variance for an unbiased estimator. So in this restricted subset, the search for an estimator with the smallest MSE boils down to that for the minimum-variance estimator and, fortunately, it can be found on many occasions. Another restricted subset that is often used for this purpose is the subset of all *consistent* estimators. Let $T_n = T(X_1, \ldots, X_n)$ be an estimator of $\eta(\boldsymbol{\theta})$. It is said to be *weakly consistent* for $\eta(\boldsymbol{\theta})$ if, for every $\varepsilon > 0$,

$$\lim_{n \to \infty} P_{\boldsymbol{\theta}}(\mid T_n - \eta(\boldsymbol{\theta}) \mid < \varepsilon) = 1$$

for each value of $\boldsymbol{\theta}$. A slightly stronger concept is that of a *mean-squared-error consistent* estimator. T_n will be called MSE-consistent for $\eta(\boldsymbol{\theta})$ if $\lim_{n \to \infty} E_{\boldsymbol{\theta}}(T_n -\eta(\boldsymbol{\theta}))^2 = 0$ for every value of $\boldsymbol{\theta}$. It is stronger in the sense that it implies weak consistency, but the reverse implication is not necessarily true. In any case, since an MSE-consistent estimator must be asymptotically unbiased (i.e., the bias must vanish in the limit as $n \to \infty$), one might search for the minimum-variance estimator in the MSE-consistent class. Sometimes that search is restricted to an even narrower subcollection—that of consistent and asymptotically normal estimators (i.e., consistent estimators whose distributions converge to a normal distribution as $n \to \infty$). Such an estimator, if existent, will be called *consistent asymptotically normal efficient* (CANE).

We conclude our discussion of comparison criteria by answering an important question. We have defined a UMVUE above, but how does one find such an estimator? If we knew the smallest variance that an unbiased estimator could achieve, an already familiar unbiased estimator might luckily turn out to have that variance or could be slightly modified to have that variance. Or, starting with any unbiased estimator, if we knew the technique of producing another one with smaller variance, then repeating this technique several times might ultimately lead us to a UMVUE. Fortunately, none of these is wishful thinking—we just need to introduce a couple of new concepts. The first one is *sufficiency*. When we have a random sample $X_1, \ldots, X_n$ from a p.d.f. or p.m.f. $f(x; \boldsymbol{\theta})$, a statistic $S = S(X_1, \ldots, X_n)$ is said to be sufficient for $\boldsymbol{\theta}$ if the conditional joint distribution of $X_1, \ldots, X_n$ given that $S = s$ does not involve $\boldsymbol{\theta}$. In other words, the sufficient statistic S contains all the 'juice' (i.e., relevant information about $\boldsymbol{\theta}$) in the data. An equivalent definition, which is more convenient in practice, says that a statistic $S(X_1, \ldots, X_n)$ is sufficient for $\boldsymbol{\theta}$ if the conditional distribution of any arbitrary statistic $T(X_1, \ldots, X_n)$ given $S = s$ is free from $\boldsymbol{\theta}$. A nice way to visualize a sufficient statistic is through the partition it induces in the sample space of all possible data-vectors. Suppose that $\mathcal{S}$ is the value-set of S. Then the induced partition consists of the cells $\{(X_1, \ldots, X_n) : S(X_1, \ldots, X_n) = s\}$ for $s \in \mathcal{S}$. Given that we are in any particular cell, the conditional joint distribution of the sample observations (or the conditional distribution of any statistic based on them) is free from $\boldsymbol{\theta}$. Three important points are worth noting before we move on. First, sufficient statistics are usually not unique. If S is sufficient for $\boldsymbol{\theta}$, then so is any one-to-one function of it, since it will induce exactly the same partition in the sample space. Secondly, if $\boldsymbol{\theta}$ is a vector of parameters, we may have a vector of statistics $(S_1, \ldots, S_m)$ that are jointly sufficient for them. Finally, a quick way of finding sufficient statistics is obtained from the following result, widely known as the *Fisher-Neyman factorization theorem*:

Result 3.13. A set of statistics $S_1(X_1, \ldots, X_n), \ldots, S_m(X_1, \ldots, X_n)$ is jointly sufficient for $\boldsymbol{\theta}$ (possibly a parameter-vector) if and only if the joint p.d.f. (or p.m.f.) of $X_1, \ldots, X_n$ can be factored as

$$\prod_{i=1}^{n} f(x_i; \boldsymbol{\theta}) = h_1(s_1(x_1, \ldots, x_n), \ldots, s_m(x_1, \ldots, x_n); \boldsymbol{\theta}) h_2(x_1, \ldots, x_n),$$

where h_1 is a nonnegative function depending on the data only through $\{s_i(x_1, \ldots, x_n)\}_{i=1}^{m}$ and h_2 is a nonnegative function free from $\boldsymbol{\theta}$.

If two statistics S and S^* are both sufficient for $\boldsymbol{\theta}$ and S^* is a function of S (not necessarily a one-to-one function), how do the two induced partitions compare? Well, the one induced by S^* will have fewer cells, because each of its cells will be the union of several cells from the other partition. If $s_1, \ldots, s_k$ are the values of S for which $S^*(s_i) = s^*$, $i = 1, \ldots, k$, the partition-cell corresponding to $S^* = s^*$ will be $\bigcup_{i=1}^{k} \{(X_1, \ldots, X_n) : S(X_1, \ldots, X_n) = s_i\}$. We say that S^* induces a *coarser* partition than S. From the viewpoint of data-condensation or dimension reduction, S^* is preferable, since it provides a more concise summary of the data while preserving all the relevant 'juice' regarding $\boldsymbol{\theta}$. What if another sufficient statistic S^{**} is available that is a function of S^*? It is even better in terms of data condensation. Continuing in this manner, we will ultimately get a sufficient statistic that is a function of every other sufficient statistic and induces the coarsest partition. It will be called a *minimal sufficient statistic* for $\boldsymbol{\theta}$. One way of quickly arriving at a minimal sufficient statistic is to look for a sufficient statistic S whose distribution belongs to a family with a special property. Suppose that the family of p.d.f.'s or p.m.f.'s $\{g(s; \boldsymbol{\theta})\}$ has the property: $\int \psi(s)g(s; \boldsymbol{\theta})ds = 0$ for all $\boldsymbol{\theta} \implies \psi(s) \equiv 0$. Then S will actually be a minimal sufficient statistic. The special property of the distribution-family of S mentioned above is known as *completeness*. So, in summary, a complete and sufficient statistic is a minimal sufficient statistic (although the converse is not necessarily true—it is possible sometimes to find a minimal sufficient statistic that is not complete).

Now we are in a position to go back to the question of how to 'hunt down' a UMVUE. First, how low can the variance of an unbiased estimator be? The answer is provided by the famous *Cramér-Rao inequality* (named after Harald Cramér and C.R. Rao):

Result 3.14. Suppose that $X_1, \ldots, X_n$ are i.i.d. observations from a p.d.f. (or p.m.f.) $f(x; \boldsymbol{\theta})$ and $T(X_1, \ldots, X_n)$ is an unbiased estimator of $\eta(\boldsymbol{\theta})$. Suppose also that $\frac{d}{d\boldsymbol{\theta}}\eta(\boldsymbol{\theta}) = \eta'(\boldsymbol{\theta})$ exists and $f(x; \boldsymbol{\theta})$ satisfies the following *regularity conditions*:

(i) $\frac{\partial}{\partial \boldsymbol{\theta}}\log f(x; \boldsymbol{\theta})$ exists for all x and all $\boldsymbol{\theta}$;

(ii) $0 < E_{\boldsymbol{\theta}}[(\frac{\partial}{\partial \boldsymbol{\theta}}\log f(X; \boldsymbol{\theta}))^2] < \infty$ for all values of $\boldsymbol{\theta}$;

(iii) $\frac{\partial}{\partial \boldsymbol{\theta}} \int \ldots \int \prod_{i=1}^{n} f(x_i; \boldsymbol{\theta})dx_1 \ldots dx_n = \int \ldots \int \frac{\partial}{\partial \boldsymbol{\theta}} \prod_{i=1}^{n} f(x_i; \boldsymbol{\theta})dx_1 \ldots dx_n$;

(iv) $\frac{\partial}{\partial \boldsymbol{\theta}} \int \ldots \int T(x_1, \ldots, x_n) \prod_{i=1}^{n} f(x_i; \boldsymbol{\theta})dx_1 \ldots dx_n$
$= \int \ldots \int T(x_1, \ldots, x_n)\frac{\partial}{\partial \boldsymbol{\theta}} \prod_{i=1}^{n} f(x_i; \boldsymbol{\theta})dx_1 \ldots dx_n$,

where the multiple integrals in (iii) and (iv) are to be replaced by multiple sums in case $f(x; \boldsymbol{\theta})$ is a discrete p.m.f. Then we must have

$$var_{\boldsymbol{\theta}}(T) \geq \frac{(\eta'(\boldsymbol{\theta}))^2}{nE_{\boldsymbol{\theta}}[(\frac{\partial}{\partial \boldsymbol{\theta}}\log f(X; \boldsymbol{\theta}))^2]}. \qquad (3.33)$$

The expression on the right-hand side is known as the *Cramér-Rao lower bound*. Regarding the issue of starting with an arbitrary unbiased estimator and creating a new unbiased estimator from it with smaller variance, we have two famous results that are summarized below:

Result 3.15. Let $X_1, \ldots, X_n$ be i.i.d. observations from a p.d.f. or p.m.f. $f(x; \boldsymbol{\theta})$ and suppose that the statistics $S_1(X_1, \ldots, X_n), \ldots, S_m(X_1, \ldots, X_n)$ are jointly sufficient for $\boldsymbol{\theta}$. If $T(X_1, \ldots, X_n)$ is an unbiased estimator of $\eta(\boldsymbol{\theta})$ and we define a new statistic T' as $T'(S_1, \ldots, S_m) = E(T \mid S_1, \ldots, S_m)$, then

T' is also unbiased for $\eta(\boldsymbol{\theta})$ and $\mathrm{var}_{\boldsymbol{\theta}}(T') \leq \mathrm{var}_{\boldsymbol{\theta}}(T)$ for all values of $\boldsymbol{\theta}$, with a strict inequality for at least one value of $\boldsymbol{\theta}$ (unless T and T' are identical). The above technique is popularly known as *Rao-Blackwellization* (after C.R. Rao and David Blackwell).

Result 3.16. Let $X_1, \ldots, X_n$ be i.i.d. observations from a p.d.f. or p.m.f. $f(x; \boldsymbol{\theta})$ and suppose that $S_1(X_1, \ldots, X_n)$ is a complete and sufficient statistic for $\boldsymbol{\theta}$. If $T(S(X_1, \ldots, X_n))$, a function of S, is also an unbiased estimator of $\eta(\boldsymbol{\theta})$, then T must be a UMVUE of $\eta(\boldsymbol{\theta})$.

This is often called the *Lehmann-Scheffé* theorem and tells us precisely what to do in order to get a UMVUE of $\eta(\boldsymbol{\theta})$. Just get hold of a complete, sufficient statistic S for $\boldsymbol{\theta}$ and an unbiased estimator T of $\eta(\boldsymbol{\theta})$. Then define $T' = E(T \mid S)$ and T' will be a UMVUE of $\eta(\boldsymbol{\theta})$. Now we will see a series of examples illustrating all these techniques and concepts.

Example 3.37. Let $X_1, \ldots, X_n$ be i.i.d. observations from a $N(\mu, \sigma^2)$ p.d.f. Denoting the sample variance $\frac{1}{n-1} \sum_{i=1}^{n} (X_i - \bar{X})^2$ by S^2, we know that $(n-1)S^2/\sigma^2 \sim \chi^2(n-1)$. From our discussion about a Gamma p.d.f. in Section 3.3, we also know that $E[(n-1)S^2/\sigma^2] = (n-1)$ and $var[(n-1)S^2/\sigma^2] = 2(n-1)$. In other words, $E(S^2) = \sigma^2$ and $var(S^2) = \frac{2\sigma^4}{n-1}$. So it follows by Chebyshev's inequality (Result 3.6) that

$$P(| S^2 - \sigma^2 |> \varepsilon) \leq \frac{var(S^2)}{\varepsilon^2} = \frac{2\sigma^4}{(n-1)\varepsilon^2} \longrightarrow 0 \text{ as } n \longrightarrow \infty,$$

which implies that S^2 is weakly consistent for σ^2. Actually, this can be generalized to any statistic $T_n = T(X_1, \ldots, X_n)$ with asymptotically vanishing variance when $\{X_1, X_2, \ldots\}$ is a random sample from a p.d.f. or p.m.f. $f(x; \boldsymbol{\theta})$. If $\lim_{n \to \infty} E_{\boldsymbol{\theta}}(T_n) = \eta(\boldsymbol{\theta})$ and $\lim_{n \to \infty} var(T_n) = 0$, then an application of Chebyshev's inequality along with the fact that MSE = variance + bias^2 will ensure the weak consistency of T_n for $\eta(\boldsymbol{\theta})$. However, in the special case involving S^2 and σ^2, a stronger result can be obtained: S^2 is in fact MSE-consistent for σ^2. That is because $E[(S^2 - \sigma^2)^2] = var(S^2) = \frac{1}{n}(\mu_4 - \frac{n-3}{n-1}\sigma^4)$, which goes to 0 as $n \to \infty$. In fact, if $a_1, a_2, \ldots$ is any sequence of positive numbers such that $\lim_{n \to \infty} na_n = 1$, then the estimator $S^{2*} = a_n \sum_{i=1}^{n}(X_i - \bar{X})^2$ will be MSE-consistent, and hence weakly consistent, for σ^2 (can you explain to yourself why?).

Example 3.38. Let $\{X_1, \ldots, X_n\}$ be a random sample from the p.d.f. $f(x; \theta) = \theta^x (1 - \theta)^{1-x}$ for $x \in \{0, 1\}$ and $\theta \in (0, 1)$. The statistic $Y = \sum_{i=1}^{n} X_i$ has the p.d.f. $f_Y(y; \theta) =_n C_y \theta^y (1 - \theta)^{n-y}$ for $y = 0, 1, \ldots, n$. The conditional probability $P(X_1 = x_1, \ldots, X_n = x_n \mid Y = y)$ is

$$\frac{\prod_{i=1}^{n} \theta^{x_i}(1-\theta)^{1-x_i}}{nC_y\theta^y(1-\theta)^{n-y}} = \frac{\theta^{\sum_1^n x_i}(1-\theta)^{n-\sum_1^n x_i}}{nC_y\theta^y(1-\theta)^{n-y}} = \frac{1}{nC_y},$$

which is free from θ and, therefore, implies that Y is sufficient for θ.

Example 3.39. If $X_1, \ldots, X_n$ is a random sample from a Gamma(α, θ) p.d.f., we know that $Y = \sum_{i=1}^{n} X_i$ has a Gamma$(n\alpha, \theta)$. The conditional joint p.d.f. of $X_1, \ldots, X_n$ given $Y = y$ is $[\Gamma(n\alpha)/(\Gamma(\alpha))^n] \frac{x_1 \ldots x_n}{(x_1 + \ldots + x_n)^{n\alpha - 1}}$. This being free from θ, Y is sufficient for θ if α is known. But if α is unknown, Y is *not* sufficient for the parameter vector (α, θ).

Example 3.40. Let $X_1, \ldots, X_n$ be i.i.d. observations from a $N(\mu, \sigma^2)$ density. Let $\boldsymbol{\theta} = (\mu, \sigma)$. The joint density of $X_1, \ldots, X_n$ can be factored as

$$\prod_{i=1}^{n} \frac{1}{\sqrt{2\pi}\sigma} \exp[-\frac{(x_i - \mu)^2}{2\sigma^2}] = \frac{1}{\sqrt{2\pi}^n \sigma^n} \exp[-\frac{1}{2\sigma^2}(\sum_{1}^{n} x_i^2 - 2\mu \sum_{1}^{n} x_i + n\mu^2)],$$

so that according to Result 3.13, the statistics $\sum_{1}^{n} X_i$ and $\sum_{1}^{n} X_i^2$ are jointly sufficient for $\boldsymbol{\theta} = (\mu, \sigma)$. Here $\frac{1}{(2\pi)^{n/2}}$ plays the role of the function $h_2(x_1, \ldots, x_n)$. Note that the mapping $\sum_{1}^{n} X_i, \sum_{1}^{n} X_i^2 \longrightarrow \bar{X}, S^2$ is one-to-one, so that $\bar{X}$ and S^2 are also jointly sufficient for $\boldsymbol{\theta} = (\mu, \sigma)$. Also note that if the parameter μ is known, the factorization theorem tells us that the vector $(\sum_{1}^{n} X_i, \sum_{1}^{n} X_i^2)$ is still jointly sufficient for σ. But is it minimally sufficient for σ? No, the statistic $\sum_{1}^{n}(X_i - \mu)^2$ is.

Example 3.41. Suppose that $X_1, \ldots, X_n$ are i.i.d. observations from a Uniform $[\theta_1, \theta_2]$ density. Then their joint p.d.f. is $\prod_{1}^{n}(\theta_2 - \theta_1)^{-1} I_{[\theta_1, \theta_2]}(x_i)$, where $I_{[a,b]}(x) = 1$ if $x \in [a, b]$ and is 0 otherwise. So the joint p.d.f. can be written as $(\theta_2 - \theta_1)^{-n} I_{[\theta_1, \theta_2]}(\min_i x_i) I_{[\theta_1, \theta_2]}(\max_i x_i)$. So, by the factorization theorem (Result 3.13), $(\min_i x_i, \max_i x_i)$ is sufficient for $\boldsymbol{\theta} = (\theta_1, \theta_2)$, with $h_2(x_1, \ldots, x_n) \equiv 1$.

Example 3.42. Recall the definition of the exponential family of densities or p.m.f.s from (3.20)-(3.21). If $\mathbf{X} = (X_1, \ldots, X_n)$ has the p.d.f. or p.m.f. in (3.20) or (3.21), the factorization theorem immediately implies that the statistics $t_1(X_1, \ldots, X_n), \ldots, t_k(X_1, \ldots, X_n)$ are jointly sufficient for $\boldsymbol{\theta}$. It can actually be shown that they are complete and, therefore, are minimally sufficient for $\boldsymbol{\theta}$.

Example 3.43. Here is an example of a statistic that is not sufficient. Let X_1, X_2 be i.i.d. observations from a Poisson(λ) p.m.f. and look at the statistic $X_1 + 2X_2$. Is the conditional joint p.m.f. of X_1 and X_2 given $X_1 + 2X_2$ free from λ? No. For example,

$$P(X_1 = 0, X_2 = 1 \mid X_1 + 2X_2 = 2) = \frac{P(X_1 = 0, X_2 = 1)}{P(X_1 = 0, X_2 = 1) + P(X_1 = 2, X_2 = 0)}$$

and this ratio is $\frac{1}{1 + \lambda/2}$, not free from λ.

Example 3.44. Let $X_1, \ldots, X_n$ be i.i.d. Bernoulli(θ) random variables. Then, clearly, the statistic $T = \sum_{1}^{n} X_i$ is sufficient for θ. Notice also that T has a

Binomial(n, θ) p.m.f. and the family of p.m.f.s $\{\text{Binomial}(n, \theta) : \theta \in (0, 1)\}$ is complete. This is because

$$E_\theta \psi(T) = \sum_{t=0}^{n} \psi(t)_n C_t \theta^t (1 - \theta)^{n-t} = 0 \implies (1 - \theta)^n \sum_{t=0}^{n} \psi(t)_n C_t \left(\frac{\theta}{1 - \theta}\right)^t = 0$$

for all θ. But the last sum above is a polynomial in $\theta/(1 - \theta)$, so it being zero for all θ must imply that the coefficients are all zero. In other words, $\psi(t) = 0$ for $t = 0, 1, \ldots, n$. Hence, T is minimal sufficient for θ.

Example 3.45. Let $\{X_1, \ldots, X_n\}$ be a random sample from a $N(\mu, \sigma^2)$ density with μ known. Then by the factorization theorem, $T = \sum_1^n (X_i - \mu)^2$ is sufficient for σ. But it is also complete, because as was mentioned in the context of the Gamma p.d.f. in (3.16), $(X_i - \mu)^2/\sigma^2 \sim \chi_1^2$ for each i and $\sum_1^n (X_i - \mu)^2/\sigma^2 \sim \chi_n^2$. So,

$$E_\sigma \psi(T) = 0 \text{ for all } \sigma > 0 \implies \int_0^\infty \psi(t) \frac{1}{\Gamma(n/2)(2\sigma^2)^{n/2}} e^{-t/(2\sigma^2)} t^{(n/2)-1} dt = 0$$

for all $\sigma > 0$, which can happen if and only if $\int_0^\infty \psi(t) t^{(n/2)-1} e^{-t/(2\sigma^2)} dt = 0$ for all $\sigma > 0$. However, by the uniqueness property of Laplace transforms (the *Laplace transform* of a function $g(t)$ at s is defined as $\int g(t) e^{-ts} ds$), only the function identically equal to zero can have a Laplace transform that is identically zero. So $\psi(t) \equiv 0$. Hence, the statistic T is minimally sufficient for σ.

Example 3.46. Suppose that $X_1, \ldots, X_n$ are i.i.d. observations from a $N(\mu, \mu^2)$ density. Clearly, by the factorization theorem, $T = (T_1, T_2) = (\sum_1^n X_i, \sum_1^n X_i^2)$ is sufficient for μ. But is T complete? No, because $\psi(t) = \psi(t_1, t_2) = 2t_1^2 - (n + 1)t_2$ is a function of t that is not identically zero, yet $E_\mu \psi(T) = 0$ for all $\mu \in \mathcal{R}$.

Example 3.47. Let $\{X_1, \ldots, X_n\}$ be a random sample from a Poisson(λ) p.m.f. We will derive the Cramér-Rao lower bound for an unbiased estimator of $\eta(\lambda) = e^{-\lambda} = P(X = 0)$ and that of $\eta(\lambda) = \lambda$. It can be shown that the regularity conditions of Result 3.14 are satisfied, but we omit the details. Since

$$\frac{\partial}{\partial \lambda} \log f(x; \lambda) = \frac{\partial}{\partial \lambda} \log \frac{e^{-\lambda} \lambda^x}{x!} = -1 + \frac{x}{\lambda},$$

so we have

$$E_\lambda \left[\frac{\partial}{\partial \lambda} \log f(X; \lambda)\right]^2 = E_\lambda [(X/\lambda) - 1]^2 = \frac{1}{\lambda^2} E_\lambda (X - \lambda)^2 = \frac{\lambda}{\lambda^2} = \frac{1}{\lambda}.$$

So the denominator of the Cramér-Rao lower bound (3.33) is n/λ. For $\eta(\lambda) = e^{-\lambda}$, the numerator of (3.33) will be $e^{-2\lambda}$. So the lower bound is $\lambda e^{-2\lambda}/n$. For instance, an unbiased estimator of $\eta(\lambda) = e^{-\lambda}$ is the sample proportion of zeroes, that is, $\hat{\eta}(\lambda) = n^{-1} \sum_{i=1}^n I_{\{0\}}(X_i)$. Since each X_i will either be zero or nonzero and $P(X_i = 0) = e^{-\lambda}$, clearly $\sum_{i=1}^n I_{\{0\}}(X_i)$ follows a

Binomial$(n, e^{-\lambda})$ p.m.f. and its variance is $ne^{-\lambda}(1 - e^{-\lambda})$. So the variance of $\hat{\eta}(\lambda) = n^{-1}\sum_{i=1}^{n} I_{\{0\}}(X_i)$ is $e^{-\lambda}(1-e^{-\lambda})/n$, which is $\geq$ the Cramér-Rao lower bound (verify it for yourself). Next, if $\eta(\lambda) = \lambda$ itself, the Cramér-Rao lower bound will be $\frac{1}{n/\lambda} = \frac{\lambda}{n}$. An unbiased estimator of λ is, of course, the sample mean $n^{-1}\sum_{i=1}^{n} X_i$. Its variance is $n\lambda/n^2 = \frac{\lambda}{n}$. So it is indeed a UMVUE of λ.

Example 3.48. Consider the Poisson(λ) example again. We will now derive a UMVUE for $\eta(\lambda) = e^{-\lambda}$ following the prescription in Result 3.16. For this purpose, we can start with *any* unbiased estimator of $e^{-\lambda}$. One such estimator is $I_{\{0\}}(X_1)$, which is a Bernoulli$(e^{-\lambda})$ random variable. Clearly, $\sum_{1}^{n} X_i$ is sufficient for λ, by the factorization theorem. It is actually minimal sufficient, since its p.m.f. belongs to the exponential family (see the remark at the end of Example 3.42). So, according to the Lehmann-Scheffe theorem, $E(I_{\{0\}}(X_1) \mid \sum_{1}^{n} X_i)$ will be a UMVUE of λ. To see what this conditional expectation actually is, notice that $P(X_1 = 0 \mid \sum_{1}^{n} X_i = r)$ is equal to

$$\frac{P(X_1 = 0 \text{ and } \sum_{2}^{n} X_i = r)}{P(\sum_{1}^{n} X_i = r)} = \frac{e^{-\lambda}e^{-(n-1)\lambda}((n-1)\lambda)^r/r!}{e^{-n\lambda}(n\lambda)^r/r!},$$

which simplifies to $(1 - n^{-1})^r$. In this derivation, we have used the fact that $\sum_{1}^{n} X_i \sim$ Poisson$(n\lambda)$ and $\sum_{2}^{n} X_i \sim$ Poisson$((n-1)\lambda)$. In any case, since $I_{\{0\}}(X_1)$ takes the values 0 and 1 only, $E(I_{\{0\}}(X_1) \mid \sum_{1}^{n} X_i = r) = (1-n^{-1})^r$. In other words, $E(I_{\{0\}}(X_1) \mid \sum_{1}^{n} X_i) = (1 - n^{-1})^{\sum X_i}$, which must be a UMVUE of $e^{-\lambda}$.

We conclude this section with testing of hypotheses. It is closely related to interval estimation or confidence-set estimation, mentioned in the previous section. Simply speaking, a *hypothesis* is a claim, assertion or conjecture about the parameter $\boldsymbol{\theta}$ of the p.d.f. or p.m.f. generating the data. Most often these claims or assertions will be made by somebody who has something to gain if they are true—perhaps a manufacturer of light-bulbs claiming the superiority of his/her product over other brands in terms of average lifetime (the parameter $\boldsymbol{\theta}$) or a drug company claiming its hypertension-lowering drug to be better than the existing brands in terms of the average duration of effect $(\boldsymbol{\theta})$ or a genomics researcher conjecturing that certain genes will be 'differentially expressed' (i.e., will have different average expression-values) under two different experimental conditions ('treatments'). Such a claim or assertion is often called the *research hypothesis* or the *hypothesis of interest*. A statistician's job is to confirm or refute it on the basis of the 'evidence' in the data $\{X_1, \ldots, X_n\}$. In order to understand how this is accomplished, we first need to formalize the setup and introduce some notations.

The research hypothesis is commonly denoted by H_1 (sometimes H_a, because another name for it is 'alternative hypothesis'). Faced with such an H_1, the statistician does not fall for it and keeps an open mind regarding the opposite statement or assertion (i.e., the one which negates or nullifies H_1). This opposite statement is usually called the *null hypothesis* and denoted by H_0.

In other words, the statistician views the whole process as a contest between H_1 and H_0—whichever is supported by more 'substantial' evidence (from the data) wins. At the end of the day, the verdict is announced as "There are strong reasons to believe H_1 (i.e., to reject H_0)" or "H_1 lacks strong enough evidence, so H_0 cannot be rejected." The question that naturally comes to mind is: What exactly is 'substantial evidence' or 'strong reasons'? Before addressing this question, we point out another important aspect of this procedure. Conventionally, while performing a hypothesis test, a statistician has always started out by being skeptical about H_1 and temporarily assuming that H_0 is true. Only if this initial assumption is 'blown away' by a 'mountain of evidence' to the contrary does he/she change his/her mind. This initial bias toward H_0 is a 'traditional hangover,' in the sense that it dates back to the time when hypothesis testing was primarily used for drug screening or comparing commercial products. For example, if a drug company comes up with a new hypertension-lowering drug and claims that it is more effective than the existing drug in some way (which will be H_1), an initial bias toward the existing one may be justified. It is simply a matter of being cautious. The existing drug is familiar, popular and time-tested and any official indication of its inferiority from a trusted authority (such as the Food and Drug Administration) may cause confusion, sudden shift of customer loyalty and loss of business for its manufacturer. If, by chance, the declaration of its inferiority was mistaken, it would have more serious health-related consequences. So there is some wisdom in one's initial bias toward the 'status quo' unless the new invention or new product is superior beyond any reasonable doubt. But not so if testing of hypotheses is being used as a tool for discovery—discovery of hitherto-unknown genes that are differentially expressed in cancer patients (compared to normal people) or of potentially beneficial chemicals that may someday become the cure for hitherto-invincible diseases. In these cases, being extra-conservative against the research hypothesis may result in one's failure to recognize talent or potential. However, in our brief treatment of the topic, we stick to the conventional philosophy.

Now, to the question of what exactly 'strong evidence' means. Notice that, if the verdict of a hypothesis-test is wrong, the mistake can be of two types. Either H_1 is indeed incorrect but the data somehow give us the opposite impression (i.e., make us believe that H_0 should be rejected in favor of H_1). Or it is the other way around. In the first case, we call it a *type I* error and in the second case, a *type II* error. Since our final verdict is based solely on the data $\{X_1, \ldots, X_n\}$ and we do not know the true value of the parameter θ, there is always a chance that we are making either a type I or a type II error whenever a hypothesis-test is performed. The probability of making a type I error is traditionally denoted by α and is called the *significance level* (or α-*level*) of the test. Similarly, P(a type II error) is denoted by β and $(1 - \beta)$ is called the *power* of the test. We can hope to carry out the test in such a way as minimizes both α and β. However, as will be clear from the examples below, it is difficult to achieve. In order to keep α low, one has to be ultra-conservative against H_1, which will inflate β. Similarly, being ultra-liberal in favor of H_1 to get a small β will increase α. So, we have to choose one of them and make sure that it is small, hoping for the best regarding the other one. As explained earlier, the tradition is to guard against high values of α. In any hypothesis-testing problem, therefore, α will be pre-specified (typical values are 0.1, 0.05 and 0.01). The art of hypothesis-testing is to devise procedures that obey the α restriction and, at the same time, minimize β (i.e., maximize the power) as much as possible.

Formally, a research hypothesis is written as $H_1 : \boldsymbol{\theta} \in \boldsymbol{\Theta}_1$ and the corresponding null hypothesis, as $H_0 : \boldsymbol{\theta} \in \boldsymbol{\Theta}_0$. Here $\boldsymbol{\Theta}_1$ and $\boldsymbol{\Theta}_0$ are subsets of the parameter-space $\boldsymbol{\Theta}$. Often $\boldsymbol{\Theta}_1 = \boldsymbol{\Theta}_0^c$, but not always. It depends on the context of the problem. For example, if the research hypothesis comes from a light-bulb manufacturer who claims that the average lifetime $(\boldsymbol{\theta})$ of his/her bulbs exceeds 2500 hours, then $\boldsymbol{\Theta} = (0, \infty)$, $\boldsymbol{\Theta}_1 = (2500, \infty)$ and $\boldsymbol{\Theta}_0 = (0, 2500]$. On the other hand, H_1 may come from a consumers' group claiming that the low-carb soft drinks produced by a certain company have 25% more sugar per bottle on an average than is indicated on the label. If the label says 40 grams, then $\boldsymbol{\Theta}_1 = \{50\}$ and $\boldsymbol{\Theta}_0$ may very well be $\{40\}$ (if H_1 and H_0 are interpreted respectively as "the consumers' group is right" and "the producer is right"). Or $\boldsymbol{\Theta}_1$ may be $\{\boldsymbol{\theta} > 0 : \boldsymbol{\theta} \neq 50\}$, and $\boldsymbol{\Theta}_0 = \{50\}$, if the producer takes the initiative and his/her primary interest lies in 'proving' the consumers' group wrong. In case it is $H_1 : \boldsymbol{\theta} = \{50\}$ versus $H_0 : \boldsymbol{\theta} = \{40\}$, we call it a 'simple null versus simple alternative' situation. If it is $H_1 : \boldsymbol{\theta} \in \{\boldsymbol{\theta} > 0 : \boldsymbol{\theta} \neq 50\}$ versus $H_0 : \boldsymbol{\theta} = \{50\}$, we say that it is a 'simple null versus composite alternative' scenario. The previous example involving light-bulbs, however, has both a composite null and a composite alternative hypotheses.

In any of these cases, having observed the data $\mathbf{X} = \{X_1, \ldots, X_n\}$, our task is to come up with a *decision rule* $\delta(\mathbf{X})$ (often called a *test function*) that is the indicator function of a region in $\mathcal{R}^n$ and H_0 will be rejected in favor of H_1 if and only if $\delta(\mathbf{X}) = 1$. The region in $\mathcal{R}^n$ associated with δ is usually called the *rejection region* or the *critical region* of the test. Of course, given α, we have to find a δ that satisfies: $E_{\boldsymbol{\theta}}\delta(\mathbf{X}) \leq \alpha$ for all $\boldsymbol{\theta} \in \boldsymbol{\Theta}_0$. This is often referred to as the *size requirement* of the test. Let us denote the collection of all test functions δ with size α by Δ_α. The quantity $E_{\boldsymbol{\theta}}\delta(\mathbf{X}) = P_{\boldsymbol{\theta}}[\delta(\mathbf{X}) = 1]$ is a function of $\boldsymbol{\theta}$ for a fixed δ (denote it by $\beta_\delta(\boldsymbol{\theta})$). For $\boldsymbol{\theta} \in \boldsymbol{\Theta}_1$, it is called the *power function* of δ. For a particular $\boldsymbol{\theta} \in \boldsymbol{\Theta}_1$, if there is a test function $\delta_0 \in \Delta_\alpha$ such that $\beta_{\delta_0}(\boldsymbol{\theta}) \geq \beta_\delta(\boldsymbol{\theta})$ for all $\delta \in \Delta_\alpha$, we call it the *most powerful* (MP) size-α test at that $\boldsymbol{\theta}$. If we are fortunate enough to find a test function $\delta^* \in \Delta_\alpha$ which is MP at every single $\boldsymbol{\theta} \in \boldsymbol{\Theta}_1$, it will be called a *uniformly most powerful* (UMP) size-α test. Two important remarks before we see some examples: (1) The actual definition of a test function δ is more general; it can be any integrable function from $\mathcal{R}^n \longrightarrow [0, 1]$. But here we restrict ourselves to indicator functions only. (2) Often the test function will depend on the data only through a sufficient statistic which has a lower dimension. In that case, δ

will be the indicator function of a region in a lower dimensional subspace of $\mathcal{R}^n$.

Example 3.49. Let $\{X_1, \ldots, X_n\}$ be i.i.d. $N(\mu, 1)$ random variables where the mean μ is unknown and suppose we are testing $H_0 : \mu = \mu_0$ vs. $H_1 : \mu = \mu_1$, with $\mu_1 > \mu_0$ and $\alpha = 0.05$. A possible test function δ would be the indicator function of the interval $(\mu_0 + \frac{z_\alpha}{\sqrt{n}}, \infty)$, that is, a test with critical region $(\mu_0 + \frac{1.645}{\sqrt{n}}, \infty)$. It clearly satisfies the size requirement, since under H_0, $\bar{X} \sim N(\mu_0, \frac{1}{n})$, so that $\sqrt{n}(\bar{X} - \mu_0)$ is a standard normal random variable. The power of this test at $\mu = \mu_1$ is $P_{\mu_1}[\bar{X} \in (\mu_0 + \frac{1.645}{\sqrt{n}}, \infty)] = P_{\mu_1}[\sqrt{n}(\bar{X} - \mu_1) > \sqrt{n}(\mu_0 - \mu_1) + 1.645]$, which is the tail-area to the right of $\sqrt{n}(\mu_0 - \mu_1) + 1.645$ under a standard normal curve. Is this test "best" in some sense? We will return to that question.

Example 3.50. Suppose X is a single observation from a Binomial $(4, \theta)$ distribution and we want to test $H_0 : \theta = \frac{1}{2}$ vs. $H_1 : \theta = \frac{4}{5}$ at a significance level of $\alpha = \frac{1}{16} = 0.0625$. In other words, we are looking for a critical region C of size $\alpha = \frac{1}{16}$. One choice for C would be $\{x = 0\}$, since the chances of seeing no 'head' in 4 tosses of a fair coin is $\frac{1}{16}$. Another is $\{x = 4\}$. There is no third choice. If we choose $\{x = 0\}$, the power of the test at $\theta = \frac{4}{5}$ will be $(1/5)^4 = 1/625$. This is not desirable, since the test will have a much smaller chance of rejecting H_0 when H_1 is indeed true than when H_0 is true. The requirement that $P_{H_1}(\text{rejecting } H_0)$ be at least as large as $P_{H_0}(\text{rejecting } H_0)$ is known as *unbiasedness*. Let us see what happens if we choose $\{x = 4\}$. Then the power at $\theta = 45$ is $\frac{256}{625}$, which is significantly larger than the size. So the test corresponding to $\{x = 4\}$ is the most powerful unbiased (MPU) test of size $\frac{1}{16}$ (as long as we are restricting ourselves to test functions δ that are indicator functions). Notice that in this problem, the MPU test is obtained by including in the critical region all those values x of X where the ratio $f_1(x)/f_0(x)$ is the largest (f_i being the p.m.f. of X under the hypothesis H_i, $i = 1, 2$). Is this a mere coincidence? Next we address this question.

The fact that the most powerful test in the above example had a critical region that corresponded to the highest values of the ratio $f_1(x)/f_0(x)$ (the *likelihood ratio*) is just an illustration of the following well-known result in hypothesis testing:

Neyman-Pearson Theorem: Let $\{X_1, \ldots, X_n\}$ be a random sample from a p.d.f. (or p.m.f.) $f(x; \theta)$, so that their joint p.d.f. (or p.m.f.) is $L(\theta; x_1, \ldots, x_n) = \prod_{i=1}^{n} f(x_i; \theta)$. Let θ_0 and θ_1 be distinct values of θ and a be a positive number. If C is a subset of the sample space such that

(1) $P_{H_0}[(X_1, \ldots, X_n) \in C] = \alpha$,

(2) $\frac{L(\theta_0; x_1, \ldots, x_n)}{L(\theta_1; x_1, \ldots, x_n)} \le a$ for each $(x_1, \ldots, x_n) \in C$, and

(3) $\frac{L(\theta_0; x_1, \ldots, x_n)}{L(\theta_1; x_1, \ldots, x_n)} \ge a$ for each $(x'_1, \ldots, x'_n) \notin C$,

then C is a most powerful critical region (i.e., the test using C^* as the critical region is a most powerful test) of size α for testing $H_0 : \theta = \theta_0$ vs. $H_1 : \theta = \theta_1$.

Intuitively, this result says that a most powerful test rejects H_0 at sample-points where it is more likely that the sample came from $L(\theta_1; x_1, \ldots, x_n)$ than $L(\theta_0; x_1, \ldots, x_n)$ (i.e., where the latter is 'small' compared to the former). Although the three conditions listed above are *sufficient* conditions (i.e., the result is true *if* they are satisfied), it can also be shown that they are necessary conditions (i.e., the result is no longer true if any of them is violated). In a real application, C may be defined as all sample points $\{x_1, \ldots, x_n\}$ such that conditions (2) and (3) hold. Often such a definition boils down to either $u^*(\theta_0, \theta_1, x_1, \ldots, x_n) \leq a^*$ or $u^{**}(\theta_0, \theta_1, x_1, \ldots, x_n) \geq a^{**}$ for some convenient functions u^* and u^{**} and appropriate constants a^* and a^{**}, as will be evident in the examples below.

Example 3.51. Suppose that $X_1, \ldots, X_n$ are i.i.d. observations from the p.d.f. $f(x; \theta) = \theta e^{-\theta x}$ on $(0, \infty)$, which is a reparameterized version of the Exponential (θ) p.d.f. in (3.15). Suppose that we are testing $H_0 : \theta = \theta_0$ vs. $H_1 : \theta = \theta_1$ for some known, fixed values θ_0 and θ_1 $(\theta_1 > \theta_0)$. Here, $L(\theta_0; x_1, \ldots, x_n) = \theta_0^n e^{-\theta_0 \sum_1^n x_i}$ and $L(\theta_1; x_1, \ldots, x_n) = \theta_1^n e^{-\theta_1 \sum_1^n x_i}$. Let us denote their ratio by λ. The Neyman-Pearson theorem says that a most powerful test will have the form: Reject H_0 if the likelihood-ratio $\lambda \leq a$ for some constant a, which boils down to

$$(\frac{\theta_0}{\theta_1})^n \exp[-(\theta_0 - \theta_1) \sum_1^n x_i] \leq a \iff \sum_1^n x_i \leq \frac{1}{\theta_1 - \theta_0} \log_e \{a(\theta_1/\theta_0)^n\}.$$

If we now denote the quantity on the right-hand side of the last inequality by a^*, the critical region of a most powerful test takes the form: $\sum_1^n x_i \leq a^*$. The significance level of such a test will be $P_{\theta_0}(\sum_1^n X_i \leq a^*)$. In this case, we know that $\sum_1^n X_i \sim \text{Gamma}(n, \theta)$, so for a pre-specified α, the appropriate choice for a^* can be determined from the integral equation

$$\int_0^{a^*} \frac{1}{\Gamma(n)} \theta_0^n x^{n-1} e^{-x\theta_0} dx = \alpha.$$

In other words, a^* will have to be the αth percentile of a $\text{Gamma}(n, \theta_0)$ p.d.f.

Example 3.52. Let $X_1, \ldots, X_n$ be i.i.d. observations from a $N(\theta, 1)$ p.d.f. and suppose that we want to test $H_0 : \theta = 0$ vs. $H_1 : \theta = 1$. In this case, the sample likelihood under H_0 is $(1/\sqrt{2\pi})^n \exp[-(\sum_1^n x_i^2)/2]$ and that under H_1 is $(1/\sqrt{2\pi})^n \exp[-\{\sum_1^n (x_i - 1)^2\}/2]$, so the likelihood-ratio λ ultimately reduces to $\exp(-\sum_1^n x_i + \frac{n}{2})$. Hence, a most powerful test will have a critical region of the form: $\exp(-\sum_1^n X_i + \frac{n}{2}) \leq a$, which is the same as saying that $\sum_1^n X_i \geq \frac{n}{2} - \log_e a = a^{**}$ (say). Since under H_0, $\sum_1^n X_i \sim N(0, n)$, the appropriate choice for a^{**} for a given significance level α is simply the $(1-\alpha)^{th}$ percentile of a $N(0, n)$ density. Equivalently, the most powerful critical region can be expressed in terms of $\bar{X} = \frac{1}{n}\sum_1^n X_i$ as: Reject H_0 if $\bar{X} \geq a^{**}/n$. In that case, using the fact that $\bar{X} \sim N(0, \frac{1}{n})$, a^{**} can be chosen as n times the $(1-\alpha)^{th}$ percentile of a $N(0, \frac{1}{n})$ density. What about the power of this test at $\theta = 1$? It is $P_{\theta=1}[\bar{X} \geq a^{**}/n]$ or the right-hand tail-area of a^{**}/n under a $N(1, \frac{1}{n})$ density curve.

Example 3.53. Let $\{X_1, \ldots, X_n\}$ be a random sample from a Bernoulli(θ) p.m.f. and suppose that we are testing $H_0 : \theta = \theta_0$ vs. $H_1 : \theta = \theta_1$ with $\theta_0 < \theta_1$. The sample likelihoods under H_0 and H_1 are respectively $\theta_0^{\sum_1^n x_i}(1-\theta_0)^{n-\sum_1^n x_i}$ and $\theta_1^{\sum_1^n x_i}(1-\theta_1)^{n-\sum_1^n x_i}$. So the likelihood-ratio λ is less than some constant a if and only if

$$\left[\frac{\theta_0(1-\theta_1)}{(1-\theta_0)\theta_1}\right]^{\sum_1^n x_i}\left[\frac{1-\theta_0}{1-\theta_1}\right]^n \leq a \Longleftrightarrow \sum_1^n x_i \geq a^{**},$$

where a^{**} is obtained from a by taking the natural logarithm of both sides of the first inequality. Do you see why the $\leq$ changes to a $\geq$ in the second inequality above? Notice that $\log_e\{(\theta_0(1-\theta_1))((1-\theta_0)\theta_1)\}$ is negative since $\theta_0 < \theta_1$. In any case, a most powerful test will have a critical region of the form: Reject H_0 if $\sum_1^n X_i \geq a^{**}$. For a given significance level α, the choice of a^{**} will come from a Binomial(n, θ_0) p.m.f. table. Since it is a discrete p.m.f., a significance level of α may not be exactly achievable. This is an example where, in order to achieve an α-level exactly, one must use a test function that is *not* an indicator function. But we choose not to get into the details and, instead, move on to composite hypotheses.

If we are testing $H_0 : \boldsymbol{\theta} \in \boldsymbol{\Theta_0}$ against $H_1 : \boldsymbol{\theta} \in \boldsymbol{\Theta_1}$ with at least one of $\boldsymbol{\Theta_0}$ and $\boldsymbol{\Theta_1}$ not singleton, the Neyman-Pearson likelihood-ratio technique described above is not directly applicable. The modified version of it for composite hypotheses is known as the *generalized likelihood ratio* (GLR) method. Of course, the optimality criterion for a test will also change in this case. Since there are many possible values of $\boldsymbol{\theta}$ in $\boldsymbol{\Theta_1}$, we now have to talk about a *uniformly most powerful* (UMP) test over $\boldsymbol{\Theta_1}$ (i.e., a test which is most powerful at every $\boldsymbol{\theta} \in \boldsymbol{\Theta_1}$). Unfortunately, the GLR method will not always yield a UMP test, but sometimes it will. However, as will be clear from its description, the GLR method makes good intuitive sense.

As before, $L(\boldsymbol{\theta}; x_1, \ldots, x_n)$ will denote the sample likelihood function. The GLR method says that we must reject H_0 if the largest value of $L(\boldsymbol{\theta}; x_1, \ldots, x_n)$ over $\boldsymbol{\Theta_0}$ is "quite small" compared to its largest value over the entire parameter space $\boldsymbol{\Theta_0} \cup \boldsymbol{\Theta_1}$, because the ratio of these two largest values is an indicator of the chances that $\boldsymbol{\theta}$ indeed came from $\boldsymbol{\Theta_0}$. In other words, the GLR method says:

$$\text{Reject } H_0 \text{ if } \lambda^* = \frac{\sup_{\boldsymbol{\theta}\in\boldsymbol{\Theta_0}} L(\boldsymbol{\theta}; x_1, \ldots, x_n)}{\sup_{\boldsymbol{\theta}\in\boldsymbol{\Theta_0} \cup \boldsymbol{\Theta_1}} L(\boldsymbol{\theta}; x_1, \ldots, x_n)} \leq c$$

for an appropriately chosen positive fraction c. Notice that the denominator of λ^* is nothing but the sample likelihood function evaluated at the MLE of θ. Clearly $\lambda^* \in [0, 1]$, but how "small" should c be chosen for a pre-specified α? To answer this, we need to know the distribution of λ^* under H_0, which unfortunately is often difficult to derive. However, for a large sample-size n, the distribution of $-2\log_e \lambda^*$ can be well approximated by a χ^2 distribution with r degrees of freedom (where r is the dimension of Θ_0). This enables us to derive a GLR test with an approximate significance level α.

Example 3.54. Suppose, as in Example 3.51, that $X_1, \ldots, X_n$ are i.i.d. observations from the Exponential (θ) p.d.f. $f(x; \theta) = \theta e^{-\theta x}$, so that the parameter space Θ is $(0, \infty)$. Let us test $H_0 : \theta \leq \theta_0$ against $H_1 : \theta > \theta_0$. So $\Theta_0 = (0, \theta_0]$ and $\Theta_1 = (\theta_0, \infty)$. Hence,

$$\sup_{\theta \in \Theta} L(\theta; x_1, \ldots, x_n) = \sup_{\theta \in (0, \infty)} \{\theta^n \exp(-\theta \sum_1^n x_i)\} = (\frac{n}{\sum_1^n x_i})^n e^{-n}.$$

At the same time,

$$\sup_{\theta \in \Theta_0} L(\theta; x_1, \ldots, x_n) = \sup_{\theta \in (0, \theta_0]} \{\theta^n \exp(-\theta \sum_1^n x_i)\} = (\frac{n}{e \sum_1^n x_i})^n \text{ if } \frac{n}{\sum_1^n x_i} \leq \theta_0,$$

whereas it is $\theta_0^n \exp(-\theta_0 \sum_1^n x_i)$ if $\frac{n}{\sum_1^n x_i} > \theta_0$. So the generalized likelihood-ratio is

$$\lambda^* = 1 \text{ if } \frac{n}{\sum_1^n x_i} \leq \theta_0 \; , \; \lambda^* = \frac{\theta_0^n \exp(-\theta_0 \sum_1^n x_i)}{(n / \sum_1^n x_i)^n e^{-n}} \text{ if } \frac{n}{\sum_1^n x_i} > \theta_0.$$

So a GLR test would be to reject H_0 if $\lambda^* \leq c$ for some $c \in (0, 1)$. This translates to rejecting H_0 if $(n / \sum_1^n x_i) > \theta_0$ and $(\theta_0 \bar{x})^n e^{-n(\theta_0 \bar{x} - 1)} \leq c$. Notice that the largest value of $(\theta_0 \bar{x})^n e^{-n(\theta_0 \bar{x} - 1)}$ is 1 and it occurs when $\theta_0 \bar{x} = 1$. So, as long as we have $\theta_0 \bar{x} < 1$ (i.e., as long as we are to the left of that maximum value), it is easy to see that $(\theta_0 \bar{x})^n e^{-n(\theta_0 \bar{x} - 1)} \leq c \iff \theta_0 \bar{x} \leq c^*$ for some appropriate constant $c^* \in (0, 1)$. Therefore, that is what the GLR test boils down to.

3.8　Some Computational Issues

We now know that in order to find the maximum-likelihood estimators of parameters or to derive a GLR test for composite hypotheses, we need to maximize the sample likelihood function $L(\theta; x_1, \ldots, x_n)$ or its natural logarithm. This maximization is analytically possible for samples coming from a handful of distributions (e.g., normal, exponential, etc.). But in general, it is necessary to carry out this maximization numerically via some iterative algorithm. Two such algorithms are the *Newton-Raphson method* (Press et al., 1986) and the *expectation-maximization algorithm* (Dempster et al.,1977). We describe them briefly below.

The Newton-Raphson method (or its multidimensional generalization for vector parameters) is based on a Taylor series expansion of the derivative of the log-likelihood function:

$$L'(\boldsymbol{\theta_0}; x_1, \ldots, x_n) = L'(\boldsymbol{\theta}; x_1, \ldots, x_n) + (\boldsymbol{\theta_0} - \boldsymbol{\theta}) L''(\boldsymbol{\theta}; x_1, \ldots, x_n) + \text{higher-order}$$

terms, where $\boldsymbol{\theta}$ is a generic parameter-vector and $\boldsymbol{\theta_0}$ is the true (unknown) parameter-vector. Ignoring the higher-order terms and setting the expression in the above equation to zero, we get

$$\boldsymbol{\theta_0} = \boldsymbol{\theta} - \frac{L'(\boldsymbol{\theta}; x_1, \ldots, x_n)}{L''(\boldsymbol{\theta}; x_1, \ldots, x_n)}, \tag{3.34}$$

which gives us an iterative algorithm starting with an initial 'guess' $\boldsymbol{\theta}$. Notice that $L'(\boldsymbol{\theta}; x_1, \ldots, x_n)$ is a vector of first derivatives (called the *gradient vector*) and $L''(\boldsymbol{\theta}; x_1, \ldots, x_n)$ is a matrix of second derivatives (also known as the *Hessian matrix*). Usually the iterations are terminated when the updated parameter-vector $\boldsymbol{\theta}$ differs very little from its previous value (in terms of their Euclidean distance, say). In practice, this method is quick and efficient for parameter-vectors of dimension 3 or less. But for higher dimensions, the increasing size of the Hessian matrix causes some computational inefficiency and, also, the iterations can become quite unstable (in the sense of producing updates of the parameter-vector that are further away from the desired maximum than their predecessors).

An alternative to the above method that does not suffer from these drawbacks is the EM algorithm. It was originally devised for problems with missing data (e.g., due to censoring or truncating). So it is particularly suitable for problems where some 'latent' or 'hypothetical' data (that are unknowable and, hence, missing) have been introduced for analytical or computational convenience. The following is an example of this algorithm in use.

Example 3.55. Suppose we have a bimodal dataset and would like to fit a mixture of two normal distributions to it. In other words, we are assuming that $X_1, \ldots, X_n$ are i.i.d., with each of them coming from either a $N(\mu_1, \sigma_1^2)$ or from a $N(\mu_2, \sigma_2^2)$ p.d.f. Of course, it is unknown which X_i belongs to which normal p.d.f., but the probability that it comes from $N(\mu_1, \sigma_1^2)$ is $p \in (0, 1)$ and that it comes from $N(\mu_2, \sigma_2^2)$ is $1 - p$. Although this is not a missing-data problem as such, do you see how it can be interpreted as one? We could introduce a bunch of i.i.d. Bernoulli(p) random variables $Z_1, \ldots, Z_n$, with $Z_i = 1$ indicating that X_i came from $N(\mu_1, \sigma_1^2)$ and $Z_i = 0$ indicating that X_i came from $N(\mu_2, \sigma_2^2)$. These Z_i's would then be latent or unobserved 'data'. If we knew the values of $Z_1, \ldots, Z_n$, the 'success' parameter p would be estimated simply by their average. Even if we do not know them, a simple application of the Bayes theorem yields the following expression for the conditional expectation of Z_i

given X_i for $i = 1, \ldots, n$:

$$E(Z_i \mid X_i) = P[X_i \text{ from } f(.; \mu_1, \sigma_1^2)] = \frac{pf(X_i; \mu_1, \sigma_1^2)}{pf(X_i; \mu_1, \sigma_1^2) + (1-p)f(X_i; \mu_2, \sigma_2^2)}. \tag{3.35}$$

This is actually the posterior probability of X_i coming from $N(.; \mu_1, \sigma_1^2)$ given X_i. Starting with some initial 'guess' for each parameter involved, once we compute this posterior probability for each $i = 1, \ldots, n$, the mixing parameter p can be updated as their average. These computations constitute the 'expectation' part of the EM algorithm. The subsequent step is 'maximization,' which is nothing but maximum-likelihood estimation using the values obtained from the 'E' step. In other words,

$$\hat{\mu}_1 = \frac{1}{np} \sum_1^n X_i E(Z_i \mid X_i) \,, \; \hat{\mu}_2 = \frac{1}{n(1-p)} \sum_1^n X_i[1 - E(Z_i \mid X_i)] \quad (3.36)$$

and

$$\hat{\sigma}_1^2 = \frac{1}{np} \sum_1^n (X_i - \hat{\mu}_1)^2 E(Z_i \mid X_i) \,, \; \hat{\sigma}_2^2 = \frac{1}{n(1-p)} \sum_1^n (X_i - \hat{\mu}_2)^2 [1 - E(Z_i \mid X_i)]. \tag{3.37}$$

Once these MLEs are obtained, they will be used in the 'expectation' step of the next iteration.

Although estimators and testing methods such as the MLE and the GLR are widely used, it is often difficult to derive analytic expressions for some of their important features such as bias, variance or power. If the underlying sample-size is large, asymptotic approximations can be made to those quantities, but most real-life problems have moderate or small sample-sizes. There is another computer-intensive alternative—a *resampling* procedure called *bootstrap* (Efron 1979). The idea behind it has some similarity with the method-of-moments estimation technique described earlier. In order to get the method-of-moments estimators of parameters, we equate the population moments (involving those parameters) with the corresponding sample moments. In the case of bootstrap, we estimate functionals of the true (unknown) distribution function (e.g., moments, quantiles, etc.) with the corresponding functionals of the empirical distribution function of our sample data. Recall that the empirical c.d.f. of the sample observations $X_1, \ldots, X_n$ is a step-function with a 'jump' of $\frac{1}{n}$ at each X_i. How can we compute its quantiles and other functionals? By drawing a random sample (of size $m \leq n$) from it, which amounts to sampling from the sample data we already have. Hence the term *resampling*. This resampling is typically done many, many times and it can be carried out with replacement or without replacement. Since the empirical c.d.f. of the original sample gives equal weight $\frac{1}{n}$ to each data-point, each of them is equally likely to be selected during the resampling process. However, there is a procedure called *weighted bootstrap* where the data-points in the original sample receive unequal weights. In any case, suppose we want a $100(1 - \alpha)\%$ confidence interval for a parameter (say, the median) of the population distribution from where our sample data came. We can draw N (say, $N = 5000$ or 10000) bootstrap-samples from the original sample (denote them by $B_1, \ldots, B_N$) and from each of them, compute the sample-median. Thus we will have N bootstrapped sample-median values (call them $M_1^b, \ldots, M_N^b$). Denoting their simple average by $\bar{M}^b$ and their standard deviation by s_M^b, we get the desired $100(1 - \alpha)\%$ confidence interval for the population median as $(\bar{M}^b - z_{\alpha/2}\, s_M^b/\sqrt{N},\, \bar{M}^b + z_{\alpha/2}\, s_M^b/\sqrt{N})$. There is an extensive literature on bootstrap discussing conditions under which bootstrapped estimators asymptotically converge to their target parameters, but those are beyond the scope of our present discussion.

3.9 Bayesian Inference

As was indicated in Section 1, while "learning" to some of us means estimating or testing hypotheses about unknown quantities or parameters (assumed to be fixed) from the data (a random sample from the population concerned), other people interpret the word as updating our existing knowledge about unknown parameters (assumed to be random variables) from the data (considered fixed in the sense that we are only interested in the conditional behavior of the parameters given the data). This is the Bayesian approach. Recall the Bayes theorem from Section 2. There were two main ingredients: a likelihood and a bunch of prior probabilities, ultimately yielding a bunch of posterior probabilities. Following the same principle, we will now have a dataset $\{X_1, \ldots, X_n\}$ which, given the parameter $\boldsymbol{\theta} = (\theta_1, \ldots, \theta_p)$, will be conditionally i.i.d. having the common p.d.f. or p.m.f. $f(x \mid \boldsymbol{\theta})$. So their conditional likelihood given $\boldsymbol{\theta}$ is $\prod_{i=1}^{n} f(x_i \mid \boldsymbol{\theta})$. The prior knowledge about $\boldsymbol{\theta}$ will be in the form of a prior p.d.f. or p.m.f. $\pi(\boldsymbol{\theta})$. Depending on the situation, the θ_i's can be assumed to be independent *a priori*, so that $\pi(\boldsymbol{\theta}) = \prod_{i=1}^{p} \pi_i(\theta_i)$ where $\theta_i \sim \pi_i$. Or the θ_i's can be dependent, in which case, $\pi(\boldsymbol{\theta}) = \pi_1(\theta_1)\pi_2(\theta_2 \mid \theta_1)\ldots\pi_p(\theta_p \mid \theta_j, j < p)$. In what follows, we will assume this latter form with the understanding that $\pi_i(\theta_i \mid \theta_j, j < i) = \pi_i(\theta_i)$ in the case of independence. In any case, by (3.22), the joint distribution of $\{X_1, \ldots, X_n, \theta_1, \ldots, \theta_p\}$ is

$$L(x_1, \ldots, x_n \mid \boldsymbol{\theta})\pi(\boldsymbol{\theta}) = \left(\prod_{i=1}^{n} f(x_i \mid \boldsymbol{\theta})\right)\pi_1(\theta_1)\pi_2(\theta_2 \mid \theta_1)\ldots\pi_p(\theta_p \mid \theta_j, j < p).$$

(3.38)

For the denominator of the Bayes formula, we need the marginal joint p.d.f. or p.m.f. of the X_i's. To get it from (3.32), we need to integrate it (or sum it) w.r.t. the θ_i's over their full ranges. Finally, the posterior joint p.d.f. or p.m.f. of $\theta_1, \ldots, \theta_p$ will be

$$\frac{\left(\prod_{i=1}^{n} f(x_i \mid \boldsymbol{\theta})\right)\pi_1(\theta_1)\pi_2(\theta_2 \mid \theta_1)\ldots\pi_p(\theta_p \mid \theta_j, j < p)}{\int_{\theta_1}\cdots\int_{\theta_p}\left(\prod_{i=1}^{n} f(x_i \mid \boldsymbol{\theta})\right)\pi_1(\theta_1)\pi_2(\theta_2 \mid \theta_1)\ldots\pi_p(\theta_p \mid \theta_j, j < p)d\theta_p\ldots d\theta_1}.$$

(3.39)

Let us denote it by $\pi_{\boldsymbol{\theta}|\mathbf{X}}(\boldsymbol{\theta})$. Once this is obtained, we could estimate the θ_i's by the values that correspond to one of the posterior modes (i.e., highest points of the posterior surface). Or we could use the set $\{(\theta_1, \ldots, \theta_p) : \pi_{\boldsymbol{\theta}|\mathbf{X}}(\boldsymbol{\theta}) \geq 0.95\}$ as a 95% confidence region for $\boldsymbol{\theta}$ (called a *highest-posterior-density credible region* or HPDCR). However, due to the high dimensions of the parameter-vectors in real-life problems, computing (3.39) is usually an up-hill task—especially the high-dimensional integral in the denominator. Over the years, a number of remedies have been suggested for this. One of them involves being "clever" with your choice of the prior p.d.f. (or p.m.f.) so that its functional form "matches" with that of the data-likelihood and as a result, the product of these two (the joint distribution (3.38)) has a familiar form. Then the denominator of (3.39) does not actually have to be computed, because it will simply be the 'normalizing constant' of a multivariate p.d.f. or p.m.f. This is the so-called *conjugacy* trick. When this cannot be done, there are some useful analytical or numerical approximations to high-dimensional integrals (such as the LaPlace approximation) which enables us to approximate the posterior. When none of these works, we can resort to *Markov chain Monte Carlo* (MCMC) techniques. See Section 1 for the basic idea behind it. There are many variants of it, such as the *Gibbs sampler*, the *Metropolis* (or *Metropolis-Hastings*) algorithm and so on.

Recall that in Section 3.7, we mentioned statistical decision theory and one of its main principles (i.e., the minimax principle) in the context of our search for a "good" estimator. Another fundamental one is the *Bayes principle*. If δ is a decision rule (e.g., an estimator, a test function, a classification rule, etc.) and $\mathcal{R}_\delta(\boldsymbol{\theta})$ is the associated risk function with respect to some loss function $\mathcal{L}(\boldsymbol{\theta}, \delta)$, the *Bayes risk* of δ under the prior distribution $\pi(\boldsymbol{\theta})$ is defined as $r(\pi, \delta) = E_\pi[\mathcal{R}_\delta(\boldsymbol{\theta})]$. According to the Bayes principle, we should prefer the decision rule δ_1 to a competing decision rule δ_2 if $r(\pi, \delta_1) < r(\pi, \delta_2)$. If we can find a decision rule δ_π that minimizes $r(\pi, \delta)$ over all decision rules, we call it the *Bayes rule*. The associated Bayes risk $r(\pi, \delta_\pi)$ is often simply called the Bayes risk of π.

For a decision rule δ, a measure of *initial precision* is a quantity computed by averaging over all possible datasets that *might be* observed. Such a measure can be computed even before observing any data. Examples are the risk $\mathcal{R}_\delta(\boldsymbol{\theta})$, the Bayes risk $r(\pi, \delta)$, the MSE of an estimator, the type I and type II error-probabilities of a test, and so on. On the other hand, a measure of *final precision* is something that can be computed only after observing the data. Therefore, final precision is a measure *conditional* on the data. In Bayesian learning, final precision is considered the most appropriate way to assess the optimality of a decision rule. To a Bayesian, datasets that might have been (but were not) observed are not relevant to the inference-drawing activity. The only relevant dataset is the one actually observed. Does it mean that the underlying philosophy of Bayesian learning is incompatible with the Bayes principle itself? To resolve this apparent contradiction, we need to distinguish between the *normal* and the *extensive* forms of a Bayes rule. The normal method of finding a Bayes rule is the one discussed above. The extensive method is based on the fact that the Bayes risk $r(\pi, \delta)$ for a continuous parameter $\boldsymbol{\theta} \in \boldsymbol{\Theta}$ is $\int_{\boldsymbol{\Theta}} \mathcal{R}_\delta(\boldsymbol{\theta}) \pi(\boldsymbol{\theta}) d\boldsymbol{\theta}$

$$= \int_{\boldsymbol{\Theta}} [\int_{\mathcal{X}} L(\boldsymbol{\theta}, \delta(\mathbf{x})) f(\mathbf{x} \mid \boldsymbol{\theta}) d\mathbf{x}] \pi(\boldsymbol{\theta}) d\boldsymbol{\theta} = \int_{\mathcal{X}} [\int_{\boldsymbol{\theta}} L(\boldsymbol{\theta}, \delta(\mathbf{x})) f(\mathbf{x} \mid \boldsymbol{\theta}) \pi(\boldsymbol{\theta}) d\boldsymbol{\theta}] d\mathbf{x},$$

where the conditions for switching the order of integration are assumed to be satisfied and $\mathcal{X}$ denotes the sample space. So, in order to minimize $r(\pi, \delta)$, all we need to do is find a decision rule that minimizes the integrand (inside square brackets) in the last expression above. If we now denote the denominator of (3.39) by $m(\mathbf{x})$, where $\mathbf{x} = \{x_1, \ldots, x_n\}$, it is easy to see that a decision rule will minimize $r(\pi, \delta)$ if it minimizes $\int_{\boldsymbol{\Theta}} L(\boldsymbol{\theta}, \delta(\mathbf{x})) \left[\frac{f(\mathbf{x}|\boldsymbol{\theta})}{m(\mathbf{x})} \right] \pi(\boldsymbol{\theta}) d\boldsymbol{\theta}$. But this last quantity is nothing but $\int_{\boldsymbol{\Theta}} L(\boldsymbol{\theta}, \delta(\mathbf{x})) \pi_{\boldsymbol{\theta}|\mathbf{x}}(\boldsymbol{\theta}) d\boldsymbol{\theta}$, which is usually called the *Bayes posterior risk* of the decision rule δ. In other words, a Bayesian always seeks a decision rule that minimizes the Bayes posterior risk (a measure of final precision) and this does not violate the Bayes principle. Next we see some examples.

Example 3.56 Let X be a single observation from a Binomial(n, θ) population, so its p.m.f. is $f(x \mid \theta) = {}_nC_x \theta^x (1 - \theta)^{n-x}$. If $\pi(\theta)$ is the prior p.d.f. of θ, its posterior p.d.f. will be

$$\pi_{\theta|\mathbf{x}}(\theta) = \frac{\theta^x (1 - \theta)^{n-x} \pi(\theta)}{\int_0^1 s^x (1 - s)^{n-x} \pi(s) ds}.$$

As was mentioned earlier, often it will be good enough to know that $\pi_{\theta|\mathbf{x}}(\theta) \propto \theta^x (1 - \theta)^{n-x} \pi(\theta)$. In any case, the integral in the denominator of $\pi_{\theta|\mathbf{x}}(\theta)$ turns out to be difficult to calculate analytically even for some simple priors $\pi(\theta)$. But one particular choice for $\pi(\theta)$ renders analytical calculation unnecessary and it is $\pi(\theta) = (1/B(a, b)) \theta^{a-1} (1-\theta)^{b-1}$, the Beta$(a, b)$ p.d.f. Then $\pi_{\theta|\mathbf{x}}(\theta) \propto \theta^{x+a-1} (1-\theta)^{n-x+b-1}$ and clearly it is a Beta$(x+a, n-x+b)$ p.d.f. So, based on the observed data, Bayesian learning in this case would simply mean updating the prior p.d.f. parameters from (a, b) to $(x + a, n - x + b)$. This is the so-called 'conjugacy trick.' The Beta family of prior p.d.f.s is a *conjugate family* for binomial data. But are we restricting our choice for a prior p.d.f. heavily for the sake of convenience? The answer is 'no,' because the Beta family of densities provides quite a bit of flexibility in modeling the prior knowledge about θ as the parameters $a > 0$ and $b > 0$ vary.

Once the posterior p.d.f. is obtained, how do we use it to draw inference on the parameter θ? Intuitively, a reasonable estimate of θ seems to be the posterior mode. Since the mode of a Beta(s_1, s_2) density is $\frac{s_1-1}{s_1+s_2-2}$ (provided that $s_1 > 1$ and $s_2 > 1$), the posterior mode in the beta-binomial scenario is $\frac{x+a-1}{n+a+b-2}$ as long as $n > 1$ and $0 < x < n$. If in addition, both the prior parameters a and b are larger than 1, we can talk about the prior mode $\frac{a-1}{a+b-2}$. Recall from Section 3.7 that, based on a single observation from Binomial(n, θ), the MLE (as well as UMVUE) of θ is $\frac{X}{n}$. It may be interesting to observe that the posterior estimate of θ (i.e., the posterior mode) is a *weighted average* of the prior mode and the MLE. In other words,

$$\text{posterior mode} = \frac{n}{n+a+b-2}\left(\frac{X}{n}\right) + \left(1 - \frac{n}{n+a+b-2}\right)\left(\frac{a-1}{a+b-2}\right).$$
$$(3.40)$$

This is not the end of the story. If, instead of just deciding to use the posterior mode as the estimator, we take a decision-theoretic approach and try to find the 'best' estimator $\hat{\theta}$ under the squared error loss (in the sense of minimizing the Bayes posterior risk $\int_0^1 (\hat{\theta} - \theta)^2 \pi_{\theta|x}(\theta) d\theta$), the answer is the posterior mean $\int_0^1 x \pi_{\theta|x}(\theta) d\theta$. In the beta-binomial scenario, the posterior mean is $\frac{x+a}{n+a+b}$. Observe once again that

$$\frac{x+a}{n+a+b} = \frac{n}{n+a+b}\left(\frac{X}{n}\right) + \left(1 - \frac{n}{n+a+b}\right)\left(\frac{a}{a+b}\right), \qquad (3.41)$$

where X/n is the data mean (also the MLE and the UMVUE) and $a/(a+b)$ is the prior mean. These are examples of what we often see in a Bayesian inference problem: (1) A Bayesian point estimate is a weighted average of a commonly used frequentist estimate and an estimate based only on the prior distribution and (2) the weight allocated to the common frequentist estimate increases to 1 as the sample-size $n \longrightarrow \infty$. It is often said that the Bayesian point estimate *shrinks* the common frequentist estimate towards the exclusively prior-based estimate.

Another method of inference-drawing based on the posterior distribution is to construct a *credible region* for θ. A $100(1-\alpha)$ % credible region for θ is a subset C of the parameter space Θ such that $P(\theta \in C \mid X = x) \geq 1-\alpha$. As is the case for frequentist confidence intervals, usually there will be many candidates for C. An 'optimal' $100(1-\alpha)$ % credible region should have the smallest volume among them (or, equivalently, we should have $\pi_{\theta|x}(\theta) \geq \pi_{\theta|x}(\theta')$ for every $\theta \in C$ and every $\theta' \notin C$). This leads us to the concept of a *highest posterior density credible region* (HPDCR). The $100(1-\alpha)$ % HPDCR for θ is a subset C^* of Θ such that $C^* = \{\theta \in \Theta : \pi_{\theta|x}(\theta) \geq k_\alpha\}$, where k_α is the largest real number satisfying $P(\theta \in C^* \mid X = x) \geq 1 - \alpha$.

If there are Bayesian analogs of classical (or frequentist) point estimation and confidence-set estimation, you would expect a Bayesian analog of classical hypothesis testing too. Suppose, as in Section 3.7, that we want to test $H_0 : \theta \in \Theta_0$ against $H_1 : \theta \in \Theta_1$, where Θ_0 and Θ_1 constitutes a partition of the parameter space Θ. If we denote $P(\theta \in \Theta_0 \mid X = x)$ by γ, the *posterior odds ratio* is defined as $\frac{\gamma}{1-\gamma}$. Notice that this is a measure of final precision. If π_0 and $1 - \pi_0$ denote respectively the prior probabilities of θ being in Θ_0 and Θ_1, the *prior odds ratio* would be $\frac{\pi_0}{1-\pi_0}$. The *Bayes factor* is defined as the ratio of these two odds ratios. In other words, it is $\frac{\gamma(1-\pi_0)}{(1-\gamma)\pi_0}$. If the Bayes factor is smaller than 1, our degree of posterior belief in H_0 is smaller than that in H_1. Before we move to the next example, here is an interesting observation about the Bayes factor. If both Θ_0 and Θ_1 are singleton (i.e., we are under the Neyman-Pearson theorem setup), the Bayes factor indeed reduces to the likelihood-ratio used for finding the MP test in the Neyman-Pearson theorem.

Example 3.57. Let $\mathbf{X} = \{X_1, \ldots, X_n\}$ be a random sample from a $N(\mu, \sigma^2)$ p.d.f., where μ is unknown but σ^2 is known. Suppose that we choose a $N(\mu_0, \sigma_0^2)$ prior density for μ. Then the posterior density $\pi_{\mu|\mathbf{X}}(\mu)$ will be proportional to

$$\left[\exp\left(-\frac{1}{2\sigma_0^2}(\mu - \mu_0)^2\right)\right]\left[\exp\left(-\frac{1}{2\sigma^2}\sum_1^n(x_i - \mu)^2\right)\right],$$

which in turn is proportional to

$$\left[\exp\left(-\frac{1}{2\sigma_0^2}(\mu^2 - 2\mu\mu_0)\right)\right]\left[\exp\left(-\frac{n}{2\sigma^2}(\mu^2 - 2\mu\bar{x})\right)\right].$$

After completing the square in the exponent, one can see that the posterior density of μ is proportional to

$$\exp\left(-\frac{1}{2\sigma_n^2}(\mu - \mu_n)^2\right), \text{ where } \mu_n = \frac{\bar{x} + \mu_0(\sigma^2/n\sigma_0^2)}{1 + \sigma^2/n\sigma_0^2} \text{ and } \frac{1}{\sigma_n^2} = \left(\frac{1}{\sigma_0^2} + \frac{n}{\sigma^2}\right).$$

$$(3.42)$$

It should now be clear that the posterior density is $N(\mu_n, \sigma_n^2)$. In other words, the normal family of priors is a conjugate family for the mean of normal data. As before, the mode (which coincides with the mean) of this density can be used as a point estimate of μ. Once again, notice that this point estimate is a weighted average of the sample mean $\bar{x}$ and the prior mean μ_0.

Next we shed a little bit of light on the controversial issue of choosing a prior. The primary concerns that guide our choice of a prior p.d.f. (or p.m.f.) are (a) its ability to adequately represent the extent and nature of the prior information available and (b) computational convenience. There are three main ways of choosing a prior p.d.f. or p.m.f.: (a) subjective, (b) objective (informative) and (c) noninformative. A subjective choice is the most controversial, since it exclusively reflects the degree of one's personal belief about $\boldsymbol{\theta}$. For example, often in a real-life scientific experiment, expert's opinion may be available regarding the unknown parameters involved from people who are highly trained or experienced in that field. The problem with eliciting a prior from this kind of opinion is that often experts don't agree and put forward conflicting or contradictory opinions. Objective and informative priors are less controversial since they are based on either historical records about the parameter-values themselves or data from previous experiments that contain information about the parameters. In the latter case, the posterior densities or p.m.f.s obtained from such older datasets can be used as priors for the current problem at hand. Or one could possibly combine the older datasets with that for the current problem and enjoy a much bigger sample-size. The question that naturally comes to mind is when (if at all) these two approaches will yield the same posterior distributions for the parameters in the current study. It will happen only if the older datasets and the current dataset can be considered statistically independent (to be more precise, conditionally independent given the parameters). A noninformative prior is so called because it is supposed to reflect the extent of ignorance about the parameter(s). Such a prior is also sometimes called a *diffuse* prior or a *vague* prior or a *reference* prior. This concept may be confusing at times, since, for example, a *flat* prior (i.e., one that is constant over the parameter space) is not necessarily noninformative just because it is flat. In general, a noninformative prior is one that is 'dominated' by the likelihood function in the sense that it does not change much over the region in which the likelihood is reasonably large, and also does not take large values outside that region. Such a prior has also been given the name *locally uniform*. Often we feel that the best reflection of our ignorance about $\boldsymbol{\theta}$ is a constant prior density over an infinite parameter space or a nonconstant one that is so heavy-tailed that it does not integrate to 1. A prior density of this sort is called an *improper prior*. Such a prior density is not automatically disallowed as invalid, because it may still lead to a proper posterior density that integrates to 1. In many complicated real-life problems, Bayesians resort to *hierarchical modeling*. Such models often provide better insights into the dependence structure of the observed data (that may consist of response variables, covariates, etc.) and the unknown parameters and also help break down the overall variability into different layers. For instance, in Example 3.57, assuming both μ and σ^2 to be unknown, we could choose a $N(\mu_0, \sigma_0^2)$ prior for μ and a $Gamma(\alpha, \beta)$ prior (independently of μ) for $1/\sigma^2$. If there is uncertainty in our mind about the parameters $\mu_0, \sigma_0^2, \alpha$ and β, we can capture this uncertainty by imposing prior densities on them (e.g., a normal prior on μ_0, uniform priors on σ_0^2, α and β). These will be called *hyperpriors* and their parameters, *hyperparameters*. We may be reasonably certain about the hyperparameters, or sometimes they are estimated from the observed data (an approach known as *empirical Bayes*). We conclude this discussion with an example of a special noninformative prior that is widely used.

Example 3.58. Suppose that $\boldsymbol{\theta} = (\theta_1, \ldots, \theta_m)$. Recall the definition of a Fisher information matrix (FIM) from Section 3.4. Let us denote it by $I(\boldsymbol{\theta})$. It can be shown that the $(i, j)^{th}$ entry of the $m \times m$ FIM is $-E\left[\frac{\partial^2 \log f(\mathbf{X}|\boldsymbol{\theta})}{\partial \theta_i \partial \theta_j}\right]$. The prior density $\pi(\boldsymbol{\theta}) \propto \det(I(\boldsymbol{\theta}))^{0.5}$ is known as the *Jeffreys' prior* (where 'det' stands for the determinant of a matrix). When our data come from a Binomial(n, θ) distribution, the FIM is just a scalar $(= \frac{n}{\theta(1-\theta)})$. So the Jeffreys' prior in this case will be $\propto [\theta(1 - \theta)]^{-0.5}$, which is nothing but a Beta$(\frac{1}{2}, \frac{1}{2})$ density. When our data $\{X_1, \ldots, X_n\}$ come from a Normal(θ_1, θ_2^2) p.d.f., the FIM is a diagonal matrix with diagonal entries n/θ_2^2 and $2n/\theta_2^2$. So the Jeffreys' prior will be $\pi(\theta_1, \theta_2) \propto \frac{1}{\theta_2^2}$ on the upper half of the two-dimensional plane.

We conclude with a brief discussion of Bayesian computation. As indicated earlier, MCMC algorithms play a central role in Bayesian inference. Before we take a closer look at them, let us try to understand the Monte Carlo principle. The idea of Monte Carlo (MC) simulation is to draw a set of i.i.d. observations $\{x_i\}_{i=1}^n$ from a target p.d.f. $f(x)$ on a high-dimensional space $\mathcal{X}$. This sample of size n can be used to approximate the target density with the empirical point-mass function $f_n(x) = \frac{1}{n}\sum_1^n \delta_{x_i}(x)$, where $\delta_{x_i}(x)$ denotes Dirac's delta function that takes the value 1 if $x = x_i$ and 0 otherwise. As a result, one can approximate integrals such as $\int_{\mathcal{X}} g(x)f(x)dx$ with sums like $\frac{1}{n}\sum_1^n g(x_i)$, because the latter is an unbiased estimator of the former and is strongly consistent for it as well (by the *strong law of large numbers*). In addition, the central limit theorem (CLT) gives us asymptotic normality of the MC approximation error. For example, if $f(x)$ and $g(x)$ are univariate functions (i.e., $\mathcal{X}$ is some subset of $\mathcal{R}$) such that the variance of g with respect to f is finite [i.e., $\sigma_g^2 = \int_{\mathcal{X}} g^2(x)f(x)dx - (\int_{\mathcal{X}} g(x)f(x)dx)^2 < \infty$], then according to the CLT,

$$\sqrt{n}\left\{\frac{1}{n}\sum_1^n g(x_i) - \int_{\mathcal{X}} g(x)f(x)dx\right\} \Longrightarrow N(0, \sigma_g^2)$$

as $n \longrightarrow \infty$, where $\Longrightarrow$ means convergence in distribution. So it should be clear that in order to take advantage of the MC principle, we must be able to *sample* from a p.d.f. If it is a standard p.d.f. (e.g., normal, exponential or something else whose CDF has a closed-form expression), there are straightforward procedures for sampling from it. However, if it is non-standard (e.g., one with an ugly, irregular shape and no closed-form CDF) or is known only upto a proportionality-constant, sampling from it will be tricky and we need some special technique. One of them is *rejection sampling*, where we sample from a *proposal density* $f^*(x)$ that is easy to sample from and satisfies: $f(x) \leq Kf^*(x)$ for some $K < \infty$. Having sampled an observation x_i^* from f^*, we use the following acceptance-rejection scheme: Generate a random variable $U \sim \text{Uniform}[0,1]$ and accept x_i^* if $U < f(x_i^*)/[Kf^*(x_i^*)]$; otherwise reject it and sample another observation from f^* to repeat this procedure. If we continue this process until we have n acceptances, it can be easily shown that the resulting sample of size n is from $f(x)$. Although it is an easy-to-implement scheme, it has serious drawbacks in the sense that it is often impossible to bound $f(x)/f^*(x)$ from above by a finite constant K uniformly over the entire space $\mathcal{X}$. Even if such a K can be found, often it is so large that the acceptance probability at each step is very low and the process therefore becomes inefficient. So an alternative procedure has been tried. It is known as *importance sampling*. Here we once again introduce a proposal density $f^*(x)$ and realize that the integral $\int_{\mathcal{X}} g(x)f(x)dx$ can be rewritten as $\int_{\mathcal{X}} g(x)w(x)f^*(x)dx$, where $w(x) = f(x)/f^*(x)$ is usually called the *importance weight*. As a result, all we need to do is to draw n i.i.d. observations $x_1^*, \ldots, x_n^*$ from f^* and evaluate $w(x_i^*)$ for each of them. Then, following the MC principle, we will approximate the integral $\int_{\mathcal{X}} g(x)f(x)dx$ by $\sum_1^n g(x_i^*)w(x_i^*)$. This is once again an unbiased estimator and, under fairly general conditions on f and f^*, is strongly consistent as well. Like rejection sampling, this procedure is easy to implement, but choosing an appropriate proposal density may be tricky. Often the criterion that is used for this choice is the minimization of the variance of the resulting estimator $\sum_1^n g(x_i^*)w(x_i^*)$. That variance, computed with respect to the proposal density f^*, is given by

$$\text{var}_{f^*}(g(x)w(x)) = \int_\mathcal{X} g^2(x)w^2(x)f^*(x)dx - \left(\int_\mathcal{X} g(x)w(x)f^*(x)dx\right)^2.$$

Clearly, minimizing this with respect to f^* is equivalent to minimizing just the first term on the right-hand side. We can apply *Jensen's inequality* (which says that $E(h(Y)) \geq h(E(Y))$ for any nonnegative random variable Y and any convex function h such that $E(Y)$ and $E(h(Y))$ are finite) to get the lower bound $(\int_\mathcal{X} | g(x) | w(x)f^*(x)dx)^2$, which is nothing but $(\int_\mathcal{X} | g(x) | f(x)dx)^2$. So in order to achieve this lower bound, we must use the proposal density $f^*(x) =| g(x) | f(x)/[\int_\mathcal{X} | g(x) | f(x)dx]$. Sometimes it is easy to sample from this proposal density, but more often it is difficult. In those cases, one may have to resort to a Markov chain Monte Carlo (MCMC) sampling scheme such as the *Metropolis-Hastings* (MH) algorithm or the *Gibbs sampler* (GS). We first describe the MH algorithm, since GS can be viewed as a special case of it.

The MH algorithm is named after N. Metropolis who first published it in 1953 in the context of the Boltzmann distribution, and W. K. Hastings who generalized it in 1970. Suppose we want to sample from the density $f(x)$. This algorithm generates a Markov chain $\{x_1, x_2, \ldots\}$ in which each x_i depends only on the immediately preceding one (i.e., x_{i-1}). Assume that the current state of the chain is x_t. To move to the next state, the algorithm uses a *proposal density* $f^*(x; x_t)$ which depends only on the current state and is easy to sample from. Once a new observation x^* is drawn from the proposal density and a random variable U is generated from the Uniform[0,1] density, x^* is accepted to be the next state of the chain (i.e., we declare $x_{t+1} = x^*$) if $U < \min\{1, f(x^*)f^*(x_t; x^*)/[f(x_t)f^*(x^*; x_t)]\}$. Otherwise we declare $x_{t+1} = x_t$. One commonly used proposal density is the normal density centered at x_t and having some known variance σ^2 (or known variance-covariance matrix $\sigma^2 I$ in the multivariate case). This proposal density will generate new sample-observations that are centered around the current state x_t with a variance σ^2. Incidentally, this choice for f^* would be allowed under the original Metropolis algorithm too, which required the proposal density to be symmetric (i.e., $f^*(x_t; x^*) = f^*(x^*; x_t)$). The generalization by Hastings removed this 'symmetry' constraint and even allowed proposal densities to be just a function of x_t (i.e., to be free from x^*). In this latter case, the algorithm is called *independence chain Metropolis-Hastings* (as opposed to *random walk Metropolis-Hastings* when the proposal density is a function of both x_t and x^*). While the 'independence chain' version can potentially offer higher accuracy than the 'random walk' version with suitably chosen proposal densities, it requires some *a priori* knowledge of the target density. In any case, once the Markov chain is initialized by a starting value x_0, it is left running for a long time (the 'burn-in' period) to ensure that it gets 'sufficiently close' to the target density. During the burn-in period, often some parameters of the proposal density (e.g., the variance(s) in the case of a normal proposal density) have to be 'fine-tuned' in order to keep the acceptance rate moderate (i.e., slightly higher than 50%). This is because the acceptance rate is intimately related to the size of the proposal-steps. Large proposal-steps would result in very low acceptance rates and the chain will not move much. Small proposal-steps would lead to very high acceptance rates and the chain will move around too much and converge slowly to the target density (in which case, it is said to be *slowly mixing*).

Now suppose that we have a p-dimensional density $f(x_1, \ldots, x_p) = f(\mathbf{x})$ as our target. Also suppose that for each i $(1 \leq i \leq p)$, the univariate conditional density of x_i given all the other variables (call it $f_i(x_i \mid x_1, \ldots, x_{i-1}, x_{i+1}, \ldots, x_p)$) is easy to sample from. In this case, it is a good idea to choose the proposal density: $f^*(\mathbf{x}^*; \mathbf{x}_t) = f_j(x_j^* \mid \mathbf{x}_{-j,t})$ if $\mathbf{x}_{-j}^* = \mathbf{x}_{-j,t}$ and $= 0$ otherwise. Here $\mathbf{x}_{-j,t}$ means the current state $\mathbf{x}_t$ with its jth coordinate removed and $\mathbf{x}_{-j}^*$ means the newly drawn sample-observation $\mathbf{x}^*$ with its jth coordinate removed. A simple calculation shows that, under this choice, the acceptance probability will be 1. In other words, after initializing the Markov chain by setting $x_i = x_{i,0}$ for $1 \leq i \leq p$, this is how we move from the current state $\mathbf{x}_t$ to the next state $\mathbf{x}_{t+1}$: First sample $x_{1,t+1}$ from $f_1(x_1 \mid x_{2,t}, \ldots, x_{p,t})$; next sample $x_{2,t+1}$ from $f_2(x_2 \mid x_{1,t+1}, x_{3,t}, \ldots, x_{p,t})$; ...; finally sample $x_{p,t+1}$ from $f_p(x_p \mid x_{1,t+1}, \ldots, x_{p-1,t+1})$. This is known as the *deterministic scan Gibbs sampler*. Since it is a special case of the MH algorithm described above, we can actually insert MH steps within a Gibbs sampler without disrupting the properties of the underlying Markov chain. As long as the univariate conditional densities (often collectively called the *full conditionals*) are familiar or easy to sample from, we will follow the Gibbs sampling scheme, but if one of them is a bit problematic, we will deal with it using the MH technique and then get back to the Gibbs sampling scheme for the other ones.

3.10 Exercises

1. Prove *Boole's inequality*: For any two sets A and B, $P(A^c \cap B^c) \geq 1 - (P(A) + P(B))$.

2. Suppose you are drawing two cards (with your eyes closed) one by one *without* replacement from a well-shuffled deck of 52 cards. How many outcomes are there in the sample space? What are the chances that the two cards are of different colors? That they belong to different suites? Given that the first card turned out to be black, what are the chances that the second card will be the queen of clubs?

3. If your company has a twelve-member management team with a third of them being women, what are the chances that a committee of four people randomly chosen from them will have no woman at all? Will have at least one man? Will have all women?

4. How many distinct ten-digit phone numbers are possible if the first digit of the area code, as well as the first digit of the three-digit exchange number, is not allowed to be zero? If you are randomly choosing such a phone number, what are the chances that it will have an area code of Chicago (i.e., 312)?

5. Suppose you are randomly permuting the letters of the alphabet. What are the chances that the vowels are together in the correct order AEIOU? That the vowels are together (in any order)? That the positions of the vowels are multiples of 5 (i.e., $5^{th}, 10^{th}, \ldots, 25^{th}$)?

6. Two points are chosen at random on a line-segment of unit length. Find the probability that each of the three parts thus formed will have length $> \frac{1}{5}$.

7. In a group of 23 randomly selected people, what are the chances that at least two will share a birthday? Answer the same question for a group of 35 people.

8. You learn from the local newspaper that there have been 5 automobile accidents in your town in the last 7 days. What are the chances that they all happened on the same day? That they happened on 5 different days? That at least two of them happened on the same day?

9. In a manufacturing plant, each product undergoes four independent inspections. The probability that a defective product is detected at each inspection is 0.9. What are the chances that a defective product escapes detection by all four inspections? That it is detected only by the very last inspection?

10. A fair coin is tossed repeatedly. How many tosses must be made so that the chances of at least one TAIL occurring is more than 90% ?

11. Show that if two different coins (both with the same $P(H) = p$) are being repeatedly tossed independently of one another and X_i is the number of tosses needed for the ith coin ($i = 1, 2$) to produce the first HEAD, the p.m.f. of $Y = X_1 + X_2$ is the same as it would be if Y counted the number of tosses of a fair coin needed to see the second head.

12. Write down $P(X = k)$ where X has a Binomial(n, p) p.m.f. Then take its limit as $n \to \infty$ and $p \to 0$ with np remaining constant. Show that you get a Poisson p.m.f. as a result.

13. Show that each of the following functions is a legitimate p.d.f. on the real line and find its mean if you can: (a) $f(x) = \frac{1}{2}e^{-|x|}$, (b) $f(x) = \frac{1}{\pi(1+x^2)}$. [Note: The first one is actually known as a *bilateral exponential* or *Laplace* p.d.f. and the second one is known as a *Cauchy* p.d.f.]

14. Suppose $X_1, \ldots, X_n$ are i.i.d. continuous random variables having a common p.d.f. $f(x)$ and c.d.f. $F(t)$. Let $Y = \max_{1 \le i \le n} X_i$ and $Y^* = \min_{1 \le i \le n} X_i$. Find the c.d.f. of Y and that of Y^*. From these, find their p.d.f.'s. Incidentally, Y and Y^* are respectively known as the largest and the smallest *order statistic* associated with the sample $\{X_1, \ldots, X_n\}$.

15. Let $X_1, \ldots, X_n$ be i.i.d. having an Exponential(β) p.d.f. Find the p.d.f. and the rth raw moment of their smallest order statistic.

16. Let Y_1 and Y_2 be i.i.d. Poisson(λ) random variables and $X = \max(Y_1, Y_2)$. Show that (i) $E(X)$ is between λ and 2λ; (ii) $E(X)$ is between $\lambda + e^{-\lambda} - e^{-2\lambda}$ and $2(\lambda + e^{-\lambda}) - 1 - e^{-2\lambda}$.

17. If X has a Normal$(0,1)$ p.d.f., find that p.d.f. of $Y = |X|$. This is known as the *folded normal* p.d.f. Compute the relative entropy of this p.d.f. with respect to an Exponential(1) p.d.f. and vice versa.

18. For $X \sim$ Normal(μ, σ^2), find the Fisher information matrix for its two parameters.

19. Let $\{X_1, \ldots, X_n\}$ be an i.i.d. sample from the Gamma(α, β) p.d.f. Can you find the maximum likelihood estimates of its two parameters?

20. Suppose $\{X_1, \ldots, X_n\}$ are i.i.d. having a *zero-inflated* Poisson p.m.f. with parameters $w \in (0, 1)$ and $\lambda > 0$, that is, $P(X_i = 0) = wI(X_i = 0) + (1-w)e^{-\lambda}$ and $P(X_i = k) = (1 - w)e^{-\lambda}\lambda^k/k!$ for $k = 1, 2, \ldots$. Find the method-of-moments estimators for its two parameters.

21. Suppose an insect is laying its eggs on tree-leaves. The number of eggs (X)

it lays on a leaf has a Poisson(λ) p.m.f. Given $X = x$, the number of eggs (Y) among them that will ultimately hatch has a Binomial(x, p) conditional p.m.f. If we have just one observation-pair x_1 and y_1, assume conjugate priors on the two parameters and find their posteriors. If we only had observation(s) on Y, can you choose a conjugate prior to keep the posterior computation simple?

22. Suppose $X_1, \ldots, X_n$ are i.i.d. observations from a Normal(μ, σ^2) p.d.f. Assume a Gamma(α, β) prior density on $1/\sigma^2$ and given σ^2, assume a Normal(μ_0, $\tau\sigma^2$) prior on μ for some known $\tau > 0$. Find out the posterior densities for the parameters.

23. Suppose $X_1, \ldots, X_n$ are i.i.d. observations from a Poisson(λ) p.m.f. Find out the Jeffreys' prior for λ. What family of prior densities is the conjugate family in this case? Assume a conjugate prior on λ and find out the posterior density.

References

[1] Bhat, U.N. (1984) *Elements of Applied Stochastic Processes.* New York: Wiley.
[2] Cramér, H. (1946) *Mathematical Methods of Statistics.* Princeton University Press.
[3] Dempster, A.P., Laird, N.M. and Rubin, D.B. (1977) Maximum Likelihood from Incomplete Data via the EM Algorithm. In *Journal of the Royal Statistical Society. Series B (Methodological)* **39:1**, pp. 1–38.
[4] Efron, B. (1979) Bootstrap Methods: Another Look at the Jackknife. In *The Annals of Statistics* **7:1**, pp. 1–26.
[5] Ewens, W.J. and Grant, G. (2001) *Statistical Methods in Bioinformatics: An Introduction.* New York: Springer-Verlag.
[6] Leon, S.J. (1998) *Linear Algebra with Applications.* Prentice-Hall.
[7] Mood, A.M.F., Graybill, F.A. and Boes, D.C. (1974) *Introduction to the Theory of Statistics.* McGraw-Hill; Kogakusha.
[8] Press, W.H., Teukolsky, S.A., Vetterling, W.T. and Flannery, B.P. (1986) *Numerical Recipes: The Art of Scientific Computing.* Cambridge University Press.
[9] Rao, C.R. (1965) *Linear Statistical Inference and its Applications.* New York: Wiley.
[10] Rohatgi, V.K. and Saleh, A.K.M.E. (2001) *An Introduction to Probability and Statistics.* New York: Wiley.
[11] Schervish, M.J. (1996) *Theory of Statistics.* Springer.
[12] Shmueli, G., Minka, T.P., Kadane, J.B., Borle, S. and Boatwright, P. (2005) A useful distribution for fitting discrete data: revival of the Conway-Maxwell-Poisson distribution. In *Journal of the Royal Statistical Society Series C (Applied Statistics).* **54:1**, pp. 127–142.

CHAPTER 4

Classification Techniques

4.1 Introduction and Problem Formulation

The area of classification also known as pattern recognition and supervised
learning has grown substantially over the last twenty years, primarily due to
the availability of increased computing power, necessary for executing sophisticated algorithms. The need for classification arises in most scientific fields,
ranging from disease diagnosis, to classifying galaxies by shape, to text and
image classification, to applications in the financial industry, such as deciding which customers are good credit risks or constructing efficient portfolios,
just to name a few. In bioinformatics, examples of classification tasks include
classification of samples to different diseases based on gene and or protein
expression data, prediction of protein secondary structure and identification
and assignment of spectra to peptides and proteins obtained from mass spectrometry. It should be noted that the emergence of genomic and proteomic
technologies, such as cDNA microarrays and high density oligonucleotide chips
and antibody arrays, gave a big impetus to the field, but also introduced a
number of technical challenges, since the availability of more features (variables) than samples represented a shift in the classical paradigm.

The field of classification originated with the work of Fisher [15] and experienced fast growth in the subsequent 40 years with the introduction of flexible
classification techniques, such as nearest neighbor classifiers, the perceptron
and neural networks. More recently, more computationally flexible classifiers
were proposed in the literature, such as classification trees, support vector
machines, as well as regularized versions of more classical classifiers. Finally,
over the last ten years ensemble methods emerged such as bagging and boosting, based in PAC learning theory [41] which established that classifiers whose
performance is slightly better than random guessing when appropriately combined can exhibit a superior performance.

On the theoretical front, some of the major developments include the development of a general statistical decision framework for classification (see below)
and the connection of the problem to learning theory. The objective of learning
theory is to study mathematical properties of learning machines. Such properties are usually expressed as those of the function class that the learning
machine can implement (see below).

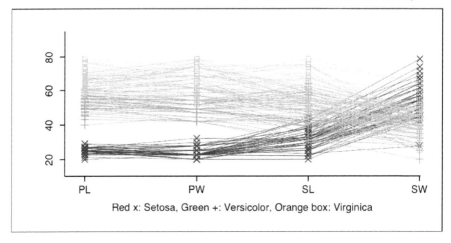

Figure 4.1 *Parallel coordinates plot for the iris data.*

The classification problem in layman's terms is as follows: given a number of measurements on a particular object, *assign* the object to one of a *prespecified* fixed number of *classes* (groups). The following examples provide some motivation about the task at hand.

Example 1: The goal is to assign an iris plant to one of the following three classes, *setosa, virginica, versicolor*, using information on the length and width of its petals and sepals. In order to achieve this goal, data on 150 plants were collected. Figure 4.1 shows the profiles (in a parallel coordinates plot) of 150 plants, 50 from each class [15]. It can be immediately seen that petal length (PL) and width (PW) separate the three classes well enough. The scatterplot of these variables given in Figure 4.2 shows a fairly clean separation of the three classes.

Example 2: This example comes from forensic testing of glass used by Evett and Spiehler from the Central Research Establishment, Home Office Forensic Science Service. The variables are the refractive index and weight percent of the following oxides: sodium, magnesium, aluminum, silicon, potassium, calcium, barium and iron. The six possible classes are: A=Building windows (float processed) (Green), B=Building windows (nonfloat processed) (Red), C=Vehicle windows (Yellow), D=Containers (Blue), E=Tableware (Pink), and F=Headlamps (Orange). Figure 4.3 shows the profiles (in a parallel coordinates plot) of 214 cases; 70 from A, 76 from B, 17 from C, 13 from D, 9 from E and 29 from F. Some discrimination between glass types is apparent even from single attributes. For example class F is high on barium and aluminum and low on magnesium, while class A is high on magnesium. Nevertheless, the goal of assigning objects to their appropriate classes seems more difficult in this case than in the iris dataset.

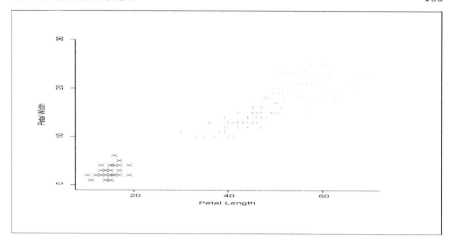

Figure 4.2 *Scatterplot of petal length and petal width for the iris data.*

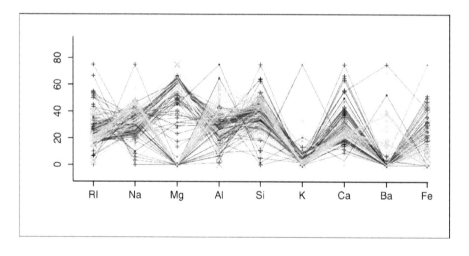

Figure 4.3 *Parallel coordinates plot for the glass data.*

4.2 The Framework

Suppose we have collected data on N objects. Each object is characterized by an attribute, or feature, vector x comprised of p elements, belonging to a suitable space (e.g., a subset of $\mathbb{R}^p$ or $\mathbb{Z}^p$). Each object needs to be classified into (assigned to) one of K prespecified classes (groups, types). In order to achieve this goal we need to *design* a **decision rule** ($\hat{c}$) that would assign object i with attribute vector x_i to class c_k, $k = 1, ..., K$

$$c_k \leftarrow \hat{c}(x_i). \tag{4.1}$$

A good classification rule would be one that minimizes "inaccuracy."

4.2.1 A decision theoretic framework

Let C denote the class label of a random feature vector $\mathbf{x} \in \mathbb{R}^p$. Assume that the prior probability in the population is given by $\pi_k = P(C = k)$, $k = 1, \cdots, K$. Let $\hat{c}(\mathbf{x}) : \mathbb{R}^p \rightarrow \{1, \cdots, K, D\}$ be a decision or classification rule, with D denoting the *in doubt* option.

In order to determine whether a classification rule is 'good' or not, we need a *loss function* that measures its quality. In classification, the most commonly used loss function is the 0/1 loss; i.e., $L(\ell, k) = 1$, if the rule assigns the object to class ℓ but the true class is k, and 0 otherwise. Further, we assume that $L(\ell, D) = d$ for all classes k (see [33]). Finally, assume that the observations from class k have a *class conditional distribution function* denoted by $p_k(\mathbf{x})$.

The risk function for classifier $\hat{c}$ is the expected loss as a function of the unknown class k:

$$R(\hat{c}, C = k) = \mathbb{E}_\mathbf{x}[L(\hat{c}, k)|C = k]$$

$$= \sum_{\ell=1}^{K} L(\ell, k)P(\hat{c} = \ell|C = k) + L(D, k)P(\hat{c} = D|C = k)$$

where the expectation is taken with respect to the distribution of the feature vector $\mathbf{x}$.

The *total risk* is the total expected loss, viewing both the class C and the vector $\mathbf{x}$ as random:

$$R(\hat{c}) = \mathbb{E}_{k,\mathbf{x}}R(\hat{c}, C) = \sum_{k=1}^{K} \pi_k P(\hat{c}(\mathbf{x}) \neq k|C = k) + d\sum_{k=1}^{K} \pi_k P(\hat{c}(\mathbf{x}) = D|C = k).$$

This gives the overall misclassification probability plus d times the overall doubt probability.

Let $p(k|\mathbf{x}) = P(C = k|\mathbf{x} = x) = \frac{\pi_k p(k|\mathbf{x})}{\sum_{\ell=1}^{K} \pi_\ell p(\ell|\mathbf{x})}$ be the *posterior probability* of class k given a feature vector $\mathbf{x} = x$.

Proposition: Suppose that both the prior probabilities π_k, $k = 1, \cdots, K$ and the class conditional densities $p(k|\mathbf{x})$ are known. Then, the classification rule which minimizes the total risk under the 0/1 loss function is given by

$$c(x) = \begin{cases} k & \text{if } p(k|\mathbf{x}) = \max_{1 \leq \ell \leq K} p(\ell|\mathbf{x}) \text{ and exceeds } 1 - d \\ D & \text{if each } p(k|\mathbf{x}) \leq 1 - d. \end{cases}$$

Proof: We have that

$$R(\hat{c}) = \mathbb{E}_\mathbf{x}[\mathbb{E}_k[L(\hat{c}(\mathbf{x}), C)|\mathbf{x} = x]] = \int \mathbb{E}[L(\hat{c}(\mathbf{x}, C))|\mathbf{x} = x]p(\mathbf{x})d\mathbf{x}$$

where $p(\mathbf{x}) = \sum_{\ell=1}^{K} \pi_\ell p_\ell(\mathbf{x})$ is the marginal density for $\mathbf{x}$. It suffices to minimize the conditional expectation for each $\mathbf{x}$ separately, which can be written as

$$\sum_{\ell=1}^{K} L(c, \ell) p(\ell|\mathbf{x})$$

with respect to c for each $\mathbf{x}$. For $c = D$ we have that

$$\sum_{\ell=1}^{K} L(\ell, D) p(\ell|\mathbf{x}) = d.$$

Under our loss function, it is easy to see that the minimum is given by

$$1 - p(1|\mathbf{x}), 1 - p(2|\mathbf{x}), ..., 1 - p(K|\mathbf{x}), d$$

for $\hat{c} = 1, 2, ..., K, D$, respectively. So, the solution is given by the $\max\{\max_{1 \leq \ell \leq K} p(k|\mathbf{x}), \ 1 - d\}$.

Under the 0/1 loss function, another way to write the optimal classification rule is to choose the class with the highest $\pi_k p(k|\mathbf{x})$, provided that it exceeds $(1 - d)p(\mathbf{x})$.

The optimal classification rule is referred to in the literature as the *Bayes rule* [33]. When two or more classes attain the maximal posterior probability, the tie can in principle be broken in an arbitrary fashion. The value $R(c)$ of the total risk for the Bayes rule is called the *Bayes risk*. This value is the best one can achieve if both the prior probabilities π_k and the class conditional distributions $p_k(\mathbf{x})$ are a priori known; hence, it provides a theoretical benchmark for all other procedures.

Some algebra shows that without the in doubt option, for a two class problem the Bayes risk is given by $\mathbb{E}\min\{p(1|\mathbf{x}), p(2|\mathbf{x})\}$, while for K classes by $\mathbb{E}[1 - \max_k p(k|\mathbf{x})]$.

Example: In order to illustrate the decision theoretic framework, we look at an example involving two normal populations with *common* covariance matrix (the one-dimensional case is illustrated in Figure 4.4). Assume that $p_k(\mathbf{x}) \sim N(\mu_k, \Sigma)$, $k = 1, \cdots, K$ and disregard the doubt option. Then, an observation with feature vector $\mathbf{x} = x$ is allocated to the class k with smallest value of $\delta(x, \mu_k) - 2 \log \pi_k$ (by calculating the posterior probability), where

$$\delta(x, \mu_k) = [(x - \mu_k)' \Sigma^{-1} (x - \mu_k)]^{1/2},$$

the Mahalanobis distance between the mean μ_k of class k and the observation x. Notice that the quadratic term $x' \Sigma^{-1} x$ is common to all classes and therefore the optimal rule can be written as

$$\text{minimize} - 2\mu_k' \Sigma^{-1} x + \mu_k' \Sigma^{-1} \mu_k - 2 \log \pi_k, \text{ over } k = 1, \cdots, K.$$

If all classes are equally likely, then an observation with feature vector x is classified to the *nearest* population in the sense of having the smallest Mahalanobis distance to its mean.

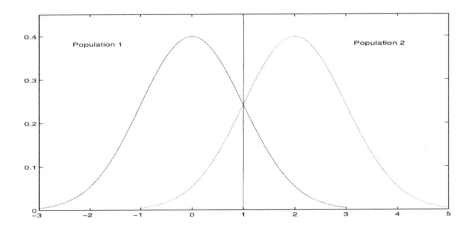

Figure 4.4 *Demonstration of decision boundary in the case of two normal populations with equal variances.*

4.2.2 Learning a classification rule $c(x)$

Notice that the theoretical framework previously defined assumes perfect knowledge of the prior distribution over classes and in particular of the class conditional distributions (or densities).

In order to make this decisions theoretical framework *operational*, a *training data set* $\mathcal{T} = \{c_i, x_i\}_{i=1}^{N}$ will be used to "learn" a classification rule $\hat{c}(x|\mathcal{T})$, by estimating $\hat{p}(k|x, \mathcal{T})$. Notice that under very mild conditions, the frequency of N_k/N in $\mathcal{T}$ (i.e., the number of observations of class k in the training data over the total size of the training data) is a consistent estimate of π_k. Two widely used approaches for learning $c(x)$ are: (i) parametric and (ii) nonparametric methods. In parametric methods, the class conditional densities are explicitly specified (e.g., multivariate normal or t distribution) and the goal becomes to learn their parameters from the training data. In nonparametric methods, the class conditional densities *themselves* are estimated from the training data. In what follows, we review these approaches, together with some more recent developments for estimating the decision rule.

4.2.3 Assessing the performance of classification rules

The performance of the classification rule $c(x)$ is usually assessed by calculating the misclassification error rate; i.e., the probability of making an incorrect classification for a future randomly sampled observation. The apparent error rate is the proportion of mistakes made, when classifying either a training or an *independent test* data set. If the training set is used, the error rate will usually be biased downwards, since the training data set was used both for constructing the classifier and also assessing its performance. Hence, an independent test data set is preferable for performance assessment purposes. The error rate is calculated by classifying each observation in the test data, counting the errors and dividing by the size of the test data set. This measure is clearly unbiased, but can be highly variable. Further, the use of an independent test set wastes data that could have been used for training purposes, especially in cases where labeled examples are expensive to acquire, which is often the case in biological applications. In order to overcome this difficulty, the idea of *cross-validation* proves useful. Suppose that the training data $\mathcal{T}$ are partitioned into M pieces. Then, one part can be used as the test set and the remaining $M - 1$ pieces for training purposes and this exercise can be repeated M times. Each estimate of the error rate is unbiased and by averaging over the M estimates the variability is reduced. The extreme version of this strategy is to take $M = N$, the so called *leave-one-out* cross validation [31]. However, this is computationally the most demanding strategy. Further, excluding one observation leads to assessment of the classifier through $\mathcal{O}(1/N)$ perturbations from $\mathcal{T}$. But the variability in the parameter estimates is usually of the order $\mathcal{O}(1/\sqrt{N})$, which implies that for large sample sizes they are calculated from smaller perturbations. Thus, cross validated estimates in this case can be rather variable; on the other hand, a smaller M may lead to larger bias, but smaller variance and consequently mean squared error.

4.2.4 Connection to statistical learning theory

In Section 2.1, a decision theoretic framework was introduced for deriving optimal classification rules. It was shown that an optimal rule minimizes the Bayes risk. In practice, since the true distribution of the data is unknown and has to be estimated, one is interested in minimizing the empirical counterpart of the Bayes risk, defined as:

$$R_n(c) = \frac{1}{n} \sum_{i \in \mathcal{T}} I(c(x_i) \neq y_i), \tag{4.2}$$

where $I(\cdot)$ denotes the indicator function and y_i the label of observation i in the training data set $\mathcal{T}$. The empirical Bayes risk corresponds to misclassification error rate in $\mathcal{T}$; i.e., the average number of misclassifications in the training data.

Since the risk $R(c)$ can not be computed directly, but only approximated by $R_n(c)$, it is not reasonable to look at classifiers (functions) that minimize $R_n(c)$ among all possible functions. The reason is that one can always construct such a function that performs perfectly on the training data and always misclassifies new observations. Hence, avoiding *over-fitting* the training data becomes an important consideration. This can be achieved in two ways: (i) restrict the class of functions (classification rules) over which the minimization takes place and (ii) modify the optimization criterion; for example, by adding a term that penalizes 'complicated' functions, or in other words *regularize* the problem under consideration.

A classification rule trained on n observations $(c_n(x))$ is designed to map inputs (variables x) to outputs (class labels). Its risk is a random variable $R(c_n)$ (since it depends on the data) and can not be computed directly. Statistical learning theory is interested in producing bounds for the following quantities:

- Error bound for the estimation of the risk from an empirical quantity.
- Error bound for the rule given the functional class it is assumed to belong to.
- Error bound relative to the Bayes risk.

These questions have attracted a lot of attention over the last decade; a good overview is given in Vapnik's book [43], in Scholkopf et al. book [35] and in the review paper by Bousquet et al. [4].

4.3 Classification Methods

In this section, we discuss a number of widely used in practice classification methods.

4.3.1 Classification rules via probability models

Linear and Quadratic Discriminant Analysis

The objects in class k are assumed to be normally distributed with mean vector μ_k and covariance matrix Σ_k. Then, the Bayes rule decision rule chooses for each observation with feature vector x the class k that minimizes

$$Q_k(x) = -2\log(p(x|k)) - 2\log(\pi_k) = (x-\mu_k)'\Sigma_k^{-1}(x-\mu_k) + \log|\Sigma_k| - 2\log(\pi_k). \quad (4.3)$$

The first term of (4.3) corresponds to the Mahalanobis distance of object i from the center of class k. This rule is known in the literature as *quadratic discriminant analysis*.

The expression in (4.3) simplifies if one is willing to assume equal covariance matrices amongst the K groups; i.e., $\Sigma_1 = \cdots = \Sigma_K \equiv \Sigma$. In that case the rule minimizes,

$$Q_k(x) = -2\log(p(x|k)) - 2\log(\pi_k) = \mu_k\Sigma^{-1}x + \mu_k'\Sigma_k^{-1}\mu_k - 2\log(\pi_k). \quad (4.4)$$

Therefore, the rule becomes *linear* in the data x. For a 2-class problem, it corresponds to Fisher's linear discriminant that was derived based on a different criterion. The above rule is known in the literature as *linear discriminant analysis*.

In practice, the following quantities need to be estimated from the training data set $\mathcal{T}$ by their sample counterparts: μ_k and Σ_k. The maximum likelihood estimates are given by $\hat{\mu}_k = \bar{X}_k$ (the multivariate mean for each class k and $\hat{\Sigma}_k = \frac{1}{n_k} \sum_{j=1}^{n_k} (x_j - \hat{\mu}_k)(x_j - \hat{\mu}_k)'$. For linear discriminant analysis, the pooled covariance estimate $\hat{\Sigma} = \sum_{k=1}^{K} n_k / N \hat{\Sigma}_k$ is used. Often the bias-corrected estimator of Σ with divisor $N - K$ is used, but makes very little difference to the linear rule (and none if the class prior probabilities π_k are the same).

It should be noted that in linear discriminant analysis $Kp + p(p+1)/2$ parameters need to be estimated from the training data, while $Kp + Kp(p+1)/2$ for quadratic discriminant analysis. This suggests that unless there are sufficient training data for each class, parameter estimates can be rather volatile, which in turn implies that the quadratic rule can be well outperformed by the linear one for moderate sample sizes. For this reason, in practice a diagonal version of quadratic discriminant analysis is often employed, where $\Sigma_k = \text{diag}(\sigma_j^k)$, $j = 1, \cdots, p$.

Another multivariate distribution used for classification purposes is the multivariate t with a moderate number of degrees of freedom, that exhibits heavier tails than the multivariate normal. It can be thought of as a flexible model for distributions with elliptical densities with heavy tails. The density for a p-variate vector x with $\nu > 2$ degrees of freedom is given by

$$p(x|k) = \frac{\Gamma(\frac{1}{2}(\nu + p))}{(\nu\pi)^{p/2}\Gamma(\nu/2)} |\Sigma|^{-1/2} \left(1 + \frac{1}{nu}(x - \mu_k)'\Sigma_k^{-1}(x - \mu_k)\right)^{-1/2(\nu+p)}. \quad (4.5)$$

The optimal classifier minimizes

$$Q_k(x) = \frac{\nu + p}{2} \log\left(1 + \frac{1}{\nu}(x - \mu_k)'\Sigma_k^{-1}(x - \mu_k)\right) + \frac{1}{2}\log|\Sigma_k| - \log(\pi_k). \quad (4.6)$$

If the covariance matrix is common to all classes, a linear rule is once again obtained. The effect of the heavier tails of the t distribution is to down-weigh observations which are far from the class mean.

An illustration of the quadratic type of decision boundaries produced by linear and quadratic discriminant analysis is given in Figures 4.5-4.6. It can be seen that the linear discriminant analysis misclassifies four observations, while the more flexible boundary of quadratic discriminant analysis results in three misclassifications.

Parametric models can be used in more complex situations, where for example the data exhibit clear multi-modality. In that case, one can employ a finite mixture of multivariate normal distributions as the conditional class density. However, the problem of estimating all the parameters involved (means, covariances and mixing coefficients) is not a trivial one and care should be taken regarding identifiability issues (see [44]).

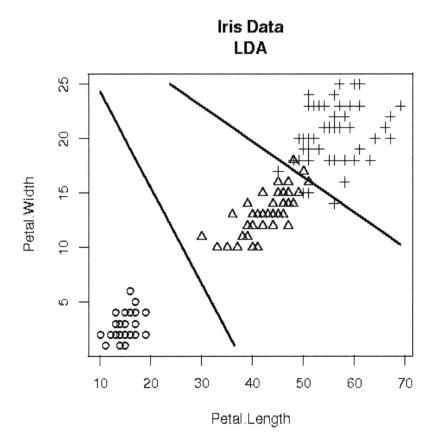

Figure 4.5 *Decision boundaries for linear discriminant analysis for the Iris data set.*

Logistic Discrimination

This model arises when one wants to model the posterior probabilities of the K classes through linear functions of x. As a motivating example, consider the normal model for the K classes with common covariance matrix. By comparing the ratio of the class k posterior probability to that of class 1, we obtain

$$\log(\frac{p(k|x)}{p(1|x)}) = (x - \mu_k)'\Sigma^{-1}(x - \mu_k) - (x - \mu_1)'\Sigma^{-1}(x - \mu_1) + \log(\pi_k/\pi_1) =$$

$$(\mu_k - \mu_1)'\Sigma^{-1}x - (\mu_k + \mu_1)'\Sigma^{-1}(\mu_k - \mu_1) + \log(\pi_k/\pi_1) = \alpha_k + \beta_k'x,$$

a linear model in x.

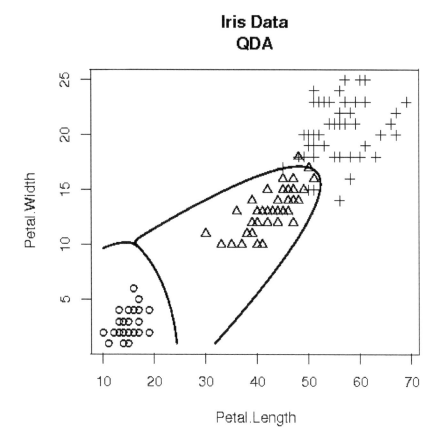

Figure 4.6 *Decision boundaries for quadratic discriminant analysis for the Iris data set.*

In general, we can model $\log(p(k|x)) - \log(p(1|x))$ by

$$p(\text{class} = k|x) = \frac{\exp(\beta'_k x)}{\sum_{k=1}^{K} \exp(\beta'_k x)}, \qquad (4.7)$$

thus modeling the log-odds by a linear function. This model implies that odds increase multiplicatively by e^{β}_{kh} for every unit increase in the value of the attribute x_h. This procedure can be generalized to allow modeling of the log-odds by more general functions.

The logistic regression model given in (4.7) is estimated by maximum likelihood methods. Since the resulting score equations are nonlinear in the parameter β_k, an iterative algorithm like Newton-Raphson needs to be employed (for more details, see [21]).

4.3.2 Nonparametric methods

In case one is not willing to assume that the underlying populations are normally distributed, we need to resort to nonparametric methods for estimating $p(x|k)$ and its sample counterpart $p(x|k,\mathcal{T})$. The naive estimate of the underlying density is the sample histogram of the training dataset $\mathcal{T}$. The main problem with this estimate is its lack of "smoothness." A possible solution is the use of *kernel* methods to smooth it. The estimate then takes the following form

$$p(x|k,\mathcal{T}) = \frac{1}{n_k} \sum_{i=1}^{n_k} \frac{1}{h} K(\frac{x - x_i}{h}), \qquad (4.8)$$

where h is the *bandwidth* and K the *kernel* that satisfies $\int_{\text{sample space}} K(x)dx = 1$.

The choice of h (a scalar parameter) controls the "smoothness" of the resulting estimate. For small values of h we get more wiggly estimates, while for large values of h fairly smooth ones. Possible choices of the kernel function K include the Gaussian, the rectangular, the triangular, the Epanechnikov, etc. [36]. There are various automatic approaches for choosing the bandwidth that have been proposed in the literature (for a discussion see [36]). In high dimensions most of the underlying sample space is empty of objects. The latter fact implies that one needs to choose either a very large value of h, or use a product estimate of $p(x|k,\mathcal{T})$.

The product estimate of $p(x|k,\mathcal{T})$ leads to the naive Bayes classifier, that has remained popular over the years. Specifically, it is assumed that conditional on the class $C = k$, the features are *independent*; i.e.,

$$p(x|k,\mathcal{T}) = \prod_{j=1}^{K} p_j(x_j|k,\mathcal{T}).$$

The individual class-conditional densities $p_j(x_j|k,\mathcal{T})$ can be estimated using one dimensional kernel density estimates. The latter is a generalization of the original implementation of naive Bayes classifiers that relied on the extra assumption of normality for the individual class conditional densities. The popularity of the naive Bayes classifier stems from its capability to exhibit a very good performance in many applications. A theoretical justification of its performance, in the case where there are more variables than observations in the training data, is given in [3].

Nearest Neighbor Methods

One simple adaptive kernel method is to choose a uniform kernel K, which takes a constant value over the nearest r objects and is zero elsewhere. As mentioned in [33] this does not in fact define a density as the estimate of $p(x|k)$, since its integral is infinite but we can nevertheless estimate the posterior distribution as the proportions of the classes amongst the nearest r objects. The resulting classification rule is known as the *nearest neighbor rule* and dates back to Fix and Hodges [16].

The basic steps to classify a new object with attribute vector x_0 are:

- Step 1: Determine the r nearest neighbors (usually with respect to the Euclidean distance) of the new object in T.
- Step 2: Classify the new object to the class that contains the majority of the r nearest neighbors.

The version with $r = 1$ corresponds to dividing the data space into the cells of the Dirichlet tessellation of the data points, and every new observation is classified according to the label of the cell it falls in. The behavior of the 1-nearest neighbor classification can be characterized asymptotically. Specifically, let E^* denote the error rate of the Bayes rule in a K-class problem. Then the error rate of the 1-nearest neighbor rule *averaged* over training sets converges to a value E_1 which is bounded above by $E^*(2 - K/(K - 1)E^*)$ (see [11]). In another development, Stone [38] showed that if the number of neighbors used $r \to \infty$, while $r/N \to 0$, the risk for the r-nearest neighbor rule converges in probability to the Bayes risk (not averaged over training data sets). However, these results are asymptotic in nature and do not necessarily apply to finite samples. In particular, these results are independent of the metric used for defining the nearest neighbor (e.g., Euclidean). Experience shows that the choice of metric can be important.

The flexible boundaries produced by nearest neighbor classifiers are illustrated in Figure 4.7, where the value $r = 3$ was chosen. It can be seen that one observation from the Viriginica class is classified as Versicolor and vice versa, thus exhibiting a slightly better apparent misclassification error rate than linear discriminant analysis.

Classification Trees

The goal of classification trees is to partition the feature space into *hypercubes* and assign a class to every hypercube.

We illustrate their use by an example. In Figure 4.8 the classification boundaries for the three classes in the iris data are shown, corresponding to the tree shown in the right panel. It can be seen that if the objects' petal length is less than 24.5 then they are assigned to class Setosa, while if it is larger than 24.5 to some other class. Then, if petal width is larger than 17.5 and petal length less than 49.5 the objects are classified to class Virginica, and so on. In Figure 4.9 the constructed tree is shown with the misclassifications in the terminal nodes also included; for example, all the observations in the Setosa class are correctly classified, while 5 observations in the Viriginica class are classified as Versicolor, etc. This is an example of a binary tree, where each node *splits* into

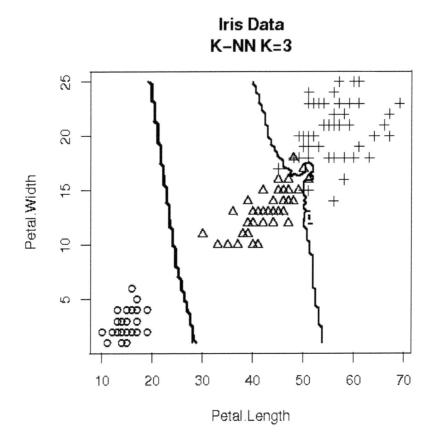

Figure 4.7 *Decision boundaries produced by r = 3 nearest neighbor classifier.*

two branches. The number of possible binary splits is $m(N-1)$ provided that the variables are real-valued without any repeated values. The main advantage of binary splits is their inherent simplicity: numerical variables are split according to whether their values are above or below some threshold value τ, and the same holds for ordinal variables; nominal variables with L levels can be split in $2^{L-1}-1$ possible ways.

Classification trees are *rooted* ones, with the root corresponding to the top node. Observations are passed down the tree along the branches, with decisions being made at each node until a terminal node, also called *leaf* node, is reached. Each nonterminal node contains a question on which a split is based, while each leaf contains the label of a classified observation.

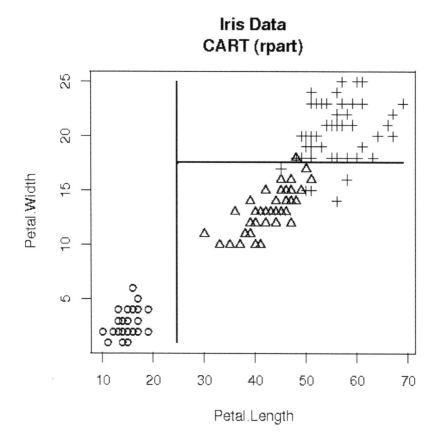

Figure 4.8 *Classification boundaries for the Iris data.*

It can be seen that the goal of the classification tree is to split the feature space in such a way that most members belonging to a particular class fall in the same hypercube. However, nothing prevents the tree to grow in such a way so that there is a single object only in each hypercube. The main issues then become on how the tree should be constructed and also pruned, in order to avoid the one object per hypercube phenomenon.

Almost all current tree-construction methods use a one-step look-ahead approach. That is, they choose the next split (which variable to use in the next split) in an optimal way with respect to this greedy strategy, without attempting to optimize the performance of the whole tree. Common measures for choosing the splits are the Gini index and the cross-entropy (also called deviance) [5]. At each node ν of a tree we have a probability distribution $p_{\nu k}$ over the classes. By conditioning on the observed attributes x_i of the objects

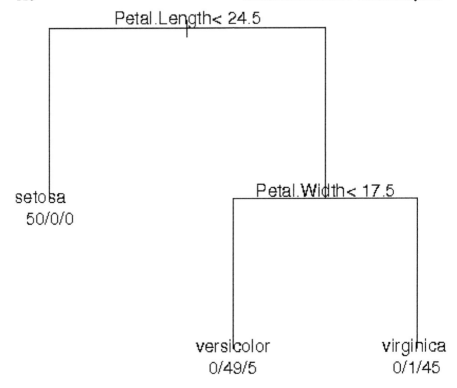

Figure 4.9 *Corresponding classification tree.*

in the training dataset $\mathcal{T}$ we learn the numbers n_ν assigned to every node of the tree. The Gini index is defined for node ν as

$$G(\nu) = \sum_{k=1}^{K} p_{\nu k}(1 - p_{\nu k}),$$

whereas the cross-entropy by

$$E(\nu) = \sum_{k=1}^{K} p_{\nu k} \log(p_{\nu k}).$$

For the tree as a whole we have $G = \sum_\nu G(\nu)$ and $E = \sum_\nu E(\nu)$. Now consider splitting a particular node ν into nodes ν_1 and ν_2. The split that maximizes the reduction in the Gini index or cross-entropy $D_\nu - D_{\nu_1} - D_{\nu_2}$ is the one that is chosen.

With noisy data there is always the danger of over-fitting the training dataset $\mathcal{T}$. The established methodology to avoid this problem is to "prune" the tree.

For any tree TR, define $R(\text{TR})$ to be the error rate (i.e., the number of misclassifications) in $\mathcal{T}$. Define an error-complexity measure

$$C_\alpha(\text{TR}) = R(\text{TR}) + \alpha \text{Size}(\text{TR}), \qquad (4.9)$$

which we are interested in minimizing over all trees. Notice that for $\alpha = 0$ we are only interested in minimizing $R(\text{TR})$ while for $\alpha \to \infty$ we are only interested in constructing the simplest possible tree. Note that by choosing a particular value of α appropriately we can strike a good compromise between the two objectives; namely minimizing the number of misclassifications vs the size of the tree. The value of $\alpha = 2(K-1)$ corresponds to the tree with minimum Akaike information criterion value, a measure of trading off the fit of particular model against its complexity in terms of number of parameters that need to be estimated.

Support Vector Machines

Support vector machines and kernel methods have proven to be powerful tools in numerous applications and have thus gained widespread popularity, e.g., in machine learning and bioinformatics. In order to motivate the idea behind the technique we start our exposition with the simplest case, that of two *separable classes by a linear boundary*. The data in the training set $\mathcal{T}$ are the pairs (y_i, x_i), $i = 1, \cdots, N$ with the response taking values in the set $\{-1, 1\}$ (positive and negative examples) and $x_i \in \mathbb{R}^p$. Suppose there exists a *hyperplane* that *separates* the positive from the negative examples. The points x that lie on the hyperplane satisfy $< w, x > +b = 0$, where w is a vector normal (orthogonal) to the hyperplane and $b/||w||$ is the perpendicular distance from the hyperplane to the origin, with $||w||$ denoting the Euclidean norm of w and $< \cdot, \cdot >$ the inner product operator. Let m_+ and m_- be the shortest distance from the separating hyperplane to the closest positive and negative example, respectively. The quantities m_+ and m_- are called the *margin* of a separating hyperplane for the two types of examples. For the linearly separable case, the support vector algorithm searches for the separating hyperplane with the largest margin. To rigorously formulate such an algorithm, notice that in the current setting all the training data satisfy: $< w, x_i > +b \geq +1$ for $y_1 = +1$ and $< w, x_i > +b \leq -1$ for $y_i = -1$. Consider the positive and negative examples for which these relations hold with equality. The positive examples lie on a hyperplane $h_+ :< w, x_i >_{+b=1}$ with a normal vector w and perpendicular distance from the origin $|1 - b|/||w||$, while the negative examples on another hyperplane $h_- :< w, x_i > +b = -1$, with the same normal vector w and perpendicular distance to the origin $|-1 - b|/||w||$. Therefore, $m_+ = m_- = 1/||w||$ and the margin is given by $2/||w||$. As shown in Figure 4.10, h_+ and h_- are parallel hyperplanes, since they have the same normal vector and no training points fall between them. Hence, we can find the pair of hyperplanes that gives the maximum margin by minimizing

$$\min ||w||^2 \text{ subject to } y_i(< w, x_i > +b) - 1 \geq 0, \quad \text{all } i \in \mathcal{T}. \qquad (4.10)$$

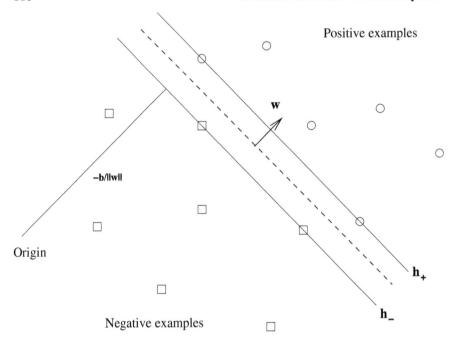

Figure 4.10 *Illustration of linear separating hyperplanes for the separable case.*

This is a convex optimization problem with a quadratic criterion and linear constraints, whose solution can be found by first switching to a Lagrangian unconstrained formulation and then exploring its corresponding dual problem.

The above formulation and the resulting quadratic programming algorithm, when applied to nonseparable data, will find no feasible solution. A relaxation of the constraints on h_+ and h_- is required, which can be achieved by the introduction of *slack* variables $\xi_i \geq 0$ for all i. The defining hyperplanes are then given by $h_+ :< w, x_i > +b \geq 1 - \xi_i$ for $y_i = +1$ and $h_- :< w, x_i > +b \geq -1 + \xi_i$ for $y_i = -1$. Hence, when mistakes occur, the corresponding ξ_i must be larger than one and so $\sum_i \xi_i$ serves as an upper bound on the number of training errors. A natural way to assign costs for errors is to change the objective function to be minimized from $||w||^2$ to $||w||^2 + a(\sum_i \xi)$, with a being a tuning parameter that controls the assigned penalty in the presence of mistakes. This leads once again to a quadratic programming problem (for more details on the Lagrangian formulation of the dual, see [21]).

The resulting classifier is given by:

$$c(x) = \text{sign} < w, x > +b, \tag{4.11}$$

where $w = \sum_{i \in S} \alpha_i y_i x_i$, α_i are obtained from the solution of the dual problem and S denotes the set of examples that define the support vectors h_+ and h_-.

In most real life examples, the two classes are not going to be approximately linearly separable and that is where the so-called 'kernel trick' comes into play. The main idea is that separation of the classes by a hyperplane may be easier in higher dimensions. The construction depends on inner products that have to be evaluated in the variables' space; the latter task may become computationally intractable, if the dimensionality of that space is too large. Kernel (generalized covariance) functions that are defined in lower dimensional space, but behave like inner products in higher (possibly infinite) dimensional space will accomplish this goal, as originally observed by Boser et al. [2]. Suppose that the data are mapped to some other Euclidean space Q through:

$$\Phi : \mathbb{R}^p \longrightarrow Q, \qquad (4.12)$$

where Q may be infinite dimensional. Notice that the training algorithm would depend on the data through inner products of the form $< \Phi(x_\ell, \Phi(x_j) > $ in Q space. The challenge is to find a bivariate function K such that $K(x_\ell, x_j) =< \Phi(x_\ell, \Phi(x_j) >$; this would require only knowledge of K in the computations. Examples of K include γ-degree polynomials of the form $K(x_\ell, x_j)(1+ < x_\ell, x_j >)^\gamma$, radial basis functions $K(x_\ell, x_j) = \exp(-||x_\ell - x_j||^2/u)$ and neural network functions $K(x_\ell, x_j) = \tanh(\eta_1 < x_\ell, x_j > +\eta_2)$. The classifier for a new point x is given by

$$c(x) = \sum_{i \in \mathcal{S}} \alpha_i y_i K(x, x_i). \qquad (4.13)$$

Remark 1: An extensive treatment of the theory of support vector machines, together with learning with kernels, is given in the book by Scolkopf and Smola [35]. Another fairly extensive presentation and connections to a regularization framework is given in [21].

Remark 2: Support vector machines have proved particularly successful in a variety of applications. Lin [27] investigated their empirical success and argued that they implement the Bayes classifier in an efficient manner. Specifically, as discussed above, the Bayes rule minimizes the misclassification error rate, given by $\mathbb{P}(y = +1|x) - 1/2$ for the positive examples class. It was also shown in [27] that the support vector machines solution targets directly this rate without estimating the posterior probability of the positive (or negative) class.

The classical support vector machine paradigm presented above primarily deals with binary (2-class) classification problems and has a nice geometric interpretation. The most common strategy adopted to solve multi-category classification problems is to treat them as a series of binary problems, with the positive examples corresponding to class k and the negative examples to all the remaining classes. Dietterich and Bakiri [14] discuss a general scheme for carrying out this task and Allwein et al. [1] proposed a unifying framework for it. More direct formulations of the multi-category problem for support vector machines are given in [12, 26].

The flexible decision boundaries produced by support vector machines are

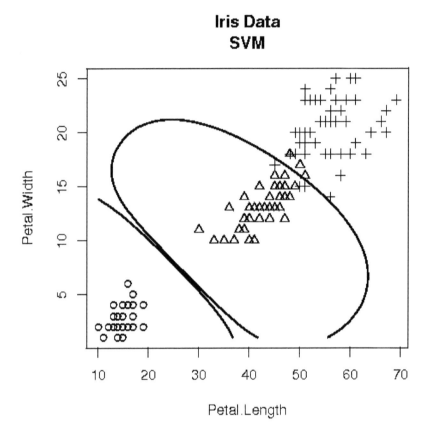

Figure 4.11 *Classification boundary produced by the Support Vector Machine classi-fier using a radial basis kernel.*

shown in Figure 4.11. It can be seen that five observations from the training data set are misclassified. It should be noted that in the presence of only two variables, support vector machines can not take full advantage of their flexibility.

Ensemble Methods: Bagging and Boosting

In this section, we discuss some techniques that do not produce a single clas-sification rule, but an *ensemble* of classifiers. We focus on bagging and its extension random forests and on boosting.

Bagging:

The main idea is to generate *perturbed* versions of the training data of the same size as the original set $\mathcal{T}$, train a classifier on each perturbed version and aggregate the results by majority voting. The perturbed versions are obtained by bootstrapping observations from $\mathcal{T}$. The bootstrap aggregation or *bagging* algorithm introduced in [6] is described next:

1. Create B bootstrapped training data sets $\mathcal{T}_1, \cdots, \mathcal{T}_B$

2. Train a classifier $c_b(x)$ on each bootstrapped training data set $\mathcal{T}_b$

3. Output: for a new observation x, classify it to the majority class predicted by the B classifiers

Since the final result is based on averaging a number of predictions, it can be shown that bagging reduces the variance of the classification procedure, but does not affect its bias. Further, empirical evidence suggests that high variance classifiers (those with the tendency to overfit the data, such as classification trees) primarily benefit from bagging. Unlike techniques that produce a single classification rule, bagging produces an ensemble of rules that although may exhibit superior performance in terms of misclassification error rate, they are nevertheless hard to interpret as a 'model.' A variation of bagging based on creating convex pseudo-sets of training data was proposed in [7]. The idea is to generate a new training set where the observations are linear combinations of two observations selected at random from the original training set $\mathcal{T}$. This procedure is repeated B times, so that the desired classifier can be trained on B perturbed training data sets and its results averaged through majority voting as in bagging.

Empirical evidence showed that bagging improved performance in practice but not by a wide margin. A variation of bagging specifically designed for classification trees as the base classifier was introduced in [8]. The main steps of the algorithm, called Random Forests, are described next:

1. Create B bootstrapped training data sets $\mathcal{T}_1, \cdots, \mathcal{T}_B$. The observations not included in the bootstrapped sample $\mathcal{T}_b$ form an *out-of-bag* sample

2. Train a classification tree on each bootstrapped training data set $\mathcal{T}_b$ as follows:
   at every node, consider *only a random set of variables* for the next split
   Note that trees are recommended to be grown to maximum size and *not pruned*

3. Output: for a new observation x, classify it to the majority class predicted by the B classifiers

Empirical evidence suggest that over-fitting due to growing the tree to maximum size is not an issue. Further, selection of a small number of variables at every node works well and the performance of random forests seems to be insensitive to it. Finally, the algorithm provides a mechanism for estimating the importance of each variable in the ensemble.

The variable importance estimate is obtained as follows: After each tree is constructed, the values of each variable in the out-of-sample sample are randomly permuted and the out-of-bag sample is run down the tree and therefore a classification for each observation is obtained. Thus, p misclassification error rates are obtained and the output is the percent increase in misclassification rate as compared to the out-of-sample rate with all variables intact.

Boosting:

Boosting has proved to be an effective method to improve the performance of base classifiers, both theoretically and empirically. The underlying idea is to combine simple classification rules (the base classifiers) to form an ensemble, whose performance is significantly improved. The origins of boosting lie in PAC learning theory [41], which established that learners that exhibit a performance slightly better than random guessing when appropriately combined can perform very well. A provably polynomial complexity boosting algorithm was derived in [34], whereas the Adaptive Boosting (AdaBoost) algorithm in various varieties [19, 20] proved to be a practical implementation of the boosting ensemble method. The basic steps of the AdaBoost algorithm are described next for a binary classifier. Let $c(x)$ be the classifier under consideration and (y_i, x_i), $i = 1, \cdots, N$ the data in the training set $\mathcal{T}$. Then,

1. Initialize weights $w_i = 1/N$

2. for $m = 1$ to M do

3. fit $y = c_m(x)$ as the base weighted classifier using weights w_i

4. let $W_{c_m} = \sum_{i=1}^{N} w_i I(y_i c_m(x_i) = -1)$ and $\alpha_m = \log[(1 - W_-(h))]/W_-$

5. $w_i = w_i \exp\{\alpha_m I(y_i \neq c_m(x_i))\}$ scaled to sum to one

6. end for

7. Output: $\hat{c}(x) = \text{sign}(\sum_{m=1}^{M} \alpha_m c_m(x))$

The popular intuition behind the success of the algorithm is that hard to classify observations receive higher weights α_i at later iterations; thus, the algorithm concentrates more on these cases and hence manages to drive the misclassification error rate on the training data to zero.

Remark 1: Friedman et al. [17] established connections between the AdaBoost algorithm and statistical concepts such as loss functions, additive models and logistic regression. Specifically, it was shown that this algorithm fits a stagewise additive logistic regression model that minimizes the expectation of the exponential loss function $\exp(-yc(x))$. A good discussion is also included in [21]. For some interesting observations on boosting see also [28].

Remark 2: It can be seen that this algorithm produces as output the new observation's predicted class. Proposals for the algorithm to produce as output the class probability estimates include the RealBoost and the GentleBoost algorithms [17].

Remark 3: The boosting algorithm performs a gradient descent for minimizing the underlying exponential loss function and consequently determine the weights and the classifier at each iteration. However, this is a greedy strategy, since the solution pays attention to reducing the misclassification error rate on the training data. A variant of the algorithm that focuses more on the misclassification error rate for future observations along with a regularized version (Stochastic Gradient Boosting) are discussed in [18].

4.3.3 Variable selection

Variable selection is an important topic in classification, especially in applications where the number of variables exceeds the number of available observations in the training data set T ($p >> n$). Two domains where $p >> n$ are text classification and gene/protein microarray data. The goal of variable selection is to improve the predictive performance of the classifier, to reduce its computational cost and finally to provide a more interpretable rule. ¿From the description of the most commonly used techniques to train classifiers, it can be seen that some algorithms have a *built-in* variable selection mechanism; examples include classification trees, random forests. On the other hand, logistic discrimination, support vector machines and especially nearest neighbor methods do not offer such mechanisms.

Various approaches have been proposed in the literature to deal with this topic. We briefly outline two popular approaches: (i) regularization and (ii) dimension reduction. The idea behind regularization is to modify the objective function that is being optimized by adding a penalty term that penalizes complex models - in this setting, models that include a large number of variables. This applies directly to logistic, support vector machines and linear and quadratic discriminant analysis. The ideal penalty would be an ℓ_0 norm that would eliminate "unnecessary" variables. However, it generally leads to a nonconvex optimization problem which in turn is hard to solve. Hence, the ℓ_1 norm has become a very popular alternative, since it shrinks the coefficients of variables to zero, thus effectively performing variable selection, while at the same time keeping the underlying optimization problem convex. Another popular penalty is the ℓ_2 norm that shrinks the coefficients of the variables, but not all the way to zero. This penalty is best suited for reducing the mean squared error of the estimated model, but does not eliminate variables. Another approach is based on the idea of reducing the dimensionality of the variable space. Possible techniques include principal components analysis and clustering. Principal components analysis creates new variables that are linear combinations of the original ones, which can then be used as inputs to the various classification techniques. Clustering of variables replaces a group of 'similar' variables by a representative one, that can be their weighted average. A more ad hoc method is variable ranking, which is attractive because of its simplicity and scalability. A classification rule is used with a single variable and the final output is a ranking of all variables according to their predictive ability. This procedure is not necessarily designed to produce a good final model, but to discard the most 'uninformative' variables.

4.4 Applications of Classification Techniques to Bioinformatics Problems

Applications of classification techniques to problems in bioinformatics abound. One of the most active areas has been the class prediction problem (e.g., different stage of cancer of patients) using primarily gene but more recently protein microarray data. Actually, the availability of data for a significantly larger number of genes (in the thousands) than observations (less than 100)– the so called large p, small n paradigm–has led researchers to incorporate variable selection in many classifiers and in addition develop variants of the methods surveyed in this chapter (see for example [25, 45]). The number of application papers is in the 100s and it is hard to do justice to all of them. Therefore, we give selected references that give a starting point to the reader. One of the earliest papers that introduced the class prediction problem in the context of microarray data is by Golub et al. [22]. Other application papers include Khan et al. [20], van't Veer et al. [42], Tibshirani et al. [39], Zhang et al. [46], just to name a few. A comprehensive comparison of the performance of various classifiers using gene microarray data is given in [13, 37].

Another interesting area of applications deals with peptide and protein identification in mass spectrometry, where variations of logistic discrimination have been used for this purpose (see Zhang et al. [47], Nesvizhskii and Aebersold, [32], Ulintz et al. [40]). Other applications in the proteomics domain include functional classification of proteins (Keck and Wetter [23], Lau and Chasman [25], Cheng et al. [10]), and prediction of protein-protein interactions [9].

4.5 Exercises

1. Consider the following two *univariate* populations. The first one is normally distributed with mean μ_1 and variance σ_1^2 and the second one is normally distributed with mean μ_2 and variance σ_2^2. Further assume that the prior probabilities are given by $\pi_1 \neq \pi_2$. Derive the optimal Bayes decision rule and calculate the misclassification error rate.

How does the optimal Bayes rule look if population 1 is exponentially distributed with parameter θ_1 and population 2 is also exponentially distributed with parameter θ_2? Calculate the misclassification error rate.

2. Suppose there are three classes in a population with equal prior probabilities; i.e., $\pi_1 = \pi_2 = \pi_3 = 1/3$. The data in class k are Poisson distributed with parameter λ_K, where $\lambda_1 = 5$, $\lambda_2 = 7.5$ and $\lambda_3 = 10$. Derive the Bayes rule; i.e., identify the class boundaries. Also, calculate the overall misclassification error rate. How does the Bayes rule and the misclassification error rate change, if two *independent* measurements are obtained; i.e., X_1, X_2 are Poisson distributed with parameters λ_k?

3. Suppose there are two classes in the population C_1 and C_2. Further, assume that a p-variate observation x comes from one of these two populations with probability π_1 and π_2, respectively. Observations from class C_1 follow a multivariate normal distribution $N(\mu_1, \Sigma)$, while those from C_2 follow another multivariate normal distribution $N(\mu_2, \Sigma)$. Assuming that the cost of misclassifying an observation is equal for the two classes, calculate the Bayes rule for assigning x to C_1 or C_2. Also calculate the corresponding misclassification probabilities.

References

[1] Allwein, E.L., Schapire, R.E. and Singer, Y. (2000), Reducing multi-class to binary: A unifying approach for margin classifiers, *Journal of Machine Learning Research*, 113-141

[2] Boser, B.E., Guyon, I.M. and Vapnik, V. (1992), A training algorithm for optimal margin classifiers, in Proceedings of ACM Workshop on Computational Learning Theory, 144-152, Pittsburgh, PA

[3] Bickel, P.J. and Levina, E. (2004), Some theory for Fisher's Linear Discriminant function, "naive Bayes," and some alternatives when there are many more variables than observations, *Bernoulli*, 10, 989-1010

[4] Bousquet, O., Boucheron, S. and Lugosi, G. (2004), Introduction to Statistical Learning Theory, Advanced Lectures on Machine Learning Lecture Notes in Artificial Intelligence 3176, 169-207. (Eds.) Bousquet, O., von Luxburg, U. and Ratsch, R. Springer, Heidelberg, Germany

[5] Breiman, L., Friedman, J.H., Olshen, R. and Stone, C. (1984), *Classification and Regression Trees*, CRC Press, Boca Raton, FL

[6] Breiman, L. (1996), Bagging predictors, *Machine Learning*, 24, 123-140

[7] Breiman, L. (1998), Arcing classifiers, *Annals of Statistics*, 26, 801-849

[8] Breiman, L. (2001), Random forests, *Machine Learning*, 45, 5-32

[9] Chen, X.W. and Liu, M. (2005), Prediction of proteinprotein interactions using random decision forest framework, *Bioinformatics*, 21, 4394-4400

[10] Cheng B.Y., Carbonell J.G., Klein-Seetharaman J. (2005), Protein classification based on text classification techniques, *Proteins*, 58, 955-970

[11] Cover, T. and Hart, P. (1967), Nearest neighbor pattern classification, *IEEE Transactions on Information Theory*, 13, 21-27

[12] Crammer, K. and Singer, Y. (2001), On the algorithmic implementation of multi-class kernel-based vector machines, *Journal of Machine Learning Research*, 2, 265-292

[13] Dudoit, S., Fridlyand, J. and Speed, T.P. (2002), Comparison of discrimination

methods for the classification of tumors using gene expression data, *Journal of the American Statistical Association*, 97, 77-87

[14] Dietterich, T.G. and Bakiri, G. (1995), Solving multi-class learning problems via error correcting output codes, *Journal of Artificial Intelligence Research*, 2, 263-286

[15] Fisher, R.A. (1936), The use of multiple measurements in taxonomic problems, *Annals of Eugenics*, 7, 179-188

[16] Fix, E. and Hodges, J.L. (1951), Discriminatory Analysis. Nonparametric Discrimination; Consistency Properties, Report Number 4, USAF School of Aviation Medicine, Randolph Field, TX

[17] Friedman, J.H., Hastie, T. and Tibshirani, R. (1998), Additive logistic regression: A statistical view of boosting, *Annals of Statistics*, 28, 337-407 (with discussion)

[18] Friedman, J.H. (2002), Stochastic gradient boosting, *Computational Statistics and Data Analysis*, 38, 367-378

[19] Freund, Y. and Schapire, R.E. (1997), Experiments with a new boosting algorithm, In Proceedings of the International Conference on Machine Learning, 148-156.

[20] Freund, Y. and Schapire, R.E. (1997), A decision-theoretic generalization of on-line learning and an application to boosting, *Journal of Computer and System Sciences*, 55, 119-139

[21] Hastie, T., Tibshirani, R. and Friedman, J. (2001), *The Elements of Statistical Learning: Data Mining, Inference and Prediction*, Springer, NY

[22] Golub, T.R., Slonim, D.K, Tamayo, P., Huard, C., Gaasenbeek, M., Mesirov, J.P., Coller, H., Loh, M.L, Downing, J.R., Caligiuri, M.A, Bloomfield, C.D. and Lander, E.S. (1999), Molecular classification of cancer: Class discovery and class prediction by gene expression monitoring, *Science*, 286, 531-537

[23] Keck, H.P. and Wetter, T. (2003), Functional classification of proteins using a nearest neighbour algorithm, *In Silico Biology*, 3, 23

[24] Khan, J., Wei, J.S., Ringner, M., Saal, L.H., Ladanyi, M., Westermann, F., Berthold, F., Schwab, M., Antonescu, C.R., Peterson, C., and Meltzer, P.S. (2001), Classification and diagnostic prediction of cancers using gene expression profiling and artificial neural networks, *Nature Medicine*, 7, 673-679

[25] Lau, A.Y. and Chasman, D.I. (2004), Functional classification of proteins and protein variants, *Proceedings of the National Academies of Science USA*, 101, 6576-6581

[26] Lee, Y., Lin, Y. and Wahba, G. (2004), Multi-category support vector machines, theory and application to the classification of microarray data and satellite radiance data, *Journal of American Statistical Association*, 99, 67-81

[27] Lin, Y. (2002), Support vector machines and the Bayes rule in classification, *Data Mining and Knowledge Discovery*, 6, 259-275

[28] Mease, D. and Wyner, A. (2008), Evidence contrary to the statistical view of boosting, *Journal of Machine Learning Research*, 9, 131-156

[29] Michailidis, G. and Shedden, K. (2003), The application of rule-based methods to class prediction problems in genomics, *Journal of Computational Biology*, 10, 689-698

[30] Middendorf, M., Ziv, E. and Wiggins, C.H. (2005), Inferring network mechanisms: The Drosophila melanogaster protein interaction network, *Proceedings of the National Academies of Science USA*, 102, 3192-3197

[31] Mosteller, F. and Wallace, D.L. (1963), Inference in an authorship problem, *Journal of the American Statistical Association*, 58, 275-309

[32] Nesvizhskii A.I., Aebersold R. (2005), Interpretation of shotgun proteomic data: The protein inference problem, *Molecular & Cellular Proteomics*, 4, 1419-1440

[33] Ripley, B.D. (1996), *Pattern Recognition and Neural Networks*, Cambridge University Press, Cambridge, UK

[34] Schapire, R. (1990), The strength of weak learnability, *Machine Learning*, 5, 197-227

[35] Scholkopf, B. and Smola, A.J. (2002), *Learning with Kernels*, MIT Press, Cambridge, MA

[36] Scott, D.W. (1992), *Multivariate Density Estimation: Theory, Practice, and Visualization*, John Wiley, New York.

[37] Statnikov, A., Aliferis, C.F., Tsamardinos, I., Hardini, D. and Levy, S. (2005), A comprehensive evaluation of multicategory classification methods for microarray gene expression cancer diagnosis, *Bioinformatics*, 21, 631 - 643.

[38] Stone, C. (1977), Consistent nonparametric regression, *Annals of Statistics*, 5, 595-620

[39] Tibshirani, R., Hastie, T., Narasimhan, B. and Chu, G. (2002), Diagnosis of multiple cancer types by shrunken centroids of gene expression, *Proceedings of the National Academies of Science USA*, 99, 6567-6572

[40] Ulintz, P.J., Zhu, J., Qin, Z.S. and Andrews, P.C. (2006), Improved classification of mass spectrometry database search results using newer machine learning approaches, *Molecular and Cellular Proteomics*, 5, 497-509

[41] Valiant, L.G. (1984), A theory of the learnable, *Communications of the ACM*, 1134-1142

[42] van't Veer, L.J., Dai, H., van de Vijver, M.J., He, Y.D., Hart, A.A., Mao, M., Peterse, H.L., van der Kooy, K., Marton, M.J., Witteveen, A.T., et al. (2002), Gene expression profiling predicts clinical outcome of breast cancer, *Nature*, 415, 530-536

[43] Vapnik, V. (1998), *Statistical Learning Theory*, Wiley, NY

[44] Venables, W. and Ripley, B. (1994), *Modern Applied Statistics with S-Plus*, Springer, New York

[45] Wang, S. and Zhu, J. (2007) Improved centroids estimation for the nearest shrunken centroid classifier, *Bioinformatics*, 23, 972-979

[46] Zhang, H., Yu, C.Y. and Singer, B. (2003), Cell and tumor classification using gene expression data: construction of forests, *Proceedings of the National Academies of Science USA*, 100, 673-679

[47] Zhang, N., Aebersold, R. and Schwikowski, B. (2002), ProbID: A probabilistic algorithm to identify peptides through sequence database searching using tandem mass spectral data, *Proteomics*, 2, 1406-1412

CHAPTER 5

Unsupervised Learning Techniques

5.1 Introduction

Technological advances have permitted scientists to monitor simultaneously thousands of genes and proteins under different experimental conditions or across different tissue types. This offers the opportunity for a systematic genome- and proteome-wide approach to understand function and regulatory mechanisms in cellular signaling pathways, or gain insight in diverse disease mechanisms.

For example, consider microarray data comprised of thousands of genes (features/variables) and dozens of samples. In this setting, one may be interested in reducing the dimensionality of the 'gene space,' by constructing 'supergenes' that simplify the underlying structure of the data. This strategy has proved useful in classification [20]. Dimension reduction aids also in visualizing the structure of the data set to uncover interesting patterns. In this case, one seeks an efficient representation of high-dimensional data in a 2-3 dimensional space. The main techniques we consider for dimension reduction are principal components analysis and multidimensional scaling.

In cluster analysis, one is interested in partitioning the data into groups, so that the objects (samples and/or variables) in a group are more 'similar' to each other than objects in different groups. Cluster analysis techniques have proved useful in identifying biologically relevant groups of both genes and samples and have also provided insight into gene function and regulation.

5.2 Principal Components Analysis

Principal components analysis (PCA) belongs to the class of *projection methods*. Such methods choose one or more *linear combinations* of the original variables (features, attributes) to *maximize* some measure of *"interestingness."* In PCA the goal is to reduce the *dimensionality* of a data set comprised of a large number of *interrelated* variables, while retaining as much as possible of the *variation* present in the data. PCA also has considerable theoretical importance because of its relationship to elliptical distributions and to standard distance.

We briefly give an intuitive introduction to the technique through a toy example. Consider a 2-dimensional data set shown in the left panel of Figure 5.1. We are interested in projecting the data along the direction that captures most of the variability, which is also shown on the plot. One could continue along this line and request for a second direction that captures the remaining variability in the data set, which in addition is *orthogonal* to the first one. These directions that are obviously linear combinations of the original variables are called principal components. The projected data set onto the space spanned by the first two principal components is shown in the right panel of Figure 5.1. It should be noted that this exercise is most useful, when few principal components suffice to capture most of the variation in the data and consequently reduce its effective dimensionality. We formalize these ideas below.

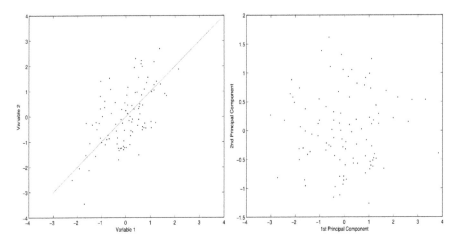

Figure 5.1 *Left panel: a two-dimensional synthetic data set and the direction of maximum variability captured by the first principal component. Right panel: the projection of the data set onto the principal components space.*

5.2.1 Derivation of principal components

Let $\mathbf{x}$ be a vector of p real random variables and let Σ denote their covariance matrix; i.e., $\Sigma = \mathbb{E}(\mathbf{x} - \mu)(\mathbf{x} - \mu)'$. Without loss of generality, we set $\mu = \vec{0}$; otherwise, the mean of the random variables can always be subtracted.

The covariance matrix summarizes all the *linear bivariate* associations in this set of random variables. In PCA, we are interested in reducing the dimensionality p of the original variables, by considering *linear combinations* of them and retaining the *"most important"* of these new variables.

Define a new variable

$$y_1 = \beta'_1 \mathbf{x} = \sum_{j=1}^{p} \beta_{j1} x_j, \qquad (5.1)$$

that is a linear combination of the x's with coefficients $\{\beta_{j1}\}$.

The question is how to choose the weights β_1. The choice depends on the measure of "interestingness" that we are going to use. For PCA as was shown with the toy example, the measure of interestingness that we want to maximize is the *variance* of the new variable $\mathbf{y}_1$.

Criterion: $\max_{y_1} \mathrm{Var}\,(y_1) = \max_{\beta_{j1}} \mathrm{Var}\,\left(\sum_{j=1}^{p} \beta_{j1} x_j\right) = \max_{\beta_1} \beta'_1 \Sigma \beta_1$.

A moment of reflection shows that the maximum would be achieved for an infinite β_1; thus, a *normalization constraint* must be imposed. The standard constraint employed is $||\beta_1|| = 1$. Hence, the problem becomes

$$\max_{\beta_1}\ \beta'_1 \Sigma \beta_1 \quad \text{s.t} \quad ||\beta_1|| = 1. \qquad (5.2)$$

The Rayleigh-Rietz theorem indicates that the solution is given by the eigenvector corresponding to the largest eigenvalue λ_1 of Σ.

If a second linear combination $y_2 = \beta'_2 x$ is desired, we require in addition to be *orthogonal* to the first one; i.e., formally, $< y_1, y_2 >= 0$, where $< \cdot, \cdot >$ denotes the inner product of two real-valued vectors. Then the problem becomes

$$\max_{\beta_2}\ \beta'_2 \Sigma \beta_2 \quad \text{s.t } ||\beta_2|| = 1 \quad \text{and} \quad < \beta_1, \beta_2 >= 0. \qquad (5.3)$$

In general, we can construct p linear combinations $\mathbf{y} = B'\mathbf{x}$, B being a $p \times p$ matrix whose k-th column corresponds to β_k.

The optimal set of weights is given by the eigenvectors of Σ;

$$\Sigma = B \Lambda B'.$$

Notice that the *Principal Components* $\mathbf{y}$ are by construction an *orthonormal linear transformation* of $\mathbf{x}$.

We discuss next the main properties of Principal Components.

(1) $\mathbb{E}(\mathbf{y}) = \mathbb{E}(B'\mathbf{x}) = B'\mathbb{E}(\mathbf{x}) = 0$, ex hypothesis.

(2) $\mathrm{Cov}(\mathbf{y}) \equiv \Sigma_y = B'\Sigma B = B'(B\Lambda) = \Lambda$. Hence, the principal components are uncorrelated and $\mathrm{Var}(y_j) = \lambda_j$.

(3) Correlations of the original random variables $\mathbf{x}$ with the principal components $\mathbf{y}$.
We have that $\mathbf{Cov}(\mathbf{x}, \mathbf{y}) = \mathbf{Cov}(\mathbf{x}, B'\mathbf{x}) = \Sigma B = B\Lambda$. This shows that for a fixed y_j we have $\mathbf{Cov}(y_j, \mathbf{x}) = \lambda_j \beta_j$. Hence, $\mathbf{Corr}(x_i, y_j) = \lambda_j \beta_{ji} / \sqrt{\lambda_j}$. The quantities $\sqrt{\lambda_j} \beta_j$ are called *factor loadings*.

(4) Proportion of variance explained by the principal components.
Notice that $\mathbf{trace}(\Sigma) = \mathbf{trace}(B\Lambda B') = \mathbf{trace}(\Lambda B'B) = \mathbf{trace}(\Lambda)$. Thus,

each principal component "explains" $\lambda_j/(\sum_{j=1}^{p}\lambda_j)$ proportion of the total variance.

(5) Consider the set of p-dimensional ellipsoids of the form

$$\mathbf{x}'\Sigma^{-1}\mathbf{x} = c,$$

Then, the principal components define the principal axes of these ellipsoids.

The set of principal components is given by $\mathbf{y} = B'\mathbf{x}$. Because of the fact that B is orthogonal, we get $\mathbf{x} = B\mathbf{y}$. Plugging this in we obtain

$$\mathbf{x}'\Sigma^{-1}\mathbf{x} = \mathbf{y}'B'\Sigma^{-1}B\mathbf{y} = c.$$

But the eigenvalue decomposition of Σ^{-1} gives

$$\Sigma^{-1} = B\Lambda^{-1}B',$$

i.e., the inverse of the covariance matrix has the same set of eigenvectors as the covariance matrix and the reciprocals of its eigenvalues (provided that Σ is strictly positive definite). Some algebra shows that $B'\Sigma^{-1}B = \Lambda^{-1}$ and therefore $\mathbf{y}'\Lambda^{-1}\mathbf{y} = c$. The last expression corresponds to the equation of an ellipsoid referred to its principal axes. It also implies that the half lengths of the principal axes are proportional to $\lambda_j^{1/2}$.

Remark: Principal Components based on the correlation matrix
Let $\tilde{\mathbf{x}} = (x_j/\sigma_j)$, where σ_j^2 is the variance of x_j. We can define the principal components of $\tilde{\mathbf{x}}$ as before; namely,

$$\tilde{\mathbf{y}} = \tilde{B}'\tilde{\mathbf{x}}, \tag{5.4}$$

where $\tilde{B}$ contains the eigenvectors of the correlation matrix of the $\tilde{x}_j$'s.

It is worth noting that the principal components derived from the covariance and the correlation matrix are not related, except in very special circumstances. This is because the principal components are *invariant* under *orthogonal* transformations of $\mathbf{x}$ but not under other transformations. However, the transformation from $\mathbf{x}$ to $\tilde{\mathbf{x}}$ is not orthogonal. In practice, the main advantage using the correlation matrix in PCA is that the results are not sensitive to the units of measurement for $\mathbf{x}$.

5.2.2 Sample principal components

We turn our attention to PCA applied to data. Let X be an $n \times p$ data matrix. Further, assume that the observations are *independent* replicates from the p-dimensional random vector $\mathbf{x}$. We will denote the columns (variables) of the data matrix by $X_1, X_2, \cdots, X_n$. Without loss of generality assume that $\text{Mean}(X_i)=0$r; in practice the data matrix X is always centered, by subtracting the sample mean of each variable. Then, the *sample covariance* matrix is given by $S = X'X/(n-1)$.

Let $S = B\Lambda B'$ be the eigenvalue decomposition of the sample covariance matrix. Then, the sample principal components are defined by

$$Y = XB, \tag{5.5}$$

where Y is a $n \times p$ matrix containing PC scores, and B a $p \times p$ orthonormal matrix containing the weights. In direct analogy to the population version of PCA the elements of the $p \times p$ diagonal matrix Λ are the variances of the p principal components; i.e., $\mathbf{Var}(Y_j) = \lambda_j$, $j = 1, \cdots, p$.

Remark: The properties derived in Section 2.1 carry over, with the sample covariance matrix S replacing the population covariance matrix Σ. One small difference arises in property (5). The ellipsoids $\mathbf{x}'S^{-1}\mathbf{x} = c$ do not represent contours of constant probability; rather, they correspond to contours of equal Mahalanobis distance from the sample mean, which it is assumed to be 0 (the data have been centered). In the literature, the following interpretation of PCA appears; namely, PCA finds successive orthogonal directions for which the Mahalanobis distance from the data set to a hypersphere enclosing all the data is minimized.

The next geometric property of sample principal components gives another interpretation of PCA. In fact, it is a generalization of an idea used by Karl Pearson back in 1901 to define PCA for 2-dimensional data.

Consider n observations denoted by $X(i, \cdot)$ ($X(i, \cdot)$ being a p-dimensional column vector). Let C be an orthogonal $p \times q$ matrix and let $Y(i, \cdot) = C'X(i, \cdot)$ be the projection of the data onto a q-dimensional subspace. A goodness-of-fit measure of this q-dimensional subspace to $X(1, \cdot), \cdots, X(n, \cdot)$ is given by the sum of *squared perpendicular distances* of the $X(i, \cdot)$'s from the subspace. This measure is *minimized* when $C = B_q$ which contains the first q columns of the coefficient matrix B. The plot in Figure 5.2 schematically illustrates this property.

5.2.3 Illustration of PCA

In this example, we illustrate PCA with a gene expression data set from a cancer study [20] that deals with small, round blue-cell tumors of childhood. There are four types of tumors: neuroblastomas (NB), rhabdomyosarcomas (RMS), Burkitt lymphomas (BL) and the Ewing family of tumors (EWS). The data were filtered and contain 2308 genes (see [20]) and 63 samples (12 NB, 20 RMS, 8 BL and 23 EWS). The projection of the data onto the first two principal components that capture about 55% of the total variance is shown in Figure 5.3. It can be seen that the classes are not well separated; however, the BL class of tumors is fairly well separated from the RMS along the first principal component and to a lesser extent from the NB one. A look at the component loadings (not shown here) indicates that the first principal component can be interpreted as an overall average of gene expression, which

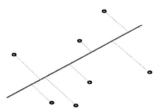

Figure 5.2 *Geometry of principal components, where the direction (line) minimizes the orthogonal projections of the data to it.*

in turn suggests that the expression level of the BL tumors is clearly different and opposite than the majority of the RMS ones. Further, the analysis reveals the presence of a RMS sample at the bottom of the plot, which strongly indicates that it is an outlier.

For further illustration purposes, we concentrated on the two largest tumor classes, EWS and RMS. In addition, the 83 most discriminating genes between these two classes were selected through t-tests. The projection of the data onto the first two principal components is shown in Figure 5.4. As expected, a very strong separation of the two classes occurs along the first principal component, which again can be interpreted as an overall average of gene expression.

5.2.4 PCA and the Singular Value Decomposition (SVD)

Given an $n \times p$ centered data matrix X of full column rank we can apply the Singular Value Decomposition and obtain

$$X = ULB', \qquad (5.6)$$

where U and B are $n \times p$ and $p \times p$ orthonormal matrices and L a $p \times p$ diagonal matrix.

Some straightforward algebra shows that

$$(n-1)S = X'X = BLU'ULB' = BL^2B'. \qquad (5.7)$$

Hence, the matrix B contains the eigenvectors of the sample covariance matrix, while a rescaled version of L its eigenvalues. Specifically $\mathbf{Var}(y_i) = \ell_i^2/(n-1)$.

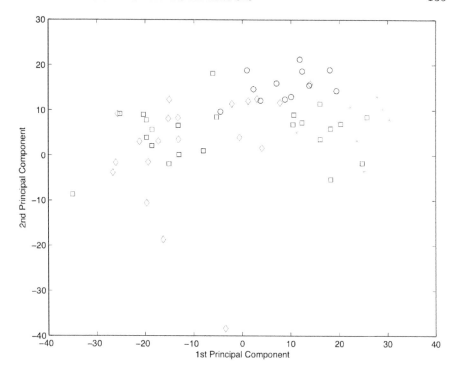

Figure 5.3 *Projection of the cancer microarray data set onto the first two principal components. The NB samples are represented by* o, *the RMS by* ⋄, *the BL by* ⋆ *and the EWS by* □.

Further, notice that the PC scores $Z = XB$ are given by $Z = XB = ULB'B = UL$.

It is also of interest the fact that $XX' = ULB'BLU' = UL^2U$; i.e., the matrix, U contains the eigenvectors of XX'. This result comes in handy in some applications where the role of 'variables' and 'observations' is reversed.

The main interest in the SVD is that in addition to providing a computationally efficient way to do PCA, it also gives additional insight into the technique. For example, let $\tilde{X}_q = U_qL_qB_q'$ be the $n \times q$ matrix formed by retaining the q largest in terms of variance principal components. Then, the Householder-Young theorem implies that $\tilde{X}_q$ gives the best q-rank approximation of the data in the Frobenius norm; i.e., $\mathbf{argmin}_Q||X - Q||_F = \tilde{X}_q$.

Biplots: The SVD also provides an efficient way of visualizing together objects (observations) and variables. The emphasis of the PCA of the city crime data was on cities (objects). However, a graphical representation of the data that contains some information about the variables is particularly useful. One such representation is given by the *biplot*, introduced by Gabriel in 1971, shown for

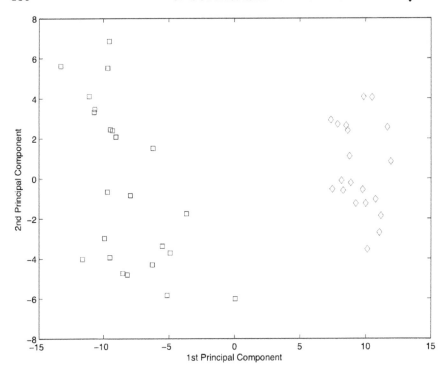

Figure 5.4 *Projection of the RMS (◇) and EWS (□) tumor classes set onto the first two principal components based on a subset of 83 genes.*

the gene expression data set comprised of the RMS and EWS tumor samples in Figure 5.5. The arrows on the biplot indicate where one can find samples with high values on a particular gene (variable). The angles between the arrows, which represent the variables, capture the correlation between them provided that the total amount of variance explained by the first two PCs is reasonably high. The biplot is constructed as follows: from the SVD of the centered data matrix X we get that $X = ULV'$. Define L^{α} and $L^{1-\alpha}$ for any $\alpha \in [0,1]$. Let $G = UL^{\alpha}$ and $H' = L^{1-\alpha}V'$. Notice that for $\alpha = 1$ we get $G = Y$ the PC scores. A biplot is a plot of the first k (usually $k = 2$ or 3) columns of G and H.

5.3 Multidimensional Scaling

Multidimensional scaling (MDS) is a method that represents measurements of *similarity* or *dissimilarity* among pairs of objects as distances between points in a low-dimensional space [7].

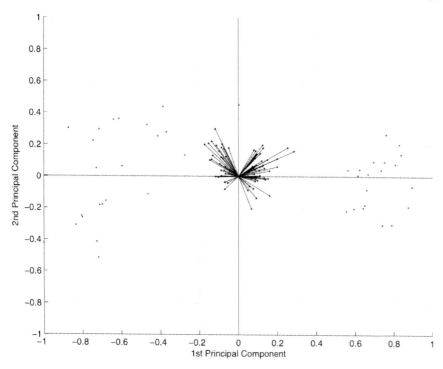

Figure 5.5 *Biplot of the gene expression data set containing only the RMS and EWS tumor samples.*

Historically, MDS was developed as a method for analyzing proximity data that arise in the social sciences (similarity ratings for pairs of stimuli such as tastes, colors, etc), classification problems, archeology (similarity in the features of artifacts). Another early use of MDS has been *dimension reduction* (Kruskal-Shepard MDS). The idea is to compute the pairwise distances between the objects in a high dimensional space, and attempt to find an arrangement of the objects in a low dimensional space so that their distances approximate as close as possible the original distances.

5.3.1 Metric scaling

The setting is as follows: given an $N \times p$ data matrix (N observations, p variables) X calculate the corresponding dissimilarity matrix D. If one is interested in the samples (observations), then D is a $N \times N$ matrix, while if one is interested in the variables then D is a $p \times p$ matrix.

A dissimilarity d_{ij} between observations i and j is a distance-like measure that satisfies the following properties:

1. Nonnegativity property: $d_{ij} \geq 0$ for all i and j, with $d_{ii} = 0$.

2. Symmetry property: $d_{ij} = d_{ji}$ for all i, j.

If, in addition, d_{ij} satisfies the triangle inequality ($d_{ij} \leq d_{ik} + d_{kj}$ for any k) then it is a proper *distance*. Euclidean, Mahalanobis and Manhattan distances are commonly used in data analysis.

The objective of MDS is to approximate dissimilarities (distances) calculated from the data in p-dimensional space by Euclidean distances defined in a $q \ll p$ dimensional space. In practice, q is usually chosen equal to two or three, since one is interested in visualizing the structure of the data. Let Z be a $N \times q$ matrix that contains the coordinates of the objects in the low dimensional space. The quality of the approximation is measured by a loss function. Common loss functions include:

$$\text{Stress}(Z) = \sum_{i=1}^{N} \sum_{j=i+1}^{N} (d_{ij}(Z) - d_{ij})^2, \tag{5.8}$$

which is invariant under rotations and translations, but not invariant to stretching and shrinking. This implies that the value of *Stress* is scale dependent and a better criterion is given by

$$\sqrt{\frac{\text{Stress}(Z)}{\sum_{i<j} d_{ij}^2(Z)}}. \tag{5.9}$$

A more general form of the Stress function incorporates weights w_{ij} that reflect variability, measurement error or missing data:

$$\text{WStress}(X) = \sum_{i=1}^{N} \sum_{j=i+1}^{N} w_{ij}(d_{ij}(Z) - d_{ij})^2. \tag{5.10}$$

If the weights are chosen so that $w_{ij} = d_{ij}^{-1}$, we then recover the *Sammon mapping*. One reason for using weights is to reduce the effect of outliers on the low dimensional configuration, which can distort the results [7].

Finding the optimal configuration Z of the observations in q-dimensional space is not a trivial problem, since the loss function is nonconvex. Iterative algorithms are discussed in [7].

An application of MDS to the gene expression data of the tumor samples is shown in Figure 5.6. Manhattan distances were calculated between the samples and a 2-dimensional representation based on the Stress loss function obtained. It can be seen that a different view of the data compared to that of PCA is obtained, although some features, such as the grouping of the BL samples and their separation from the RMS and the NB samples, are still preserved. In Figure 5.7, the MDS representation of the gene expression data for the RMS and EWS samples based on Manhattan distances is shown. In this case a clear separation of the two groups of tumors is observed, as was the case with the principal components analysis.

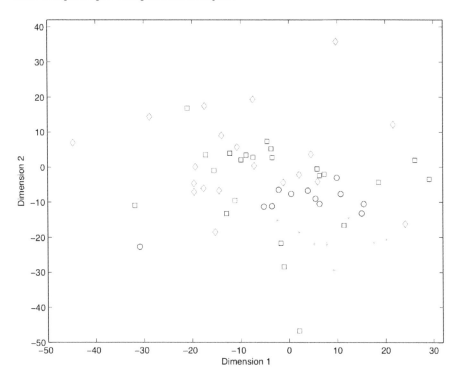

Figure 5.6 *MDS representation of the gene expression data based on Manhattan distances. The NB samples are represented by* ∘, *the RMS by* ◇, *the BL by* ⋆ *and the EWS by* □.

5.4 Other Dimension Reduction Techniques

We briefly outline extensions of principal components analysis and multidimensional scaling that have proved useful in various data analytic tasks. All of these approaches attempt to deal with the presence of strong nonlinearities in the data. Finally, in the presence of categorical data a somewhat different approach is required (for a comprehensive review see [25]).

Local PCA: If one would like to retain the conceptual simplicity of PCA, together with its algorithmic efficiency in the presence of nonlinearities in the data, one could resort to applying PCA locally. One possibility is to perform a cluster analysis, and then apply different principal components analyses to the various clusters [8]. Some recent advances on this topic are discussed in [24].

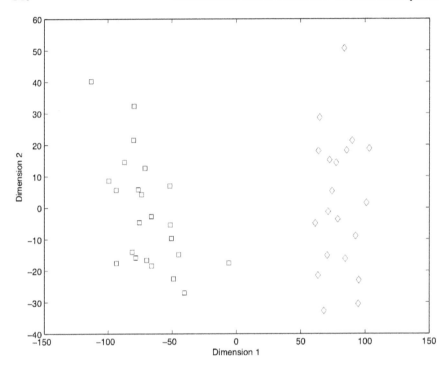

Figure 5.7 *MDS representation of the gene expression data of the RMS (◇) and EWS (□) samples, based on Manhattan distances.*

Kernel PCA: The idea behind kernel PCA [29] is that nonlinear patterns present in low dimensional spaces can be linearized by projecting the data into high dimensional spaces, at which point classical PCA becomes effective. Although this concept is contrary to the idea of using PCA for data reduction purposes, it has proved successful in some application areas like hand-written digit recognition. In order to make this idea computationally tractable, kernels (for a brief discussion see the chapter on Classification Techniques) are used that essentially calculate inner products of the original variables.

Manifold learning methods: These methods assume that the data have a nonlinear structure that is captured by a smooth connected manifold. These techniques such as Local Linear Embedding (LLE) [27], Hessian LLE [10], Laplacian Eigenmaps [5] and Isomap [31] construct a low dimensional data representation using a cost function that retains local properties of the data. For example, the Isomap technique constructs a dissimilarity matrix for only locally neighboring points and then uses multidimensional scaling for deriving the final configuration. On the other hand, LLE uses local linear approximations of the manifold and then pieces them together; it can be viewed as defining a graph-based kernel to be used in conjunction with kernel PCA. For a comprehensive review of these methods, see also [28].

5.5 Cluster Analysis Techniques

5.5.1 Problem formulation and algorithms

The main idea behind cluster analysis is to investigate relationships between the objects in order to establish whether or not the data can validly be summarized by a small number of groups (clusters) of similar objects. A good outcome manages to assign the N objects in the dataset into a small number of groups in such a way that members in the same group are as 'similar' as possible, while members of different groups are as 'dissimilar' as possible.

Formally we have: let $\mathcal{O} = \{o_1, o_2, \ldots, o_N\}$ be a set of N objects and let $\mathcal{C} = \{C_1, \ldots, C_K\}$ be collection of subsets of $\mathcal{O}$; i.e., $C_k \subset \mathcal{O}$ for all $k = 1, \cdots, K$. Each subset is called a *cluster*, and $\mathcal{C}$ is a clustering solution.

The input data for a clustering problem is typically given in one of two forms:

- Profile data matrix $X_{N \times p}$, where each object o_i is associated with a real-valued vector, called its profile, which contains measurements on p variables, e.g., gene expression levels at different experimental conditions.

- Dissimilarity matrix $\Delta_{N \times N}$, that contains the pairwise dissimilarities (or similarities) that are usually computed from the profile data.

When dealing with a cluster analysis problem the following issues need to be studied [15]:

- what is the criterion used (e.g., homogeneity vs separation).
- what type of clustering should be considered.
- how difficult it is to perform the clustering (issue of complexity).
- how should the clustering be done (issue of algorithmic design).
- is the resulting clustering meaningful (i.e., how should the results be interpreted).

Remark:

In most clustering problems few assumptions, usually posed in set-theoretic terms, are made. However, in some cases the problem is posed as one where the clusters correspond to mixtures of distributions whose number and parameters need to be estimated (learned) from the data [26].

We examine next some of the most popular algorithmic approaches for cluster analysis. Primarily they can be categorized into two groups: (i) *partition
methods* and (ii) *tree-type methods*.

Partition methods create a family of clusters where each object belongs to just
a single member of the partition. Formally, we require $C_i \cap C_j = \emptyset$ for clusters
i and j with $i \neq j$ and $\cup_k C_k = \mathcal{O}$. To generate such partitions the requirement
is that distances between pairs of objects belonging to the same cluster are
smaller than distances between pairs of objects in different clusters. Formally,
suppose that objects i and j belong to cluster A, while object k belongs to
cluster B. We then require that $d_{ij} < d_{ik}$ and $d_{ij} < d_{jk}$.

Tree methods build a tree of clusters that includes all the objects and for which
any two clusters are either disjoint or one cluster is a superset of the other.
In tree methods it is usually required that the distance be an *ultrametric*;
i.e., $d_{ij} \leq \max\{d_{ik}, d_{jk}\}$. Equivalently we have that for every triple of objects
(i, j, k) we have that the 2 largest values in the set $\{d_{ij}, d_{jk}, d_{ik}\}$ are equal.

Partition methods:
Points are assigned to clusters with the objective of optimizing some criterion.
We can decompose the variation as

$$T = W + B, \tag{5.11}$$

where

$$T = \sum_{\text{all objects} i} (x_i - \bar{x})(x_i - \bar{x})' \tag{5.12}$$

and

$$W = \sum_{\text{all clusters} k} n_k(x_i - \bar{x}_k)(x_i - \bar{x}_k)' \tag{5.13}$$

where n_k is the number of objects in cluster k, $\bar{x}$ is the overall mean of the data
and $\bar{x}_k$ is the mean of cluster k. It is worth noting that the above definitions
are identical to the ones in MANOVA. The only difference is that we *don't
know a priori* which group an object belongs to.

Given that T is fixed a good clustering algorithm seeks to minimize some
measure of W (the within groups variation) and hence maximize some measure
of B (the between groups variation). Criteria that have been suggested in the
literature include:

1. Minimize trace(W) (not invariant to changes in scale).
2. Minimize det(W).
3. Maximize the sum of the eigenvalues of $W^{-1}B$.

Remark: This optimization problem is a hard one from a computational point
of view. Just notice that there are 10^{30} ways to allocate 100 objects to 2 groups.
Hence, exhaustive search is not a viable option. Therefore, various heuristic

methods have been proposed in the literature to deal with the problem at hand. One of the most popular ones is the K-means clustering algorithm.

The K-means algorithm:
The K-means algorithm optimizes the following objective function

$$H(K) = \sum_{i=1}^{K} \sum_{X_k \in C_i} ||X_k - \mathbf{m}_i||^2, \tag{5.14}$$

where $\mathbf{m}_i$ are the centroids or means of the clusters C_i, and $|| \cdot ||$ denotes the Euclidean norm. The algorithm proceeds by partitioning N objects into K nonempty subsets. Note that the value of K is determined by the user. During each partition, the centroids or means of the clusters are computed. The main steps of the K-means algorithm are as follows:

1. Assign initial means $\mathbf{m}_i$ to each of the K clusters.
2. Assign each data object o_j with profile X_j to the cluster C_i for the closest mean.
3. Compute new mean for each cluster using

$$\mathbf{m}_i = \frac{\sum_{X_k \in C_i} X_k}{|C_i|}, \tag{5.15}$$

   where $|C_i|$ is the number of objects in cluster C_i.
4. Iterate until criterion function converges, i.e., there are no more new assignments.

The K-means algorithm is simple and has been widely used in the analysis of microarray data. Typically, the algorithm converges fairly fast to a solution. However, it requires as input parameters the cluster number K and initial centroids. Randomly chosen initial centroids may lead to poor results. Further, due to its nature the algorithm favors spherically shaped clusters.

Model based clustering:
In this approach, we try to turn the problem into a density estimation one based on statistical *mixture* models. They provide a probabilistic alternative to deterministic partition algorithms. The idea is that the objects are independent samples from an unknown number of group populations, but the group labels have been lost. If we knew that the vector α gave the group labels, and each group had class-conditional probability density functions $f_{\alpha_k}(x_k; \theta)$ then the likelihood would be given by

$$\Pi_k f_{\alpha_k}(x_k; \theta). \tag{5.16}$$

Since the labels are unknown, these are regarded as parameters, and hence we need to maximize the likelihood function over (θ, α). For arbitrary densities $f_{\alpha_k}(x_k; \theta)$ the problem is almost computationally intractable. However, for normal densities the expectation-maximization algorithm allows one to estimate the parameters of interest.

In case we choose the conditional probability density functions to be $N(\mu_k, \sigma^2 I)$ (different means, common covariance matrix), then this procedure is equivalent to K-means. Other possibilities allow clusters of different sizes, shapes and orientations, by imposing constraints on the covariance matrix.

In [4], a general framework was proposed based on parameterizing covariance matrices through the eigenvalue decomposition as follows:

$$\Sigma_k = \lambda_k D_k A_k D_k', \quad k = 1, \ldots, K, \qquad (5.17)$$

where D_k is the orthonormal matrix of eigenvectors, A_k is a diagonal matrix whose elements are proportional to the eigenvalues and λ_k is an associated constant of proportionality. D_k governs the orientation of the kth cluster, A_k its shape and λ_k its volume, which is proportional to $\lambda_k \det(A_k)$. For example, if the largest eigenvalue of Σ_k is much larger in magnitude than the remaining ones, then the k-th cluster will be concentrated close to a line, which will correspond to the first PC of the k-th group. If the first two eigenvalues are much larger than the rest, then the k-th cluster will be concentrated close to a plane. If finally all the eigenvalues of the k-th group are about the same, then the k-th cluster will be roughly spherical.

One of the advantages of the model based clustering formulation is that it provides a principled mechanism to determine the number of clusters and compare the parameterizations of different models (volume, orientation, shape). Given the likelihood formulation in (5.16) one can use the Bayesian Information Criterion (BIC) for model selection purposes in terms of number of clusters and parameterizations.

A review of model based clustering together with extensions of the framework that incorporates more complex parameterizations with applications to genomic data are discussed in [17, 18].

Self-Organizing Maps:
The self-organizing map (SOM) is a neural network algorithm that has been used in a wide variety of applications (for microarray data, see [14]). For a comprehensive treatment of the topic consult Kohonen's book [22].

This particular algorithm is motivated by ideas in *competitive learning* (see [3, 22]), that can be described as the adaptive process in which neurons in a neural network gradually become more sensitive to different input samples (data). 'A kind of division of labor emerges in the network when different neurons specialize to represent different types of input' [19]. The specialization is enforced by competition among the neurons: when a new input x arrives (e.g., a new data point), the neuron that is best able to represent it wins the competition and is allowed to learn it even better.

If there exists an ordering between the neurons, e.g., the neurons are located on a discrete lattice, then a competitive learning algorithm can be generalized to allow not only for the winning neuron, but also for its closest neighbors to learn. Thus, neighboring neurons will eventually specialize to represent similar inputs, and the resulting representation will become *ordered* on the map lattice. We formulate these heuristic ideas next.

Consider the self-organizing network given in Figure 5.8. Let M input signals be simultaneously incident on each of an $N \times N$ array of neurons. The output of the ith neuron is defined as

$$\eta_i(t) = \boldsymbol{\sigma} \left[[\boldsymbol{m}_i(t)]^T \, \boldsymbol{x}(t) + \sum_{k \in S_i} w_{ki} \, \eta_k(t - \triangle t) \right], \qquad (5.18)$$

where $\boldsymbol{x}$ is the M-dimensional input vector incident on it along the connection weight vector $\boldsymbol{m}_i$, k belongs to the subset S_i of neurons having interconnections with the ith neuron, w_{ki} denotes the fixed feedback coupling between the kth and ith neurons, $\boldsymbol{\sigma}[.]$ is a suitable sigmoidal output function, t denotes a discrete time index and T stands for the transpose.

Input points that are close in p-dimensional space are also mapped closely on the q-dimensional lattice. Each lattice cell is represented by a neuron that has a p-dimensional adaptable weight vector associated with it. With every input the match with each weight vector is computed. Then the best matching weight vector and some of its topological neighbors are adjusted to match the input points a little better. Initially, the process starts with a large neighborhood; with passage of time (iteration), the neighborhood size is reduced gradually. At a given time instant, within the neighborhood, the weight vector associated with each neuron is not updated equally. The strength of interaction between the winner and a neighboring node is inversely related to the distance (on the lattice) between them.

Initially the components of $\boldsymbol{m}_i$ are set to small random values lying in the range $[0, 0.5]$. If the best match between vectors $\boldsymbol{m}_i$ and $\boldsymbol{x}$ occurs at neuron c, then we have

$$\|\boldsymbol{x} - \boldsymbol{m}_c\| = \min_i \|\boldsymbol{x} - \boldsymbol{m}_i\|, \quad i = 1, 2, \dots, N^2, \qquad (5.19)$$

where $\|.\|$ indicates the Euclidean norm.

The weight updating is given as [22]

$$\boldsymbol{m}_i(t+1) = \begin{cases} \boldsymbol{m}_i(t) + \alpha(t) \left(\boldsymbol{x}(t) - \boldsymbol{m}_i(t) \right) & \text{for } i \in N_c \\ \boldsymbol{m}_i(t) & \text{otherwise}, \end{cases} \qquad (5.20)$$

where $\alpha(t)$ is a positive constant that decays with time and N_c defines a topological neighborhood around the maximally responding neuron c, such that it also decreases with time. Different parts of the network become selectively

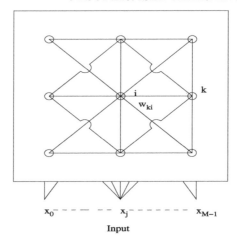

Figure 5.8 *Depiction of Kohonen's neural network.*

sensitized to different inputs in an ordered fashion so as to form a continuous map of the signal space. After a number of sweeps through the training data, with weight updating at each iteration obeying Eq. (5.20), the asymptotic values of m_i cause the output space to attain proper topological ordering. This is basically a variation of *unsupervised* learning.

Vector quantization can be seen as a mapping from an n-dimensional Euclidean space into a finite set of prototypes. Based on this principle, Kohonen proposed an unsupervised learning algorithm, which is a special case of SOFM and is known as LVQ [22]. In LVQ, only the weight vector associated with the winner node is updated with every data point by Eq. (5.20). The topological neighborhood is not updated here. Such a learning scheme, where all nodes compete to become the winner, is termed *competitive learning*. It is essentially a clustering network that does not care about preserving the topological order. Its main uses are for clustering, classification and image data compression.

There exists a family of LVQs, termed LVQ1 and LVQ2 [22]. These algorithms are supervised learning schemes, essentially used as classifiers. The basic idea behind LVQ1 is as follows. If the winner prototype m_i has the same class label as that of the data point x, then bring m_i closer to x; otherwise, move m_i away from x. Nonwinner nodes are not updated. LVQ2, a modified form of LVQ1, is designed to make the learning scheme comply better with Bayes' decision-making philosophy. This algorithm considers the winner along with the runner-up (second winner).

Agglomerative Hierarchical Algorithms:
In general hierarchical methods create an iterated partition. The clustering solution is represented by a *dendrogram*, which is a rooted weighted tree

$\mathcal{T}$, with leaves corresponding to the objects. A tree satisfies the following conditions:

$$\text{(i) } \mathcal{O} \in \mathcal{T}, \text{ (ii) } \emptyset \notin \mathcal{T}. \text{ (iii) } o_i \in \mathcal{T} \text{ all } i \in \mathcal{O} \tag{5.21}$$

$$\text{(iv) if } A, \ B \in \mathcal{T}, \text{ then } A \cap B \in \{\emptyset, A, B\}. \tag{5.22}$$

Each edge defines the cluster of objects contained in the subtree below that particular edge. The edge's length (or weight) reflects the dissimilarity between that cluster and the remaining clusters.

Although hierarchical clustering of objects determines a set of ultrametric distances, practitioners rarely make use of the values themselves. Attention is most often paid only in the ordering of the lengths. More specifically, such methods start with one object per cluster and merge them to form progressively larger groups. Since groups are merged to form larger groups we need a definition of distance between groups. The three most common possibilities are:

Single linkage: The dissimilarity between two clusters is defined by

$$d(C_i, C_j) = \min_{s,t}\{d(s,t)|o_s \in C_i, \ o_t \in C_j\}. \tag{5.23}$$

Complete linkage: In this case, the dissimilarity is defined by

$$d(C_i, C_j) = \max_{s,t}\{d(s,t)|o_s \in C_i, \ o_t \in C_j\}. \tag{5.24}$$

Average linkage: In this case, the dissimilarity is defined by

$$d(C_i, C_j) = \frac{\sum_{s \in C_i} \sum_{t \in C_j} d(i,j)}{n_i n_j}. \tag{5.25}$$

Other possibilities to perform the join operation include:

1. *Ward's method* that merges clusters that minimize $\sum_{\text{all clusters}}$(SS about the mean in cluster).

2. *The weighted average linkage.*

We illustrate next the complete linkage algorithm by using it on the complete gene expression data set for tumors and that based on the RMS and EWS samples only. In the first case, dissimilarities were computed based on all 2308 genes using the Manhattan metric, while in the latter, they were based on 83 selected genes. The results are shown in Figure 5.9. In both cases, the clustering algorithm reveals some substructure of the samples, with some of them being very similar. Further, for the second data set, a clear clustering pattern emerges, as expected.

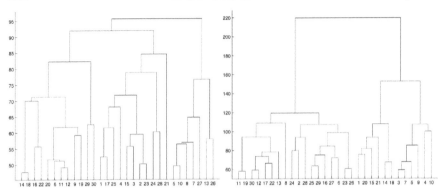

Figure 5.9 *Left panel: Complete linkage based clustering of tumor samples based on 2308 genes. Right panel: Complete linkage based clustering of RMS and EWS tumor samples based on 83 selected genes.*

5.5.2 Some practical considerations

Cluster analysis is based on the distance between points, so variables need to be scaled appropriately. If we decide to standardize all variables it implies that we deem all variables to be equally important. On the other hand if we scale only some of the variables, it implies that we consider some of them more important than others. As data analysts you should be aware that scaling will affect the results of many clustering algorithms, since they are not robust to scale changes.

Another important issue is which variables to use in cluster analysis. You would notice that the various clustering techniques do not have built-in variable selection mechanisms. Moreover, more variables are not necessarily better. A good idea is to run a PCA in order to assess the importance of variables before applying cluster analysis.

5.5.3 Biclustering

Although cluster analysis techniques have a long history in statistics and computer science, biclustering algorithms originated in bioinformatics problems, especially in trying to find structure in gene expression data. To make things concrete consider a gene expression matrix X with rows corresponding to genes and columns corresponding to samples (e.g., experimental conditions, different a priori defined classes, etc). Cluster analysis techniques can be applied *separately* either on the genes, or on the samples, with the goal being to discover *homogeneous* groups of the former or the latter (e.g., groups of co-regulated groups, groups of similarly behaving patients, etc). As already seen in this section, clustering algorithms usually partition the genes/samples

Samples

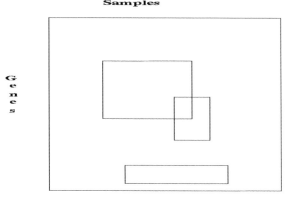

Figure 5.10 *Illustration of biclusters*

into disjoint sets. However, in many situations overlapping clusters of genes and samples may be appropriate, thus requiring a more flexible framework.

A *bicluster* is defined as a submatrix containing a set of genes and a set of samples, as shown in Figure 5.10. It is important to note that in principle, there are no a priori constraints in the organization of biclusters, which can be overlapping, with the same genes/samples being members of multiple ones. The downside of this flexibility is over-fitting, which needs to be countered by a statistical model or scoring method that identifies *significant* biclusters. Finally, since the most active area of applications of biclustering is gene expression data that are inherently noisy, the resulting algorithms must exhibit a fairly high level of robustness to noise.

Biclustering algorithms have been successfully applied to problems in genomics, but also to clinical studies. In the former case, the objective is to understand the functions of genes as members of a biological system. By collecting gene expression data on the genes under a number of biological conditions and subsequently identifying joint patterns, one can characterize transcriptional programs and assign gene function [12, 16]. Similarly in clinical studies, by collecting gene expression profiles on a large number of patients, one can identify subsets of patients associated with specific clinical conditions and/or treatment outcomes [1]. Biclustering allows the assignment of patients to multiple clinical groups.

We discuss next some of the proposed biclustering algorithms in the literature. The set of rows (genes) in the expression matrix X would be denoted by G and the set of columns (samples) by S.

Cheng and Church Algorithm:
This was the first paper [9] to consider biclustering in the context of gene expression data. Define for a subset of genes $\tilde{G} \subset G$ and a subset of samples $\tilde{S} \subset S$, the grand mean $\bar{X}_{.,.} = \sum_{i \in \tilde{G}, j \in \tilde{S}} X_{i,j}/(|\tilde{G}||\tilde{S}|)$, the row mean $\bar{X}_{i,.} =$

$\sum_{j \in \tilde{S}} X_{i,j}/|\tilde{S}|$ and the column mean $\bar{X}_{.,j} = \sum_{i \in \tilde{G}} X_{i,j}/|\tilde{G}|$. The residual of each element of submatrix $X_{\tilde{G},\tilde{S}}$ is then defined as $e_{ij} = X_{i,j} + \bar{X}_{.,.} - \bar{X}_{.,j} - \bar{X}_{i,.}$ and the mean squared residual by $M(\tilde{G}, \tilde{S}) = \sum_{i \in \tilde{G}, j \in \tilde{S}} e_{i,j}^2/(|\tilde{G}||\tilde{S}|)$. It can be seen that any submatrix that can be written as the sum of mean and column effects in analysis of variance parlance would have a mean squared residual equal to zero.

The biclustering problem is defined as identifying a bicluster of maximum size amongst all biclusters with mean squared residual not exceeding a prespecified threshold δ. The size may correspond to the number of elements or the number of columns plus the number of rows. A greedy search algorithm was introduced in [9] that guarantees convergence to a local maximum and discovers a single bicluster. In order to discover multiple biclusters, repeated application of the algorithm is suggested on modified matrices that are obtained through randomization of the entries in the values of previously discovered biclusters.

Spectral Biclustering:
Spectral biclustering is based on the singular value decomposition of the expression matrix $X = U\Lambda V'$, with Λ being a diagonal matrix. It is best suited for matrices with a 'checkerboard' structure, which in turn would be reflected in the stepwise structure of the singular vectors in U and V (namely, the eigenvectors of $X'X$ and XX', respectively). For each pair of left and right singular vectors (columns in U and V, respectively) corresponding to the *same* singular value Λ_i, one checks whether they can be approximated by a piecewise constant vector [21].

Plaid Models:
The main idea is to represent the expression matrix as s superposition of biclusters, where entries in each bicluster take a particular set of values [23]. Formally, the expression matrix X can be represented as

$$X_{ij} = \mu_0 + \sum_{k=1}^{K} \theta_{ijk} a_{ik} b_{jk},$$

where μ_0 captures the uniform background and $\theta_{ijk} = \mu_k + \alpha_{ik} + \beta_{jk}$ describe mean, row and column effects of each bicluster. The parameters a_{ik}, $b_{jk} \in \{0,1\}$ are gene/sample bicluster membership indicator variables. Hence, similar to the Cheng-Church algorithm, a bicluster is assumed to be the sum of a mean background level plus row and column specific effects. The biclustering problem is formulated as estimating the parameters of interest so that the following sum of squares criterion is minimized:

$$\sum_{i \in G, j \in S} [X_{ij} - (\mu_0 + \sum_{k=1}^{K} \theta_{ijk}] a_{ik} b_{jk},$$

subject to the identifiability constraints $\sum_{\in G} a_{ik} \alpha_{ik} = 0$ or $\sum_{\in G} b_{ik} \beta_{ik} = 0$. An iterative algorithm and a stopping criterion are discussed in [23].

The Statistical-Algorithmic Method for Bicluster Analysis Algorithm:
This algorithm uses a combination of probabilistic modeling of the data and
graph theoretic techniques to identify biclusters [30]. The expression data
are represented as a bipartite graph, whose two disjoint vertex (node) sets
correspond to the set of genes and the set of samples, respectively. Edges
between genes and samples capture significant changes in expression levels,
whose weights are assigned according to a probabilistic model, so that highly
weighted subgraphs correspond to the biclusters with high probability. The
proposed algorithm in [30] is a heuristic one, motivated by a combinatorial
algorithm for finding highly weighted bicliques.

Some Other Algorithms:
Other proposals in the literature include the coupled two-way clustering algorithm [13] and the iterative signature algorithm [6]. The former iterates
between clustering the genes and clustering the samples in identifying *stable* biclusters. This algorithm requires that the one-way clustering algorithms
used on the genes and the samples are capable of identifying significant (stable)
clusters; hence, popular clustering algorithms, such as k-means or agglomerative algorithms, can not be directly used as plug-ins. The iterative signature
algorithm tries to identify a subset of genes and samples that exhibit the following property: the average expression profile over samples and the average
expression profiles of genes in the bicluster should be 'significantly' high or
low (over- or under-expressed, respectively).

5.6 Exercises

1. (a) Let X be a $n \times p$ data matrix in which each row corresponds to a p-
variate measurement on one of n individuals. Assuming that the p variates are
numerical variables, describe three possible measures of dissimilarity of pairs
of individuals. Comment on their relative advantages and disadvantages.

(b) What are the properties that a dissimilarity function must satisfy? What
about a distance function?

(c) The values of four binary variables are measured for each of four individuals as follows:

Individual	Variable			
	V1	V2	V3	V4
1	1	1	1	0
2	0	0	1	1
3	1	1	1	1
4	0	1	0	1

Construct a dissimilarity matrix for the four individuals using *two* appropriate dissimilarity measures. Show your work.

(d) Six subjects were tested on a number of standardized tests. The scores for each subject were recorded and the Euclidean distances between each pair of subjects were calculated as follows:

	Subject					
	A	B	C	D	E	F
A	0					
B	4.2	0				
C	5.9	7.6	0			
D	1.2	7.0	10.3	0		
E	6.1	2.6	5.4	7.8	0	
F	1.3	3.6	4.5	3.2	8.2	0

Using single linkage clustering, cluster the six subjects. Sketch the dendrogram and interpret the results. How would your dendrogram change if you used a complete linkage clustering algorithm?

2. Let $x_i = (x_i^1, x_i^2, \cdots, x_i^p)$ denote the vector containing the p measurements for the i-th observation. A measure of heterogeneity of a cluster C of size n_C is given by

$$H_C = \sum_{i=1}^{n_C} d^2(x_i, \bar{x}_C),$$

where the index i runs over the observations in the cluster, $d(u_i, u_j)$ denotes the Euclidean distance between observations u_i and u_j and $\bar{x}_C$ the p-dimensional multivariate mean of cluster C. Show that when two clusters C_1 and C_2 are merged, the heterogeneity of the merged cluster, denoted by $H_{C_1+C_2}$, is increased. Provide an expression for the increase relative to the sum of the heterogeneity measures of the two original clusters. Can you suggest a clustering algorithm?

3. For the clustering of objects with p *binary* variables, a possible dissimilarity measure between two observations is based on the following contingency table:

with $\delta_1(i,j) = (b+c)/p$.

A researcher proposes an alternative dissimilarity measure $\delta_2(i,j) = 2(b+c)/(a+d+2(b+c))$; i.e., it doubles the weight of the mismatch.

(a) Show that for observations i, j, k, ℓ if

$$\delta_1(i,j) \geq \delta_1(k,\ell) \implies \delta_2(i,j) \geq \delta_2(k,\ell).$$

	Object j		
	1	0	Sum
Object i 1	a	b	a+b
Object i 0	c	d	c+d
Sum	a+c	b+d	p

(b) Two researchers decided to use single-linkage hierarchical clustering to cluster a data set comprised of binary variables. The first researcher decides to use δ_1 as the dissimilarity measure, while the second one δ_2. What can you say about the results that the two researchers will get? Explain.

References

[1] Alizadeh, A.A. et al. (2000), Distinct types of diffuse large B-cell lymphoma identified by gene expression profiling, *Nature*, 403, 503-511

[2] Alon, U., Barkai, N., Notterman, D.A., Gish, K., Ybarra, S., Mack, D. and Levine, A.J. (1999), Broad patterns of gene expression revealed by clustering analysis of tumor and normal colon tissues probed by oligonucleotide array, *Proceedings of the National Academies of Science USA*, 96, 6745-6750

[3] Amari, S.I, (1991), Dualistic geometry of the manifold of higher-order neurons, *Neural Networks*, 4, 443-451

[4] Banfield, J.D. and Raftery, A.E. (1993), Model based Gaussian and non-Gaussian clustering, *Biometrics*, 49, 803-821

[5] Belkin, M. and Nyogi, P. (2003), Laplacian eigenmaps for dimensionality reduction and data representation, *Neural Computation*, 15, 1373-1396

[6] Bergmann, S., Ihmels, J. and Barkai, N. (2003), Iterative signature algorithm for the analysis of large-scale gene expression data, *Physical Review E*, 67, 201-218

[7] Borg, I. and Groenen, P. (1997), *Modern Multidimensional Scaling: Theory and Applications*, Springer, NY

[8] Bregler, C. and Omohundro, M. (1994), Surface learning with applications to lipreading, in Cowan et al. (eds), *Advances in Neural Information Processing Systems*, Morgan Kaufman, San Mateo, CA

[9] Church, G.M. and Cheng, Y. (2000), Biclustering of expression data, in Proceedings of ISMB 2000, 93-103, AAAI Press

[10] Donoho, D.L. and Grimes, C. (2003), Hessian eigenmaps: locally linear embedding techniques for high-dimensional data, *Proceedings of the National Academies, USA*, 100, 5591-5596

[11] Eisen, M.B., Spellman, P.T., Brown, P.O. and Botstein, D. (1998), Cluster analysis and display of genome-wide expression patterns, *Proceedings of the National Academies of Science USA*, 95, 14863-14868

[12] Gasch, A.P. et al. (2001), Genomic expression responses to DNA-damaging agents and the regulatory role of the yeast ATR homolog mec1p, *Molecular Biology of the Cell*, 12, 2987-3003

[13] Getz, G., Levine, E., Domany, E. and Zhang, M.Q. (2000), Super-paramagnetic clustering of yeast gene expression profiles, *Physica A*, 279, 457

[14] Golub, T.R., Slonim, D.K, Tamayo, P., Huard, C., Gaasenbeek, M., Mesirov, J.P., Coller, H., Loh, M.L, Downing, J.R., Caligiuri, M.A, Bloomfield, C.D. and Lander, E.S. (1999), Molecular classification of cancer: Class discovery and class prediction by gene expression monitoring, *Science*, 286, 531-537

[15] Hansen, P. and Jaumard, B. (1997), Cluster analysis and mathematical programming, *Mathematical Programming*, 79, 191-215

[16] Hughes, J.D., Estep, P.E., Tavazoie, S. and Church, G.M. (2000), Computational identification of cis-regulatory elements associated with groups of functionally related genes in Saccharomyces Cerevisiae, *Journal of Molecular Biology*, 296, 1205-1214

[17] Jornsten, R. (2007), Simultaneous subset selection via rate distortion theory, submitted to *Journal of Computational and Graphical Statistics*

[18] Jornsten, R. and Keles, S. (2008), Mixture models with multiple levels, with application to the analysis of multifactor gene expression data, to appear in *Biostatistics*

[19] Kaski, S. (1997), Data exploration using self-organizing maps, *Acta Polytechnica Scandinavica*, Mathematics, Computing and Management in Engineering Series No. 82

[20] Khan, J., Wei, J.S., Ringnér, M., Saal, L.H., Ladanyi, M., Westermann, F., Berthold, F., Schwab, M., Antonescu, C.R., Peterson, C. and Meltzer, P.S. (2001), Classification and diagnostic prediction of cancers using gene expression profiling and artificial neural networks, *Nature Medicine*, 7, 673-679.

[21] Kluger, Y., Barsi, R., Cheng, J.T. and Gerstein, M (2003), Spectral biclustering of microarray data: co-clustering genes and conditions, *Genome Research*, 13, 703-716

[22] Kohonen, T. (2001), *Self-organizing Maps*, Springer, Berlin

[23] Lazzeroni, L. and Owen, A. (2002), Plaid models for gene expression data, *Statistica Sinica*, 12, 61-86

[24] Liu, Z.Y. and Xu, L. (2003), Topological local principal components analysis, *Neurocomputing*, 55, 739-745

[25] Michailidis, G. and de Leeuw, J. (1998), The Gifi system of descriptive multivariate analysis, *Statistical Science*, 13, 307-336

[26] Mirkin, B. (1996), *Mathematical Classification and Clustering*, Kluwer Academic Publishers, Dordrecht, The Netherlands

[27] Roweis, S.T. and Saul, L.K. (2000), Nonlinear dimensionality reduction by locally linear embedding, *Science*, 290, 2323-2326

[28] Saul, L.K. and Roweis, S.T. (2003), Think globally, fit locally: unsupervised learning of low dimensional manifolds, *Journal of Machine Learning Research*, 4, 119-155

[29] Scholkopf, B., Smola, A. and Muller, K.R. (1998), Nonlinear component analysis as a kernel eigenvalue problem, *Neural Computation*, 4, 1299-1309

[30] Tanay, A., Sharan, R., Kupiec, M. and Shamir, R. (2004), Revealing modularity and organization in the yeast molecular network by integrated analysis of highly heterogeneous genome-wide data, *Proceedings of the National Academies of the USA*, 101, 2981-2986

[31] Tenenbaum, J.B., de Silva, V. and Langford, J.C. (2000), A global geometric framework for nonlinear dimensionality reduction, *Science*, 290, 2319-2323

CHAPTER 6

Computational Intelligence in Bioinformatics

6.1 Introduction

In addition to the machine learning and probabilistic approaches, computational intelligence (or soft computing) is gradually opening up several possibilities in Bioinformatics – especially by generating low-cost, low-precision, good solutions. This is a consortium of methodologies that works synergistically and provides flexible information processing capability for handling real life ambiguous situations [1].

In this chapter we introduce the different soft computing paradigms, like fuzzy sets, artificial neural networks (ANNs), evolutionary computation and rough sets, and outline their role along this direction. Each of them contributes a distinct methodology for addressing problems in its domain, in a cooperative rather than a competitive manner. The result is a more intelligent and robust system providing a human-interpretable, low-cost, approximate solution, as compared to traditional techniques.

The term soft computing had its origin in the concept of fuzzy sets, and pertained to the inherent elasticity associated with the membership functions. Most biological systems behave in a fuzzy manner, with the interaction and activity of various genes attaining different levels. A single gene may be involved in different biological processes. For example, fuzzy clustering allows genes to simultaneously *(softly)* belong to multiple clusters and participate in multiple pathways. This is a more natural reflection of the biological reality of cellular metabolism. It is unlike the *hard* categorization of objects into crisp, nonoverlapping sets. Rough sets, unlike crisp sets, also provide a suitable representation for manipulation of the uncertainty prevalent in everyday life.

Artificial neural networks constitute another nature-inspired architecture, that mimics the learning or adaptivity of the biological nervous system in an attempt to incorporate intelligence. It is this pliability to adapt towards changes in environment and the capacity to correct itself from errors in judgement, that allows us to categorize the ANN under soft computing.

Genetic algorithms employ evolution-inspired operators like crossover, mutation and selection on a population of "chromosomes." The efficient search strategy represents a solution in terms of a string, which is analogically called a chromosome. It may be noted that this is not the chromosome that encompasses the DNA in biological literature. This adaptive algorithm optimizes a fitness function, based on the survival-of-the-fittest principle, while evolving towards its goal by generating the best set of chromosomes. It is the inherent softness that lets it adaptively direct the search under environmental influence.

Hybridization [2] exploiting the characteristics of the paradigms includes *neuro-fuzzy, rough-fuzzy, neuro-genetic, fuzzy-genetic, neuro-rough, rough-neuro-fuzzy, evolutionary-rough-neuro-fuzzy* approaches, to mention a few. However, among these, *neuro-fuzzy* computing is the oldest and most visible.

The major pattern recognition and data mining tasks considered here are clustering, classification, feature selection and rule mining. While classification pertains to supervised learning, in the presence of known targets, clustering corresponds to unsupervised self-organization into homologous partitions. Feature selection techniques aim at reducing the number of irrelevant and redundant variables in the dataset. Rule mining enables efficient extraction and representation of mined knowledge in human-understandable form.

Genomic sequence, protein structure, gene expression microarrays and gene regulatory networks are some of the application areas described. Since the work entails processing huge amounts of incomplete or ambiguous biological data, we can utilize the learning and generalization capability of artificial neural networks for adapting, uncertainty handling capacity of fuzzy sets and rough sets for modeling ambiguity, and the searching potential of genetic algorithms for efficiently traversing large search spaces [3].

This chapter provides an overview of the available literature on Bioinformatics, in the soft computing framework [4]. Introduction to fuzzy sets (FS), artificial neural networks (ANNs), evolutionary computing (EC) [including GAs], rough sets (RS), and their hybridizations, are provided in Sections 6.2-6.6. We describe in Section 6.7 the role of these paradigms and their hybridizations, in different application domains of Bioinformatics. It may be mentioned that there is no universally best technique; choosing particular soft computing tool(s) or some combination with traditional methods is entirely dependent on the particular application, and it requires human interaction to decide on the suitability of an approach. Finally, Section 6.8 concludes the chapter.

6.2 Fuzzy Sets (FS)

Typically, real-life data must not only be cleaned of errors and redundancy, but must also be organized in a fashion that makes sense to the problem. There can exist imperfections in raw input data needed for knowledge acquisition, mainly due to uncertainty, vagueness and incompleteness. While incompleteness arises due to missing or unknown data, uncertainty (or vagueness) can be caused by errors in physical measurements due to incorrect measuring devices or by a mixture of noisy and pure signals.

Fuzzy sets were introduced by Zadeh [5] in 1965 as a new way of representing vagueness in everyday life. They constitute the earliest and most widely reported constituent of soft computing. The theory provides an approximate and yet effective means for describing the characteristics of a system that is too complex or ill-defined to admit precise mathematical analysis [6]. Much of the logic behind human reasoning is not the traditional two-valued or even multivalued logic, but logic with fuzzy truths, fuzzy connectives and fuzzy rules of inference.

Fuzzy sets are able to handle, to a reasonable extent, uncertainties (arising from deficiencies of information) in various applications particularly in decision-making models under different kinds of risks, subjective judgment, vagueness and ambiguity. Since this theory is a generalization of the classical set theory, it has greater flexibility to capture various aspects of incompleteness or imperfection in information about a situation. The use of linguistic variables may be viewed as a form of data compression, which can be termed *granulation* [1]. The same effect can also be achieved by conventional quantization. However, in the case of quantization the values are intervals, whereas in the case of granulation the values are overlapping fuzzy sets.

The uncertainty in classification or clustering of patterns may arise from the overlapping nature of the various classes. This may be due to fuzziness or randomness. In the conventional classification technique, it is usually assumed that a pattern belongs to only one class. However, this is not necessarily realistic – either physically or mathematically. A pattern can have degrees of membership in more than one class, and both supervised and unsupervised learning should be able to accommodate this.

A fuzzy set A in a space of points $R = \{r\}$ is a class of events with a continuum of grades of membership, and it is characterized by a membership function $\mu_A(r)$ that associates with each element in R a real number in the interval $[0, 1]$ with the value of $\mu_A(r)$ at r representing the grade of membership of r in A. Formally, a fuzzy set A with its finite number of supports $r_1, r_2, \ldots, r_n$ is defined as a collection of ordered pairs

$$\begin{aligned} A &= \{(\mu_A(r_i), r_i), i = 1, 2, \ldots, n\} \\ &= \{(\tfrac{\mu_A(r_i)}{r_i}), i = 1, 2, \ldots, n\}, \end{aligned}$$

where the support of A is an ordinary subset of R and is defined as

$$Supp(A) = \{r | r \in R \text{ and } \mu_A(r) > 0\}.$$

Here μ_i, the grade of membership of r_i in A, denotes the degree to which an event r_i may be a member of A or belong to A. Note that $\mu_i = 1$ indicates the strict containment of the event r_i in A. If, on the other hand, r_i does not belong to A, then $\mu_i = 0$.

If the support of a fuzzy set is only a single point $r_1 \in R$, then $A = \frac{\mu_1}{r_1}$ is called a *fuzzy singleton*. A fuzzy set A, with its finite number of supports $r_1, r_2, \ldots, r_n$, can also be expressed in the union form of its constituent singletons as

$$A = \frac{\mu_1}{r_1} + \frac{\mu_2}{r_2} + \cdots + \frac{\mu_n}{r_n}, \qquad (6.1)$$

where the $+$ sign denotes the union.

Let us consider an example of the crisp set of ages of a group of people, defined as

$$A_c = \{5, 10, 20, 30, 40, 50, 60, 70, 80\}.$$

Let the membership grades μ_{young} of these elements (people) in the fuzzy set *young* A_{young} be given by $\{1, 1, .8, .5, .2, .1, 0, 0, 0\}$. We express

$$Supp(A_{young}) = \{5, 10, 20, 30, 40, 50\},$$

with

$$A_{young} = \left\{ \frac{1}{5} + \frac{1}{10} + \frac{.8}{20} + \frac{.5}{30} + \frac{.2}{40} + \frac{.1}{50} \right\}.$$

Fuzzy logic is based on the theory of fuzzy sets and, unlike classical logic, aims at modeling the imprecise (or inexact) modes of reasoning and thought processes (with linguistic variables) that play an essential role in the remarkable human ability to make rational decisions in an environment of uncertainty and imprecision. This ability depends on our ability to infer an approximate answer to a question based on a store of knowledge that is inexact, incomplete, or not totally reliable. In fuzzy logic, everything, including truth, is a matter of degree [8]. Zadeh has developed a theory of approximate reasoning based on fuzzy set theory.

Most biological systems behave in a fuzzy manner, with the interaction and activity of various genes attaining different levels. A single gene may be involved in different biological processes. Fuzzy clustering allows genes to simultaneously belong to multiple clusters and participate in multiple pathways. This is a more natural reflection of the biological reality of cellular metabolism.

Information integration from multiple sources is utilized to generate biologically meaningful results. Fuzzy systems enable incorporation of user-friendly domain knowledge about some genes, in the form of linguistic *If-Then* rules, into the network. Fuzzy aggregation may be used to combine this information from databases of genes and their products, along with their interactions, in a natural framework. Fuzzy measures can be employed to compute the similarity between gene products annotated with GO terms.

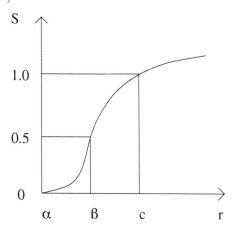

Figure 6.1 *Standard S function.*

6.2.1 Membership functions

Assignment of membership functions of a fuzzy subset is subjective in nature and reflects the context in which the problem is viewed. Note that fuzzy membership function and probability density function are conceptually different. It is often convenient to employ standardized functions with adjustable parameters (e.g., the S and π functions) which are defined in the following equations (see also Fig. 6.1):

$$
\begin{aligned}
S(r;\ \alpha, c) &= 0 & \text{for } r \leq \alpha \\
&= 2(\tfrac{r-\alpha}{c-\alpha})^2 & \text{for } \alpha \leq r \leq \beta \\
&= 1 - 2(\tfrac{r-c}{c-\alpha})^2 & \text{for } \beta \leq r \leq c \\
&= 1 & \text{for } r \geq c.
\end{aligned}
\tag{6.2}
$$

$$
\begin{aligned}
\pi(r;\ c, \lambda) &= S(r; c - \lambda, c - \tfrac{\lambda}{2}, c) & \text{for } r \leq c \\
&= 1 - S(r; c, c + \tfrac{\lambda}{2}, c + \lambda) & \text{for } r \geq c.
\end{aligned}
\tag{6.3}
$$

In eqn. (6.2) we have $\beta = (\alpha + c)/2$ as the *crossover point*, that is, the value of r at which S takes the value 0.5. In $\pi(r; c, \lambda)$, λ is the *bandwidth*, that is, the distance between the crossover points of π, while c is the central point at which π is unity.

Let us consider the linguistic variable age (x). Here the linguistic values *young* and *old* play the role of primary fuzzy sets which have a specified meaning, for example,

$$
\mu_{young} = 1 - S(20, 40),
\tag{6.4}
$$

$$
\mu_{old} = S(50, 70),
\tag{6.5}
$$

where the S and π functions are defined by Eqs. (6.2) and (6.3), and μ_{young} and μ_{old} denote the membership functions of *young* and *old*, respectively.

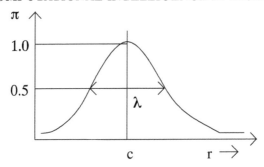

Figure 6.2 π function.

In pattern recognition problems we often need to represent a class with fuzzy boundary in terms of a π function. It is also expressed as

$$
\pi(r; c, \lambda) = \begin{cases} 2\left(1 - \frac{|r-c|}{\lambda}\right)^2, & \text{for } \frac{\lambda}{2} \leq |r - c| \leq \lambda \\ 1 - 2\left(\frac{|r-c|}{\lambda}\right)^2, & \text{for } 0 \leq |r - c| \leq \frac{\lambda}{2} \\ 0, & \text{otherwise.} \end{cases} \tag{6.6}
$$

This is shown in Fig. 6.2. Note that the membership value is maximum when $r = c$, that is, $\pi(c; c, \lambda) = 1$. The membership value is 0.5, at the crossover point, where $|r - c| = \lambda/2$.

6.2.2 Some basic operations

Basic operations related to fuzzy subsets A and B of R having membership values $\mu_A(r)$ and $\mu_B(r)$, $r \in R$ respectively, are summarized here [10].

1. A is equal to B (i.e., $A = B$) $\Rightarrow \mu_A(r) = \mu_B(r)$, for all $r \in R$.
2. A is a complement of B (i.e., $A = \overline{B}$) $\Rightarrow \mu_A(r) = \mu_{\overline{B}}(r) = 1 - \mu_B(r)$ for all $r \in R$.
3. A is contained in B ($A \subseteq B$) $\Rightarrow \mu_A(r) \leq \mu_B(r)$ for all $r \in R$.
4. The union of A and B ($A \cup B$) $\Rightarrow \mu_{A \cup B}(r) = \vee(\mu_A(r), \mu_B(r))$ for all $r \in R$, where $\vee$ denotes maximum.
5. The intersection of A and B ($A \cap B$) $\Rightarrow \mu_{A \cap B}(r) = \wedge(\mu_A(r), \mu_B(r))$ for all $r \in R$, where $\wedge$ denotes minimum.

We also have the modifiers *very* and *more or less*. These are expressed as

$$
\mu_{very\ young} = (\mu_{young})^2, \tag{6.7}
$$

$$
\mu_{more\ or\ less\ young} = (\mu_{young})^{0.5}. \tag{6.8}
$$

6.2.3 Clustering

The Fuzzy c-means (FCM) algorithm is a fuzzification of the c-means algorithm, and was proposed by Bezdek [138]. It allows a pattern to be assigned to more than one partition depending on its membership value.

The algorithm partitions a set of N patterns $\{\mathbf{X}_k\}$ into c clusters by minimizing the objective function

$$J = \sum_{k=1}^{N} \sum_{i=1}^{c} (\mu_{ik})^{m'} ||\mathbf{X}_k - \mathbf{m}_i||^2, \tag{6.9}$$

where $1 \le m' < \infty$ is the fuzzifier, $\mathbf{m}_i$ is the ith cluster center, $\mu_{ik} \in [0,1]$ is the membership of the kth pattern to it and $||.||$ is the distance norm, such that

$$\mathbf{m}_i = \frac{\sum_{k=1}^{N} (\mu_{ik})^{m'} \mathbf{X}_k}{\sum_{k=1}^{N} (\mu_{ik})^{m'}} \tag{6.10}$$

and

$$\mu_{ik} = \frac{1}{\sum_{j=1}^{c} \left(\frac{d_{ik}}{d_{jk}}\right)^{\frac{2}{m'-1}}}, \tag{6.11}$$

$\forall i$, with $d_{ik} = ||\mathbf{X}_k - \mathbf{m}_i||^2$, subject to $\sum_{i=1}^{c} \mu_{ik} = 1, \forall k$, and $0 < \sum_{k=1}^{N} \mu_{ik} < N, \forall i$. The algorithm proceeds as follows.

1. Pick the initial means $\mathbf{m}_i$, $i = 1, \ldots, c$. Choose values for fuzzifier m' and threshold ϵ. Set the iteration counter $t = 1$.

2. Repeat Steps 3-4, by incrementing t, until $|\mu_{ik}(t) - \mu_{ik}(t-1)| > \epsilon$.

3. Compute μ_{ik} by eqn. (6.11) for c clusters and N data objects.

4. Update means $\mathbf{m}_i$ by eqn. (6.10).

Note that for $\mu_{ik} \in [0,1]$ the objective function of eqn. (6.9) boils down to the hard c-means case, whereby a *winner-take-all* strategy is applied in place of membership values in eqn. (6.10).

6.3 Artificial Neural Networks (ANN)

ANNs [14, 16, 17] are signal processing systems that form a massively parallel interconnection of simple (usually adaptive) processing elements, and interact with objects of the real world in a manner similar to biological systems. The origin of ANNs can be traced to the work of Hebb [20], where a local learning rule was proposed. This rule assumed that correlations between the states of two neurons determined the strength of the coupling between them. Subsequently, a synaptic connection that was very active grew in strength and vice versa.

ANNs can typically perform tasks like pattern classification, clustering or categorization, function approximation, prediction or forecasting, optimization, retrieval by content and control. They can be viewed as weighted directed graphs in which artificial neurons are nodes and directed edges (with weights) are connections between neuron outputs and neuron inputs. On the basis of the connection pattern (architecture), ANNs can be grouped as (i) *feedforward*, in which graphs have no loops – like single-layer perceptron, multilayer perceptron (MLP), radial basis function (RBF) networks, Kohonen self-organizing map (SOM), and (ii) *recurrent* (or *feedback*), in which loops occur because of feedback connections – like Hopfield network, adaptive resonance theory (ART) models.

The adaptability of a neural network comes from its capability of learning from "environments." Broadly, there are three paradigms of learning: supervised, unsupervised (or self-organized) and reinforcement. Sometimes, reinforcement is viewed as a special case of supervised learning. Under each category there are many algorithms. In supervised learning, adaptation is done on the basis of direct comparison of the network output with known correct or desired answer. Unsupervised learning tunes the network to the statistical regularities of the input data to form categories (or partitions) by optimizing, with respect to the free parameters of the network, some task-independent measure of quality of the representation. The reinforcement learning, on the other hand, attempts to learn the input–output mapping through trial and error with a view to maximizing a performance index called *reinforcement signal*. Here the system only knows whether the output is correct, but not what the correct output is.

ANNs are natural classifiers having resistance to noise, tolerance to distorted images or patterns (ability to generalize), ability to recognize partially occluded or degraded images or overlapping pattern classes or classes with highly nonlinear boundaries and potential for parallel processing. They use nonparametric adaptive learning procedures, learn from examples and discover important underlying regularities in the task domain.

There has been widespread activity aimed at extracting the embedded knowledge in trained ANNs in the form of symbolic rules [22, 23]. This serves to identify the attributes that, either individually or in a combination, are the most significant determinants of the decision or classification. Since all information is stored in a distributed manner among the neurons and their connectivity, any individual unit cannot essentially be associated with a single concept or feature of the problem domain.

Generally ANNs consider a fixed topology of neurons connected by links in a predefined manner. These connection weights are usually initialized by small random values. *Knowledge-based networks* [24, 25] constitute a special class of ANNs that consider crude domain knowledge to generate the initial network architecture, which is later refined in the presence of training data. The use of knowledge-based nets helps in reducing the searching space and time while the network traces the optimal solution.

6.3.1 Single-layer perceptron

The *perceptron* [26, 27] was one of the most exciting developments during the early days of pattern recognition. It consists of a single neuron with adjustable weights, $w_j, j = 1, 2, \ldots, n$, and threshold θ. Given an input vector $\boldsymbol{x} = [x_1, x_2, \ldots, x_n]^T$, the net input to the neuron is

$$v = \sum_{j=1}^{n} w_j x_j - \theta. \tag{6.12}$$

The output y of the perceptron is $+1$ if $v > 0$ and is 0 otherwise. In a two-class classification problem, the perceptron assigns an input pattern to one class if $y = 1$ and to the other class if $y = 0$. The linear equation $\sum_{j=1}^{n} w_j x_j - \theta = 0$ defines the decision boundary (a hyperplane in the n-dimensional input space) that halves the space. Rosenblatt proved that when training patterns are drawn from two linearly separable classes, the perceptron learning procedure converges after a finite number of iterations. Here learning occurs only when the perceptron makes an error. The perceptron learning algorithm is outlined as follows:

1. Initialize the weights and threshold to small random numbers.

2. Present a pattern vector $[x_1, x_2, \ldots, x_n]^T$ and evaluate the output of the neuron.

3. Update the weights according to

$$w_j(t + 1) = w_j(t) + \varepsilon(d - y)x_j, \tag{6.13}$$

where d is the desired output, t is the iteration number and ε ($0.0 < \varepsilon < 1.0$) is the learning rate (step size).

However if the pattern space is not linearly separable, then the perceptron fails [28]. A single-layer perceptron is thus inadequate for situations with multiple classes and nonlinear separating boundaries.

6.3.2 Multilayer perceptron (MLP) using backpropagation of error

The multilayer perceptron (MLP) [16] consists of multiple layers of simple, two-state, sigmoid processing elements (nodes) or neurons that interact using weighted connections. There exist one or more intermediate *hidden* layers, between the input and output layers. Typically all neurons in a layer are fully connected to the neurons in the adjacent layers.

In a standard classification task, the number of units in the output layer H corresponds to the number of output classes. During training, the appropriate output node is clamped to state 1 while the others are clamped to state 0.

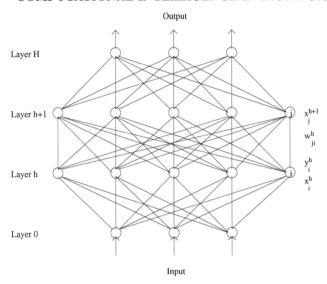

Figure 6.3 *MLP with two hidden layers.*

Let us consider the network given in Fig. 6.3. The total input x_j^{h+1} received by neuron j in layer $h+1$ is defined as

$$x_j^{h+1} = \sum_i y_i^h w_{ji}^h - \theta_j^{h+1},$$
(6.14)

where y_i^h is the state of the ith neuron in the preceding hth layer, w_{ji}^h is the weight of the connection from the ith neuron in layer h to the jth neuron in layer $h+1$ and θ_j^{h+1} is the threshold of the jth neuron in layer $h+1$.

The output of a neuron in any layer other than the input layer ($h > 0$) is expressed as

$$y_j^h = \frac{1}{1 + e^{-x_j^h}}.$$
(6.15)

For nodes in the input layer $y_j^0 = x_j^0$, where x_j^0 is the jth component of the input vector clamped at the input layer. Learning consists of minimizing the error by updating the weights in a very large parameter space.

The least mean square (LMS) error in output vectors, for a given network weight vector $\boldsymbol{w}$, is defined as

$$E = \frac{1}{2} \sum_{j,p} (y_{j,p}^H - d_{j,p})^2,$$
(6.16)

where $y_{j,p}^H$ is the state obtained for output node j in layer H for input–output pattern p and $d_{j,p}$ is its desired state specified by the teacher. One method for minimization of E is to apply the method of gradient descent by starting with any set of weights and repeatedly updating each weight by an amount

$$\Delta w_{ji}^h(t) = -\varepsilon \frac{\partial E}{\partial w_{ji}} + \alpha \Delta w_{ji}^h(t-1), \tag{6.17}$$

where the positive constant ε controls the descent, $0 \le \alpha \le 1$ is the damping coefficient or momentum and t denotes the number of the iteration currently in progress. Generally ε and α are set at constant values, but there exist approaches that vary these parameters. Initially the connection weights w_{ji}^h between each pair of neurons i in layer h and j in layer $h+1$ are set to small random values lying in the range $[-0.5, 0.5]$.

¿From Eqs. (6.14)–(6.16), we have

$$\frac{\partial E}{\partial w_{ji}} = \frac{\partial E}{\partial y_j} \frac{dy_j}{dx_j} \frac{\partial x_j}{\partial w_{ji}} = \frac{\partial E}{\partial y_j} y_j^h (1 - y_j^h) y_i^{h-1}. \tag{6.18}$$

For the output layer ($h = H$), we substitute in Eq. (6.18)

$$\frac{\partial E}{\partial y_j} = y_j^H - d_j. \tag{6.19}$$

For the other layers, using Eq. (6.14), we substitute in Eq. (6.18)

$$\frac{\partial E}{\partial y_j} = \sum_k \frac{\partial E}{\partial y_k} \frac{dy_k}{dx_k} \frac{\partial x_k}{\partial y_j} = \sum_k \frac{\partial E}{\partial y_k} \frac{dy_k}{dx_k} w_{kj}^h, \tag{6.20}$$

where units j and k lie in layers h and $h+1$, respectively.

During training, each pattern of the training set is used in succession to clamp the input and output layers of the network. After a number of iterations over the training data, the error E in Eq. (6.16) may be minimized. At this stage the network is supposed to have discovered (learned) the relationship between the input and output vectors in the training samples.

In the testing phase the neural net is expected to be able to utilize the information encoded in its connection weights to assign the correct output labels for the test vectors that are now clamped only at the input layer. Determination of the optimal number of hidden layers and/or nodes is another interesting problem.

6.3.3 Radial basis function network (RBF)

A radial basis function (RBF) network [29, 30] consists of two layers as shown in Fig. 6.4. Let the connection weight vectors of the input and output layers be denoted as $\mathbf{m}$ and $\mathbf{w}$, respectively. The basis (or kernel) functions in the hidden layer produce a localized response to the input stimulus. The output nodes form a weighted linear combination of the basis functions computed by the hidden nodes.

Let $\mathbf{x} = (x_1, \ldots, x_i, \ldots, x_n) \in R^n$ and $\mathbf{y} = (y_1, \ldots, y_i, \ldots, y_l) \in R^l$ be the

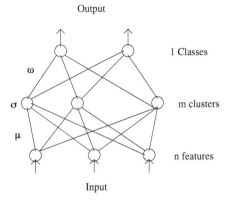

Figure 6.4 *Radial basis function network.*

input and output, respectively, and let m be the number of hidden nodes. Here the hidden nodes represent the number of clusters (specified by the user) that partition the input space.

The output u_j of the jth hidden node, using the *Gaussian kernel* function as a basis, is given by

$$u_j = \exp\left[-\frac{(\boldsymbol{x} - \boldsymbol{m}_j)^T (\boldsymbol{x} - \boldsymbol{\mu}_j)}{2\sigma_j^2}\right], \quad j = 1, 2, \ldots, m, \quad (6.21)$$

where $\boldsymbol{x}$ is the input pattern, $\boldsymbol{\mu}_j$ is its input weight vector (i.e., the center of the Gaussian for node j) and σ_j^2 is its width, such that $0 \leq u_j \leq 1$ (the closer the input is to the center of the Gaussian, the larger the response of the node).

The output y_j of the jth output node is

$$y_j = \boldsymbol{w}_j^T \boldsymbol{u}, \quad j = 1, 2, \ldots, l, \quad (6.22)$$

where $\boldsymbol{w}_j$ is the weight vector for this node, and $\boldsymbol{u}$ is the vector of outputs from the hidden layer. The network performs a linear combination of the nonlinear basis functions of Eq. (6.21).

The problem is to minimize the error

$$E = \frac{1}{2} \sum_{p=1}^{N} \sum_{j=1}^{l} (y_{j,p} - d_{j,p})^2, \quad (6.23)$$

where $d_{j,p}$ and $y_{j,p}$ are desired and computed output at the jth node for the pth pattern, N is the size of the data set and l is the number of output nodes.

Learning in the hidden layer, typically, uses the c-means clustering algorithm. Let the cluster centers be denoted as $\boldsymbol{\mu}_j$, $j = 1, \ldots, m$. The parameter σ_j represents a measure of the spread of data associated with each node. Learning in the output layer is performed after the parameters of the basis functions have been determined. The weights are typically trained online, from the presented patterns, using the LMS algorithm. It is given by

$$\Delta \boldsymbol{w}_j = -\varepsilon(y_j - d_j)\boldsymbol{u}, \qquad (6.24)$$

where ε is the learning rate.

6.4 Evolutionary Computing (EC)

The term evolutionary computing encompasses genetic algorithms (single-objective and multi-objective), evolutionary algorithms and genetic programming. It provides efficient search techniques for nonlinear optimization in fairly large and arbitrary solution spaces. In this section we provide an introduction to some of these evolutionary strategies.

6.4.1 Genetic algorithms (GAs)

Genetic algorithms (GAs) [31, 32] are adaptive and robust computational search procedures that employ evolution-inspired operators like selection, crossover and mutation while being controlled by a fitness function. The components of a GA consist of a population of individuals represented as chromosomes, their encoding or decoding mechanism, probabilities to perform the genetic operations, a replacement technique for the pool of possible solutions and the termination conditions.

Let us consider, as an example, the optimization of a function

$$y = f(x_1, x_2, \ldots, x_p).$$

A binary vector is used as a chromosome to represent real values of the variables x_i, with the length of the vector constraining the allowed precision in bits. A population is a set of individuals (chromosomes) representing the concatenated parameter set $x_1, x_2, \ldots, x_p$, where each member refers to a coded *possible* solution. For example, a sample chromosome

$$00001|01000| \ldots |11001$$

could correspond to $x_1 = 00001$, $x_2 = 01000$ and $x_p = 11001$. The *Schema theorem* provides a complete guidance on the feasible solutions in the search space.

The chromosomes can be of fixed or variable size. Selection obeys the Darwinian survival of the fittest strategy, with the objective function playing the role of Nature or environment. Variation is introduced in the population through the genetic operations like recombination (crossover) and mutation. Normally the initial population is chosen randomly.

Encoding is used to convert parameter values into chromosomal representation. In case of continuous-valued parameters, a decimal-to-binary conversion is used. For example, using a 5-bit representation, 13 is encoded as 01101. In case of parameters having categorical values, a particular bit position in the chromosomal representation is set to 1 if it comes from a certain category. For example, the gender of a person can have values from {male, female}, such that male/female is represented by the bit 1/0. These bits or strings (representing the parameters of a problem) are then concatenated, and termed a chromosome.

Decoding is the reverse of encoding. For a continuous-valued parameter the binary representation is converted to a continuous value by the expression

$$lower\_bound + \frac{\sum_{i=0}^{bits\_used-1} bit_i * 2^i}{2^{bits\_used} - 1} * (upper\_bound - lower\_bound).$$

Hence 01101 in five bits ($bits\_used$) is decoded back to 13, using $lower\_bound = 0$ and $upper\_bound = 31$. In case of categorical parameters, the value is found by referring to the original mapping.

The fitness function provides a measure on the performance of the chromosome. Selection gives more chance to better-fitted individuals, thereby mimicking the natural selection procedure. Some of the popular selection techniques include roulette wheel selection, stochastic universal sampling, linear normalization selection, and tournament selection. The roulette wheel selection procedure initially sums the fitness values (f_is) of all the N chromosomes in the population, and it stores them in slots sized accordingly. Let this sum be given by $total\_fitness$. The probability of selection p_i for the ith chromosome is expressed as

$$p_i = \frac{f_i}{total\_fitness}, \tag{6.25}$$

while the cumulative probability q_i after inclusion of the ith chromosome is given by $q_i = \sum_{j=1}^{i} p_j$. Selection is made by spinning the roulette wheel N times, on each occasion generating a random number n_r in $[0, total\_fitness]$. In rule form, we have

IF $n_r < q_1$ THEN select the first chromosome,
        ELSE select the ith chromosome such that $q_{i-1} < n_r \leq q_i$.

Crossover is modeled by choosing mating pairs from the selected chromosomes, with the crossover probability p_c determining whether or not a pair should be crossed over such that the corresponding chromosome segments are interchanged. A random number n_{rc} is generated in the range $[0, 1]$. If $n_{rc} < p_c$, the corresponding chromosome pair is selected for crossover. Again, crossover can be one point, two point, multi-point or uniform. Let us consider, as an example, two parent chromosomes $xyxyxyxy$ and $abababab$ where x, y, a, b are binary. In one-point crossover at the 4th bit involving the parent chromosomes

$$xyx|yxyxy$$

$$aba|babab,$$

one generates the children

$$xyx|babab$$

$$aba|yxyxy.$$

Here the segment involving bits 4 to 8 is interchanged between the parents. In case of two-point crossover at the 4th and 6th bits, involving parent chromosomes

$$xyx|yx|yxy$$

$$aba|ba|bab,$$

we obtain the children chromosomes

$$xyx|ba|yxy$$

$$aba|yx|bab.$$

Here the segment constituting bits 4 to 5 is swapped between the parents to generate the pair of offsprings.

The mutation operator is used to introduce diversity in the population, with the mutation probability p_m determining whether or not a bit should be mutated such that the corresponding location is flipped. For example, a mutation at the 4th bit would transform the chromosome $001|0|00$ to $001|1|00$. Probabilities p_c and p_m can be fixed or variable, and they typically have values ranging between 0.6 to 0.9, and 0.001 to 0.01, respectively.

The replacement techniques can be

1. Generational, where all the n individuals are replaced at a time by the n children created by reproduction. *Elitism* is often introduced to retain the best solution obtained so far.

2. Steady state, where $m < n$ members are replaced at a time by the m children reproduced.

The terminating criterion for the algorithm can be on the basis of (i) execution for a fixed number of generations or iterations, (ii) a bound on the fitness value of the generated solution or (iii) acquiring of a certain degree of homogeneity by the population.

Let us consider a simple example to illustrate the working principle of GAs. It is related to minimizing the surface area A of a solid cylinder, given radius r and height h. Here the fitness function can be expressed as $A = 2\pi rh + 2\pi r^2$. We need to encode the parameters r and h in a chromosome. Using a 3-bit representation, we demonstrate encoding, crossover and mutation. For $r_1 = 3$, $h_1 = 4$ and $r_2 = 4$, $h_2 = 3$, we generate parent chromosomes $011|100$ and $100|011$ with $A_1 = 132$, $A_2 = 176$, respectively. Let there be one-point crossover at bit 4, producing the children chromosomes $011|011$ and $100|100$. This is decoded as $r_{1c} = 3$, $h_{1c} = 3$ and $r_{2c} = 4$, $h_{2c} = 4$, with $A_{1c} = 16.16$

and $A_{2c} = 28.72$, respectively. Now, let there be mutation at bit 5 of the first child. This generates the chromosome $0110|0|1$, for $r_{1cm} = 3$ and $h_{1cm} = 1$, with $A_{1cm} = 10.77$. This is the minimum value of fitness obtained thus far. Consecutive applications of the genetic operations of selection, crossover and mutation, up to termination, enable the minimization (optimization) of the chosen fitness function.

GAs have been applied in diverse problems including optimization, pattern recognition, image processing, data mining and bioinformatics. Note that, unlike GAs, evolutionary algorithms [33] rely only on mutation and do not perform crossover. Applications pertaining to bioinformatics involve sequence alignment, protein tertiary structure prediction, docking, microarray clustering and genetic network extraction.

In protein structure prediction and folding, GAs try to generate a set of *native-like* conformations of a protein based on a force field, while minimizing a fitness function. The choice of the fitness function is governed by factors such as interatomic bond lengths, bond angles, electrostatic forces, potential energy– and poses challenging applications to the pharmaceutical industry in drug design.

Often classification or gene regulatory effects depend on the influence of a combination of genes. This leads to an enhancement in the potential search space, as the number of such possible combinations increases. Here lies the utility of intelligent search techniques like GAs. An optimal set of biclusters cannot be guaranteed, since it is an NP-hard problem. Therefore, the quality of biclustering is often considered to be more important than the computation time required to generate it. Hence GAs provide an alternative efficient search technique in a large solution space, based on the theory of evolution.

6.4.2 Multi-objective GA (MOGA)

Unlike single-objective optimization problems, the multiple-objective GA tries to optimize two or more conflicting characteristics represented by fitness functions. Modeling this situation with a single-objective GA would amount to a heuristic determination of a number of parameters involved in expressing such a scalar-combination-type fitness function. MOGA, on the other hand, generates a set of *Pareto-optimal* solutions [34] which simultaneously optimize the conflicting requirements of the multiple fitness functions.

In order to maintain diversity in the population, a measure called crowding distance is used. The multi-objective algorithm NSGA-II is characterized as follows.

1. Initialize the population randomly.

2. Calculate the multi-objective fitness function.

3. Rank the population using the criterion of dominance and crowding distance.

4. Do selection, using crowding selection operator, followed by crossover and mutation (as in conventional GA) to generate children population.

5. Combine parent and children population.

6. Replace the parent population by the best members of the combined population. Initially, members of lower fronts replace the parent population. When it is not possible to accommodate all the members of a particular front, then that front is sorted according to the crowding distance. Selection of individuals is done on the basis of higher crowding distance. The number selected is that required to make the new parent population size the same as the size of the old one.

6.4.3 Other evolutionary strategies

Another evolutionary scheme, often used in Bioinformatics, is *Genetic programming* (GP). This invokes exertion of evolutionary pressure on a program to make it evolve, thereby discovering optimal computer programs resulting in innovative solutions to problems [36]. The principle of operation is similar to GAs, with the focus shifting to evolving programs rather than candidate solutions. GP solutions are computer programs represented as tree structures, that are probabilistically selected according to their fitness in solving the candidate problem. These are then modified with genetic operators (crossover and mutation) to generate new solutions.

6.5 Rough Sets (RS)

Rough sets provide another formalism to handle existing uncertainty in the domain of discourse. These are also found to be useful in tasks like dimensionality reduction, and present considerable promise in the mining of high dimensional microarray data for extracting meaningful knowledge. The concept of *reducts*, from rough set theory, enables extraction of the minimal set of attributes. We present here some requisite preliminaries of rough set theory, in terms of formal definitions.

Definition 1 *An Information System $\mathcal{A} = (U, A)$ consists of a nonempty, finite set U of objects (cases, observations, etc.) and a nonempty, finite set A of attributes a (features, variables), such that $a : U \rightarrow V_a$, where V_a is a value set. We shall deal with information systems called decision tables, in which the attribute set has two parts $(A = C \cup D)$ consisting of the condition and decision attributes (in the subsets C, D of A respectively). In particular, the decision tables we take will have a single decision attribute d, and will be consistent, i.e., whenever objects x, y are such that for each condition attribute a, $a(x) = a(y)$, then $d(x) = d(y)$.*

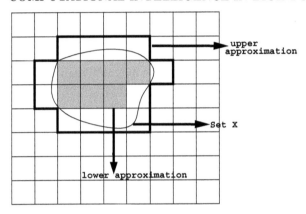

Figure 6.5 *Lower and upper approximations in a rough set.*

Definition 2 *Let $B \subset A$. Then a B-indiscernibility relation $IND(B)$ is defined as*

$$IND(B) = \{(x, y) \in U : a(x) = a(y), \ \forall a \in B\}. \tag{6.26}$$

It is clear that $IND(B)$ partitions the universe U into equivalence classes

$$[x_i]_B = \{x_j \in U : (x_i, x_j) \in IND(B)\}, \ x_i \in U. \tag{6.27}$$

Definition 3 *The B-lower and B-upper approximations of a given set $X(\subseteq U)$ are defined, respectively, as follows:*
$$\underline{B}X = \{x \in U : [x]_B \subseteq X\},$$
$$\overline{B}X = \{x \in U : [x]_B \cap X \neq \phi\}.$$
The B-boundary region is given by $BN_B(X) = \overline{B}X \setminus \underline{B}X$.

Figure 6.5 provides a schematic diagram of a rough set in the F_1-F_2 feature space. The effectiveness of RS has been investigated in the domains of artificial intelligence and cognitive sciences, especially for representation of and reasoning with vague and/or imprecise knowledge, data classification and analysis, machine learning, and knowledge discovery [37].

6.5.1 Reducts

In a decision table $\mathcal{A} = (U, C \cup D)$, one is interested in eliminating redundant *condition* attributes, and actually *relative* (D)-reducts are computed.

Let $B \subseteq C$, and consider the B-positive region of D, viz., $POS_B(D) = \bigcup_{[x]_D} \underline{B}[x]_D$. An attribute $b \in B(\subseteq C)$ is D-dispensable in B if $POS_B(D) = POS_{B \setminus \{b\}}(D)$, otherwise b is D-indispensable in B. Here B is said to be D-independent in $\mathcal{A}$, if every attribute from B is D-indispensable in B.

Definition 4 $B(\subseteq C)$ *is called a D-reduct in* $\mathcal{A}$, *if* B *is D-independent in* $\mathcal{A}$ *and* $POS_C(D) = POS_B(D)$.

Notice that, as decision tables with a single decision attribute d are taken to be consistent, $U = POS_C(d) = POS_B(D)$, for any d-reduct B.

6.5.2 Discernibility matrix

D-reducts can be computed with the help of D-discernibility matrices [38]. Let $U = \{x_1, \cdots, x_m\}$. A *D-discernibility matrix* $M_D(\mathcal{A})$ is defined as an $m \times m$ matrix of the information system $\mathcal{A}$ with the (i, j)th entry c_{ij} given by

$$c_{ij} = \{a \in C : a(x_i) \neq a(x_j), \text{and } (x_i, x_j) \notin IND(D)\}, \quad i, j \in \{1, \cdots, m\}. \tag{6.28}$$

A variant of the discernibility matrix, *viz.*, *distinction table* [39] is often used to enable faster computation.

Definition 5 *A distinction table is a binary matrix with dimensions* $\frac{(m^2-m)}{2} \times N$, *where* N *is the number of attributes in A. An entry* $b((k, j), i)$ *of the matrix corresponds to the attribute* a_i *and pair of objects* (x_k, x_j), *and is given by*

$$b((k, j), i) = \begin{cases} 1 & \text{if } a_i(x_k) \neq a_i(x_j), \\ 0 & \text{if } a_i(x_k) = a_i(x_j). \end{cases} \tag{6.29}$$

The presence of a '1' signifies the ability of the attribute a_i to discern (or distinguish) between the pair of objects (x_k, x_j).

6.5.3 Clustering

In the rough c-means clustering algorithm, the concept of c-means is extended by viewing each cluster as an interval or rough set [166]. A rough set Y is characterized by its lower and upper approximations $\underline{B}Y$ and $\overline{B}Y$ respectively. This permits overlaps between clusters. Here an object $\mathbf{X}_k$ can be part of at most *one* lower approximation. If $\mathbf{X}_k \in \underline{B}Y$ of cluster Y, then simultaneously $\mathbf{X}_k \in \overline{B}Y$. If $\mathbf{X}_k$ is not a part of any lower approximation, then it belongs to two or more upper approximations.

6.6 Hybridization

There has been a lot of research on the hybridization aspect of the different soft computing paradigms. Of these, *neuro-fuzzy (NF) computing* is the oldest and most widely reported in literature.

The concept of ANNs was inspired by *biological neural networks*, which are inherently nonlinear, adaptive, highly parallel, robust and fault tolerant. Fuzzy logic, on the other hand, is capable of modeling vagueness, handling uncertainty and supporting human-type reasoning. These are integrated in the NF framework, by augmenting each other to generate a more intelligent information system [2, 40, 41]. The integration of neural and fuzzy systems leads to a symbiotic relationship, in which fuzzy systems provide a powerful linguistic framework for expert knowledge representation, while neural networks provide learning capabilities and suitability for computationally efficient hardware implementations.

It has been proved [42] that (i) any rule-based fuzzy system may be approximated by a neural net, and (ii) any neural net (feedforward, multilayered) may be approximated by a rule-based fuzzy system. Jang and Sun [43] have shown that fuzzy systems are functionally equivalent to a class of RBF networks, based on the similarity between the local receptive fields of the network and the membership functions of the fuzzy system. Extraction of rules from neural nets enables humans to understand their prediction process in a better manner. This is because rules are a form of knowledge that human experts can easily verify, transmit and expand. Representing rules in *natural* form aids in enhancing their comprehensibility for humans. This aspect is suitably handled using fuzzy set-based representations.

Neuro-fuzzy hybridization [2] is done broadly in two ways: a neural network equipped with the capability of handling fuzzy information [termed *fuzzy-neural network* (FNN)], and a fuzzy system augmented by neural networks to enhance some of its characteristics like flexibility, speed and adaptability [termed *neural–fuzzy system* (NFS)].

In an FNN either the input signals and/or connection weights and/or the outputs are fuzzy subsets or a set of membership values to fuzzy sets (e.g., Refs. [44, 45, 46]). Usually, (i) linguistic values (such as *low, medium* and *high*) or (ii) fuzzy numbers or (iii) intervals are used to model these. A neural–fuzzy system (NFS), on the other hand, is designed to realize the process of fuzzy reasoning, where the connection weights of the network correspond to the parameters of fuzzy reasoning (e.g., Refs. [47, 48]). Using the backpropagation-type learning algorithms, the NFS can identify fuzzy rules and learn membership functions of the fuzzy reasoning. Typically, the NFS architecture has distinct nodes for antecedent clauses, conjunction operators and consequent clauses.

The state of the art for the different techniques of judiciously combining neuro-fuzzy concepts involves synthesis at various levels.

1. Incorporating fuzziness into the neural net framework: fuzzifying the input data, assigning fuzzy labels to the training samples, fuzzifying the learning procedure and obtaining neural network outputs in terms of fuzzy sets [46, 49, 45].
2. Designing neural networks guided by fuzzy logic formalism: designing neural networks to implement fuzzy logic and fuzzy decision-making, and to realize membership functions representing fuzzy sets [50, 47, 48].

3. Changing the basic characteristics of the neurons: neurons are designed to perform various operations used in fuzzy set theory (like fuzzy union, intersection, aggregation) instead of the standard multiplication and addition operations [51, 52, 53].

4. Using measures of fuzziness as the error or instability of a network: the fuzziness or uncertainty measures of a fuzzy set are used to model the error or energy function of the neural network-based system [54].

5. Making the individual neurons fuzzy: the input and output of the neurons are fuzzy sets and the activity of the networks, involving the fuzzy neurons, is also a fuzzy process [44].

Fusion of fuzzy systems and GAs can be done, under *fuzzy–genetic* hybridization, for tuning of fuzzy sets [55]. For example, this can be for membership function selection and tuning. While integrating neural networks with EC, the two components may interact in many ways [56]. For example, GAs can help to avoid the tedious backpropagation algorithm for an MLP, thereby overcoming some limitations of ANNs [57]. The optimal topology of an ANN can also be evolved using GAs [58, 59, 60]. Such an integrated approach may be termed *genetic–neural*. We can have approaches that exploit the benefits of fuzzy sets, ANNs and GAs. Such systems may be termed neuro-fuzzy–genetic (NFG) [61, 62, 63]. For example, a fuzzy reasoning system may be implemented using a multilayer network, where the free parameters of the system may be learned using GAs. Similarly, the parameters of an FNN may also be learned using GAs.

Hybridizations, exploiting the characteristics of rough sets, include the *rough–fuzzy* [64, 65] and *rough–neuro-fuzzy* [66, 67] approaches. The primary role of rough sets here is in managing uncertainty and extracting domain knowledge. Other recent investigations concern the modular *evolutionary–rough–neuro-fuzzy* integration [68, 69] for classification and rule mining. Here the use of EC helps in generating an optimal NF architecture, that is initially encoded using RS for extracting crude domain knowledge from data.

6.7 Application to Bioinformatics

In this section we highlight the role of different soft computing paradigms [3, 4, 70, 71, 72] like fuzzy sets, ANNs, GAs, rough sets and their hybridizations, in different areas of Bioinformatics. The major problems covered here include primary genomic sequence, protein structure, microarray and gene regulatory networks. We categorize the applications based on the different paradigms used.

6.7.1 Primary genomic sequence

Eukaryotic genes are typically organized as exons (coding regions) and introns (noncoding regions). Hence the main task of gene identification, from the primary genomic sequence, involves coding region recognition and splice junction detection. Sequence data are typically dynamic and order dependent. A protein sequence motif is a signature or consensus pattern that is embedded within sequences of the same protein family. Identification of the motif leads to classification of an unknown sequence into a protein family for further biological analysis.

Sequence motif discovery algorithms can follow (i) string alignment, (ii) exhaustive enumeration and (iii) heuristic methods. String alignment algorithms detect sequence motifs by minimizing a cost function which is related to the edit distance between the sequences. Multiple alignment of sequences is an NP-hard problem, with its computational complexity increasing exponentially with sequence size. Local search algorithms may lead to local optima instead of the best motif. Exhaustive enumeration algorithms, though guaranteed to find the optimal motif, are computationally too expensive. Here lies the utility of using soft computing techniques for faster convergence.

FS

Imprecise knowledge of a nucleic acid or a protein sequence of length N has been modeled by a fuzzy biopolymer [73]. This is a fuzzy subset of kN elements, with $k = 4$ bases for nucleic acids and $k = 20$ amino acids for proteins. Profiles considered in the study were a class of biopolymers generated by multiple alignment of a group of related sequences based on matrices of frequencies. A sequence is represented as a vector in a unit hypercube (corresponding to a fuzzy set) that assigns to each position-monomer pair the possibility with which the monomer (base or amino acid) appears in this position. The midpoint of a pair of fuzzy biopolymers of the same length is interpreted as an average of the knowledge of the sequences represented by them.

A systematic verification and improvement of underlying profiles has been undertaken [74], using fuzzy c-means clustering for contextual analysis. Here the authors investigate the recognition of potential transcription factor binding sites in genomic sequences.

ANN

The popularity of ANNs in genomic sequence analysis is mainly due to the involvement of high dimensional space with complex characteristics, which is difficult to model satisfactorily using parameterized approaches. We describe here the role of different models, like self-organizing map (SOM), multilayer perceptron (MLP), recurrent network, counterpropagation, radial basis function network (RBF) and adaptive resonance theory (ART), in gene identification.

Perceptron

Perceptrons were used to predict coding regions in fixed-length windows [75] with various input encoding methods, including binary encoding of codon and dicodon frequency, and the performance was found to be superior to Bayesian statistical prediction. Perceptrons have also been employed to identify cleavage sites in protein sequences [76], with the physicochemical features (of 12 amino acid residues) like hydrophobicity, hydrophilicity, polarity and volume serving as the input. However, single-layer perceptrons are limited to linearly separable classification problems.

MLP

The MLP has been employed for both classification as well as rule generation.
Classification.
An MLP, with backpropagation learning, was used to identify exons in DNA sequences in GRAIL [77]. Thirteen input features used include sequence length, exon GC composition, Markov scores, splice site (donor/acceptor) strength, surrounding intron character, etc., calculated within a fixed 99-nucleotide sequence window and scaled to lie between 0 and 1. A single output indicated whether a specific base, central to the said window, was either coding or noncoding.

A three-layered MLP, with binary encoding at input, was employed to predict acceptor and donor site positions in splice junctions of human genomic DNA sequences [78]. A joint assignment, combining coding confidence level with splice site strength, was found to reduce the number of false positives.

Prediction of the exact location of transcription initiation site has been investigated [79] in mammalian promoter regions, using MLP with different window sizes of input sequence. MLPs were also employed to predict the translation initiation sites [80], with better results being generated for bigger windows on the input sequence. Again, some of the limitations of MLPs, like convergence time and local minima, need to be appropriately handled in all these cases.

Protein classification into 137 to 178 *superfamilies* with a modular architecture involving multiple independent MLPs [81] included 400 to 1356 input features like counts of amino acid pairs, counts of exchange group pairs and triplets, and other encoded combinations using singular value decomposition. Multiple network modules run in parallel to scale up the system. This sort of divide-and-conquer strategy facilitates convergence.

Rule generation
Identification of important binding sites, in a peptide involved in pain and depression, has been attempted [82] using feedforward ANNs. Rules in M-of-N

form are extracted by detecting positions in the DNA sequence where changes in the stereochemistry give rise to significant differences in the biological activity. Browne et al. also predict splice site junctions in human DNA sequences that has a crucial impact on the performance of gene finding programs. Donor sites are nearly always located immediately preceding a GT sequence, while acceptor sites immediately follow an AG sequence. Hence GT and AG pairs within a DNA sequence are markers for potential splice junction sites, and the objective is to identify which of these sites correspond to real sites followed by prediction of likely genes and gene products. The resulting rules are shown to be reasonably accurate and roughly comparable to those obtained by an equivalent $C5$ decision tree*, while being simpler at the same time.

Rules were also generated from a pruned MLP [83], using a penalty function for weight elimination, to distinguish donor and acceptor sites in the splice junctions from the remaining part of the input sequence. The pruned network consisted of only 16 connection weights. A smaller network leads to better generalization capability as well as easier extraction of simpler rules. Ten rules were finally obtained in terms of AG and GT pairs.

SOM

Kohonen's SOM has been used for the analysis of protein sequences [84], involving identification of protein families, aligned sequences and segments of similar secondary structure, with interactive visualization. Other applications of SOM include prediction of cleavage sites in proteins [85], prediction of beta-turns [86], classification of structural motifs [87] and feature extraction [88].

Clustering of human protein sequences into families were investigated [89] with a 15×15 SOM, and the performance was shown to be better than that using statistical nonhierarchical clustering. The study demonstrated that hidden biological information contained in sequence protein databases can be well organized using SOMs.

The Self-Organizing Tree Algorithm (SOTA) is a dynamic binary tree that combines the characteristics of SOMs and divisive hierarchical clustering. SOTA has been employed for clustering protein sequences [90] and amino acids [91]. However, if the available training data are too small to be adequately representative of the actual dataset then the performance of the SOM is likely to get affected.

An unsupervised growing self-organizing ANN [92] has been developed for the phylogenetic analysis of a large number of sequences. The network expands itself following the taxonomic relationships existing among the sequences being classified. The binary tree topology of this model enables efficient classification of the sequences. The growing characteristic of this procedure allows termination at the desired taxonomic level, thereby overcoming the necessity of waiting for the generation of a complete phylogenetic tree. The time for

---

* http://www.spss.com/spssbi/clementine

convergence is approximately a linear function of the number of sequences being modeled.

RBF

A novel extension to the RBF is designed by using the concept of biological similarity between amino acid sequences [93]. Since most amino acid sequences have preserved local motifs for specific biological functions, the numerical radial basis functions are replaced here by certain such nonnumerical (bio-) basis functions. The neural network leads to reduced computational cost along with improved prediction accuracy. Applications are provided on prediction of cleavage sites as well as the characterization of site activity in the Human Immunodeficiency Virus (HIV) protease. The knowledge of these sites can be used to search for inhibitors (antiviral drugs) that block the cleavage ability of the enzyme. The prediction accuracy is reported to be 93.4%.

ART

Multiple layers of adaptive resonance theory 2 (ART2) network have been used to categorize DNA fragments [94] at different resolution levels, similar to a phylogenetic (evolutionary) analysis. The ART network trains fast, and incrementally adapts to new data without needing to review old instances. However, the ability to generalize is limited by the lack of a hidden layer.

Integration with other techniques

Benefits often accrue from using a combination of different learning strategies. A modified counter-propagation network, with supervised learning vector quantization (LVQ) performing nearest-neighbor classification, was used for molecular sequence classification [95].

Dynamic programming has been combined with MLP in GeneParser [96] to predict gene structure. Sequence information is weighted by the MLP to approximate the log-likelihood that each subinterval exactly represents an intron or exon. Dynamic programming is then applied to determine the combination of introns and exons that maximizes the likelihood function. Input to the network consists of the differences for each statistic between the correct and incorrect solutions, and the difference in the number of predicted sequence types. The output maximizes the difference between correct and incorrect solutions.

Extreme Learning Machine (ELM), a new machine learning paradigm with sigmoidal activation function and Gaussian RBF kernel for single hidden layer feedforward neural network, has been used to classify protein sequences from ten classes of superfamilies [97]. The classification accuracy is reported to be better, along with a shorter training time, as compared to that of an MLP of similar size using backpropagation. Since the ELM does not involve any control parameters like learning rate, learning epochs, stopping criteria, that require to be tuned as in MLP, this promises an added advantage.

Investigations with GAs and GP have been made on primary genomic sequences for functions involving their alignment, reconstruction and detection. These are described below.

GA

The simultaneous alignment of many amino acid sequences is one of the major research areas of Bioinformatics. Given a set of homologous sequences, multiple alignments can help predict secondary or tertiary structures of new sequences. GAs have been used for this purpose [98]. Fitness is measured by globally scoring each alignment according to a chosen objective function, with better alignments generating a higher fitness. The cost of multiple alignment A_c is expressed as

$$A_c = \sum_{i=1}^{N-1} \sum_{j=1}^{N} W_{i,j} cost(A_i, A_j), \qquad (6.30)$$

where N is the number of sequences, A_i is the aligned sequence i, $cost(A_i, A_j)$ is the alignment score between two aligned sequences A_i and A_j, and $W_{i,j}$ is the weight associated with that pair of sequences. The cost function includes the sum of the substitution costs, as given by a substitution matrix, and the cost of insertions/deletions using a model with affine gap (gap-opening and gap-extension) penalties. Roulette wheel selection is carried out among the population of possible alignments, and insertion/deletion events in the sequences are modeled using a *gap insertion* mutation operator.

Given N aligned sequences $A_1 \ldots A_N$ in a multiple alignment, with $A_{i,j}$ being the pairwise projection of sequences A_i and A_j, $length(A_{i,j})$ the number of ungapped columns in this alignment, $score(A_{i,j})$ the overall consistency between $A_{i,j}$ and the corresponding pairwise alignment in the library, and $W'_{i,j}$ the weight associated with this pairwise alignment, the fitness function was modified [99] to

$$F = \frac{\sum_{i=1}^{N-1} \sum_{j=1}^{N} W'_{i,j} * score(A_{i,j})}{\sum_{i=1}^{N-1} \sum_{j=1}^{N} W'_{i,j} * length(A_{i,j})}. \qquad (6.31)$$

The main difference with eqn. (6.30) is the use of a library that replaces the substitution matrix and provides position-dependent means of evaluation.

The generation of accurate DNA sequence is a challenging and time-consuming problem in genomics. A widely used technique in this direction is *hybridization*, which detects all oligonucleotides—short sequences of the four nucleotide bases, A, C, T, G, of a given length k—that make up the corresponding DNA fragment. (This terminology is different from the hybridization of soft computing paradigms, which we elucidate in this chapter.) The oligonucleotide library is very large, containing 4^k elements, with microarray chip technology being often used in its implementation. However, the hybridization experiment introduces both negative (missing oligonucleotides) and positive (erroneous oligonucleotide) errors in the spectrum of elements. The reconstruction of the DNA sequence, from these errors, is an NP-hard combinatorial problem. GAs have been successfully applied to difficult instances of sequence reconstruction [72], with a fitness function maximizing the number of elements chosen from the spectrum (subject to a restriction on the maximum length n) of the sequence of nucleotides. The representation of a candidate solution is in terms of a permutation of indices of oligonucleotides from the spectrum.

Phylogenetic inference has been attempted using GA [100] and parallel GA [101]. An individual in a population is a hypothesis consisting of the tree, branch lengths and parameters values for the model of sequence evolution, while the fitness is the likelihood score of the hypothesis. In the parallel version [101] each individual in a population is handled by one processor or node which computes its corresponding likelihood. This operation being extremely time consuming, the parallelization at this level causes a nearly linear order search time improvement for large data. The number of processors used is equal to the size of the evolving population, plus an additional processor for the control of operations. Selection is accomplished on the maximum-likelihood score, migration and recombination is permitted between subpopulations, and mutation can be branch-length based or topological. Results are provided on 228 taxa of DNA sequence data.

GP

Finite State Automata (FSA) has been combined with GP, to discover candidate promoter sequences in primary sequence data [102]. FSAs are directed graphs that can represent grammars in the Chomsky hierarchy and Turing machines. In the *GP-Automata*, a GP tree structure is associated with each state of the FSA. The method is able to take large base pair jumps, thereby being able to handle very long genomic sequences in order to discover gene specific *cis*-acting sites as well as genes which are regulated together. It is to be noted that an aim of drug discovery is to identify *cis*-acting sites responsible for co-regulating different genes.

The training dataset[†] consists of known promoter regions, while nonpromoter examples constitute samples from the coding or intron sequences. The objective of the GP-tree structure, in each state of the GP-Automata, is to find motifs within the promoter and nonpromoter regions. The terminal set includes A, C, T and G. The method automatically discovers motifs of various lengths in automata states, and combines motif matches using logical functions to arrive at a *cis*-acting region identification decision.

[†] http://www.fruitfly.org/

Hybridization

Extraction of motif from a group of related protein sequences has been investigated in a neuro-fuzzy framework [103], using data from PROSITE. A statistical method is first used to detect short patterns occurring with high frequency. Fuzzy logic enables the design of approximate membership functions and rules about protein motifs, as obtained from domain experts. A radial basis function (RBF) neural network is employed to optimize the classification by tuning the membership functions.

Evolving ANNs for discriminating between functional elements associated with coding nucleotides (exons) and noncoding sequences of DNA (introns and intragenic spacer) has been reported [72] in the *genetic–neural* framework. The connection weights of a fixed MLP architecture are evolved for classification, using evolutionary computation, with practical application to gene detection. Performance of the evolved network is compared to that of GRAIL [77] and GeneParser [96].

6.7.2 Protein structure

Protein structure prediction typically uses experimental information stored in protein structural databases, like the Brookhaven National Laboratory Protein Data Bank (PDB) [104]. A common approach is based on sequence alignment with structurally known proteins. The experimental approach involving X-ray crystallographic analysis and nuclear magnetic resonance (NMR) being expensive and time-consuming, soft computing techniques offer an innovative way to overcome some of these problems.

FS

A *contact map* is a concise representation of a protein's native 3D structure. It is expressed as a binary matrix, where each entry is a '1' if the corresponding protein residue pair are in "contact" (with Euclidean distance being within a threshold). When represented graphically, each contact between two residues corresponds to an edge. An alignment between two contact maps is an assignment of residues in one to those of the equivalent other. A pair of contacts is equivalent when the pairs of residues that define their end-points are also equivalent. The number of such equivalent contacts determine the overlap of the contact maps for a pair of proteins, with a higher overlap indicating increased similarity between them. A generalization of the maximum contact map overlap has been developed [105] using one or more fuzzy thresholds and membership functions. This enables a more biological formulation of the optimization problem. Investigations are reported on three datasets from the PDB. Clustering of protein structures is done to validate the results.

ANN

A step on the way to a prediction of the full 3D structure of protein is predicting the local conformation of the polypeptide chain, called the *secondary* structure. The whole framework was pioneered by Chou and Fasmann [106]. They used a statistical method, with the likelihood of each amino acid being one of the three (alpha, beta, coil) secondary structures estimated from known proteins.

In this section we highlight the enhancement in prediction performance of ANNs, with the use of ensembles and the incorporation of alignment profiles. The data consist of proteins obtained from the PDB. A fixed size window constitutes the input to the feedforward ANN. The network predicts the secondary structure corresponding to the centrally located amino acid of the sequence within the window. The contextual information about the rest of the sequence, in the window, is also considered during network training. A comparative study of performance of different approaches, for secondary structure prediction on this data, is provided in Table 6.1.

MLP

Around 1988 the first attempts were made by Qian and Sejnowski [107], to use MLP with backpropagation to predict protein secondary structure. Three output nodes correspond to the three secondary structures. Performance is measured in terms of overall correct classification Q (64.3%) and Matthews Correlation Coefficient (MCC). We have

$$Q = \sum_{i=1}^{l} w_i Q_i = \frac{C}{N} \qquad (6.32)$$

for an l-class problem, with Q_i indicating the accuracy for the ith class, w_i being the corresponding normalizing factor, N representing the total number of samples and C being the total number of correct classifications.

$$MCC = \frac{(TP * TN) - (FP * FN)}{\sqrt{(TP + FP)(TP + FN)(TN + FP)(TN + FN)}}, \qquad (6.33)$$

where TP, TN, FP and FN correspond to the number of true positive, true negative, false positive and false negative classifications, respectively. Here $N = TP + TN + FP + FN$ and $C = TP + TN$, and $-1 \leq MCC \leq +1$ with $+1$ (-1) corresponding to a perfect (wrong) prediction. The values for MCC for the α-helix, β-strand and random coil were found to be 0.41, 0.31, 0.41, respectively.

The performance of this method was improved by Rost and Sander [108, 109], by using a cascaded three-level network with multiple-sequence alignment. The three levels correspond to a sequence-to-structure net, a structure-to-structure net and a jury (combined output) decision, respectively. Correct classification increased to 70.8%, with the MCC being 0.60, 0.52, 0.51, respectively, for the three secondary classes.

Supersecondary structures (folding units), like $\alpha\alpha$- and $\beta\beta$-hairpins, and $\alpha\beta$- and $\beta\alpha$-arches, serve as important building blocks for protein tertiary structure. Prediction of supersecondary structures was made from protein sequences [110] using MLP. The size of the input vector was the same as the length of the sequence window. There were 11 networks, each with one output, for classifying one of the 11 types of frequently occurring motifs. A test sequence was assigned to the motif category of the winning network, having the largest output value. Results demonstrated more than 70% accuracy.

Protein structure comparison is often used to identify set of residue equivalencies between proteins based on their 3D coordinates, and has a wide impact on the understanding of protein sequence, structure, function and evolution. This is because it can identify more distantly related proteins, as compared to sequence comparison, since protein structures are more conserved than amino acid sequences over evolution.

The determination of an optimal 3D conformation of a protein corresponds to folding, and has manifold implications to drug design. An active site structure determines the functionality of a protein. A ligand (enzyme or drug) docks into an active site of a protein. Many automated docking approaches have been developed, and can be categorized as (i) rigid docking: both ligand and protein are rigid, (ii) flexible-ligand docking: ligand flexible and protein rigid and (iii) flexible-protein docking: both ligand and protein are flexible (only a limited model of protein variation allowed, such as side-chain flexibility or small motions of loops in the binding site).

One of the earliest ANN-based protein tertiary structure prediction in the backbone [111] used MLP, with binary encoding for a 61-amino acid window at the input. There were 33 output nodes corresponding to the three secondary structures, along with distance constraints between the central amino acid and its 30 preceding residues. A large scale ANN was employed to learn protein tertiary structures from the PDB [112]. The sequence-structure mapping encoded the entire protein sequence (66-129 residues) into 140 input units. The amino acid residue was represented by its hydrophobicity scale, normalized between -1 and +1. The network produced good prediction of distance matrices from homologous sequences, but suffered from a limited generalization capability due to the relatively small size of the training set.

Interatomic C^α distances between amino acid pairs, at a given sequence separation, were predicted [113] to be above (or below) a given threshold corresponding to contact (or noncontact). The input consisted of two sequence windows, each with 9 or 15 amino acids separated by different lengths of sequence, and a single output indicated the contact (or noncontact) between the central amino acids of the two sequence windows.

Instead of using protein sequence at input, a protein structure represented by a side-chain-side-chain contact map was employed at the input of an ANN to evaluate side-chain packing [114]. Contact maps of globular protein structures in the PDB were scanned using 7×7 windows, and converted to 49 binary numbers for the input. One output unit was used to determine whether the contact pattern is prevalent in the structure database.

Information obtained from secondary structure prediction is incorporated to improve structural class prediction using MLP [115]. The 26 input nodes include the 20-amino acid composition, sequence length and five secondary structure characteristics of the protein. Four outputs correspond to four tertiary super classes. Prediction of 83 folding classes in proteins has been attempted [116] using multiple two-class MLPs. The input was represented in terms of major physicochemical amino acid attributes, like relative hydrophobicity (hydrophobic, neutral or polar), predicted secondary structure, predicted solvent accessibility (buried or exposed), along with certain global descriptors like composition, transition and distribution of different amino acid properties along the protein sequence.

A single layer feedforward ANN, trained with scaled conjugate gradient algorithm, is used to identify catalytic residues found in enzymes [117] based on an analysis of the structure and sequence. Structural parameters like the solvent accessibility, type of secondary structure, depth and cleft that the residue lies in, along with the conservation score and residue type, are used as inputs for the ANN. Performance is measured in terms of the MCC. The network output is spatially clustered to determine the highly scoring residues, and thereby predict the location of most likely active sites.

RBF

Radial basis function (RBF) network, a supervised feedforward ANN, has been employed [118] to optimally predict the free energy contributions of proteins due to hydrogen bonds, hydrophobic interactions and the unfolded state, with simple input measures.

Ensemble networks

Prediction of protein secondary structure has been further developed by Riis and Krogh [119], with ensembles of combining networks, for greater accuracy in prediction. The *Softmax* method is used to provide simultaneous classification of an input pattern into multiple classes. A normalizing function at the output layer ensures that the three outputs always sum to one. A logarithmic likelihood cost function is minimized, instead of the usual squared error. An adaptive weight encoding of the input amino acid residues reduces the overfitting problem. A window is selected from all the single structure networks in the ensemble. The output is determined for the central residue, with the prediction being chosen as the largest of the three outputs normalized by Softmax.

The use of ensembles of small, customized subnetworks is found to improve

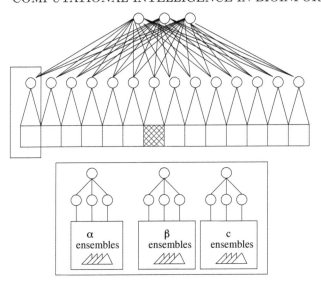

Figure 6.6 *Secondary protein structure prediction using ensemble of ANNs.*

predictive accuracy. Customization involves incorporation of domain knowledge into the subnetwork structure for improved performance and faster convergence. For example, the helix-network has a built-in period of three residues in its connections in order to capture the characteristic periodic structure of helices. Fig. 6.6 provides the schematic network structure. Overall accuracy increased to 71.3%, with the MCC becoming 0.59, 0.50, 0.41, respectively, for the three secondary classes.

Use of alignment profile

The alignment profile generated by Psi-BLAST has been incorporated by Jones [120] to design a set of cascaded ANNs. These profiles enable finding more distant sequences, use a more rigorous statistical approach for computing the probability of each residue at a specific position and properly weigh each sequence with respect to the amount of information it carries.

Prediction of segments in protein sequences containing aromatic-backbone NH interactions has been attempted [121]. (An NH interaction is a nonconventional hydrogen bonding interaction involving side-chain aromatic ring and backbone NH group.) Such interactions help in the stabilization of protein secondary and tertiary structures as well as folding, on the basis of their spatial distribution. Incorporation of evolutionary information in the form of multiple alignment, by Psi-BLAST, enhances the performance in terms of MCC. Two consecutive three-layered feedforward sequence-to-structure and structure-to-structure networks, trained by backpropagation, are employed. It is observed that a segment (window) of seven residues provides sufficient input information for prediction of these aromatic-NH interactions. The actual position of donor aromatic residue within the *potential* predicted fragment is also identified, using a separate sequence-to-structure neural network. The implementation was made on a nonredundant dataset of 2298 protein chains extracted from the Protein Data Bank (PDB).

Ensembles of bidirectional recurrent neural network architectures are used in conjunction with profiles generated by Psi-BLAST to predict protein secondary structure for a given amino acid sequence [122]. The classification decision is determined by three component networks. In addition to the standard central component associated with a local window at location t of the current prediction (as in feedforward ANNs), there exist contribution by two similar recurrent networks corresponding to the left and right contexts (like wheels rolling from the two ends of the polypeptide sequence). An ensemble of 11 networks are trained, using backpropagation. Two output categorizations are followed, *viz.*, (i) three classes (α helix, β-strand, random coil), as in SSpro, and (ii) eight classes as in DSSP[‡] programs. The output error is the relative entropy between the output and target probability distributions. At the alignment level the use of Psi-BLAST, with the ability to produce profiles that include increasingly remote homologs, enhances performance as compared to that employing only BLAST [123]. The system was implemented on proteins from the PDB, which are at least 30 amino acids long, have no chain breaks, produce a DSSP output and are obtained by X-ray diffraction methods with high resolution. The accuracy of secondary structure prediction is thereby enhanced to about 75%.

Table 6.1 *Comparative performance for protein secondary structure prediction.*

Approach	Reported overall per-residue accuracy (%)	Reported MCC
MLP [107]	64.3	0.41, 0.31, 0.41
MLP + multiple sequence alignment [109]	70.8	0.60, 0.52, 0.51
MLP ensemble + Softmax [119]	71.3	0.59, 0.50, 0.41
Recurrent network ensemble + Psi-BLAST [122]	about 75	–

[‡] http://www.cmbi.kun.nl/gv/dssp/

GAs have been mainly applied to tertiary protein structure prediction, folding, docking and side-chain packing problems. Structure alignment has been attempted in proteins using GAs [124], by first aligning equivalent secondary structure element (SSE) vectors while optimizing an elastic similarity score S. This is expressed as

$$S = \begin{cases} \sum_{i=1}^{L} \sum_{j=1}^{L} \left(\theta - \frac{d_{ij}^A - d_{ij}^B}{\bar{d}_{ij}} \right) e^{-(\bar{d}_{ij}/a)^2}, & i \neq j, \\ \theta, & i = j, \end{cases} \quad (6.34)$$

where d_{ij}^A and d_{ij}^B are the distances between equivalent positions i and j in proteins A and B respectively, $\bar{d}_{ij}$ is the average of d_{ij}^A and d_{ij}^B, and θ and a are constant parameters, with the logic implying that equivalent positions in two proteins should have similar distances to other equivalent positions. Secondly, amino acid positions are optimally aligned within the SSEs. This is followed by superposition of protein backbones, based on the position equivalencies already determined. Finally, additional equivalent positions are searched in the non-SSE regions.

Tertiary protein structure prediction and folding, using GAs, has been reported in Ref. [125, 72, 126, 127]. The objective is to generate a set of *native-like* conformations of a protein based on a force field, while minimizing a fitness function depending on its potential energy. Proteins can be represented in terms of (a) three-dimensional Cartesian coordinates of its atoms and (b) the torsional angle Rotamers, which are encoded as bit strings for the GA. The Cartesian coordinates representation has the advantage of being easily convertible to and from the 3D conformation of a protein. Bond lengths, b, are specified in these terms. In the torsional angles representation, the protein is described by a set of angles under the assumption of constant standard binding geometries. The different angles involved are the (i) bond angle θ, (ii) torsional angle ϕ, between N (amine group) and C_α, (iii) angle ψ, between C_α and C' (carboxyl group), (iv) peptide bond angle ω, between C' and N and (v) side-chain dihedral angle χ.

The potential energy $U(r_1, \ldots, r_N)$ between N atoms is minimized, being expressed as $U(r_1, \ldots, r_N) = \sum_i K_b (b_i - b_0^i)^2 + \sum_i K_\theta (\theta_i - \theta_0^i)^2 + \sum_i K_\phi [1 - \cos(n\phi_i - \delta)] + \sum_{i,j} \frac{q_i q_j}{4\pi\varepsilon_0\varepsilon_r r_{ij}} + \sum_{i,j} \varepsilon \left[\left(\frac{\sigma_{ij}}{r_{ij}} \right)^{12} - 2 \left(\frac{\sigma_{ij}}{r_{ij}} \right)^6 \right]$. Here the first three harmonic terms on the right-hand side involve the bond length, bond angle and torsional angle of covalent connectivity, with b_0^i and θ_0^i indicating the down-state (low energy) bond length and bond angle respectively, for the ith atom. The effects of hydrogen bonding and that of solvents (for nonbonded atom pairs i, j, separated by at least four atoms) are taken care of by the electrostatic Coulomb interaction and Van der Waals' interaction, modeled by the last two terms of the expression. Here K_b, K_θ, K_ϕ, σ_{ij} and δ are constants, q_i and q_j are the charges of atoms i and j, separated by distance r_{ij}, and ε indicates the dielectric constant. Two commercially available software packages, containing variations of the potential energy function, are Chemistry at HARvard Molecular Mechanics (CHARMm) and Assisted Model Building with Energy Refinement (AMBER).

Additionally, a protein acquires a folded conformation favorable to the solvent present. The calculation of the entropy difference between a folded and un-folded state is based on the interactions between a protein and solvent pair. Since it is not yet possible to routinely calculate an accurate model of these interactions, an *ad hoc* pseudo-entropic term E_{pe} is added to drive the protein to a globular state. E_{pe} is a function of its actual diameter, which is defined to be the largest distance between a pair of C_α carbon atoms in a conformation. We have

$$E_{pe} = 4^{(actual\_diameter - expected\_diameter)} \text{ kcal/mol}, \qquad (6.35)$$

where $expected\_diameter/m = 8 * \sqrt[3]{len/m}$ is the diameter in its native con-formation and *len* indicates the number of residues. This penalty term ensures that extended conformations have larger energy (or lower fitness) values than globular conformations. It constitutes the conformational entropy constituent of potential energy, in addition to the factors involved in the expression for U.

Genetic Optimization for Ligand Docking (GOLD) [128] is an automated flexible-ligand docking program, employing steady-state GA involving the is-land model (which evolves several small, distinct populations, instead of one large population). It evaluates nonmatching bonds while minimizing the po-tential energy (fitness function), defined in terms of Van der Waals' internal and external (or ligand-site) energy, torsional (or dihedral) energy and hy-drogen bonds. However, (i) an enforced requirement that the ligand must be hydrogen-bonded to the binding site and (ii) an underestimation of the hy-drophobic contribution to binding sometimes lead to failures in docking in certain cases over here.

Each chromosome in GOLD encodes the internal coordinates of both the lig-and and active protein site, and a mapping between the hydrogen-bonding sites. Reproduction operators include crossover, mutation and a migration operator to share genetic material between populations. The output is the ligand and protein conformations associated with the fittest chromosome in the population, when the GA terminates. The files handled are the Cam-bridge Crystallographic Database, Brookhaven PDB and the Rotamer library (providing the relationship between side-chain dihedral angles and backbone conformation).

AutoDock [129] works on a genome composed of a string of real-valued genes encoding the 3D coordinates and different angles. Mutation of the real-valued parameters is accomplished through the addition of a Cauchy-distributed ran-dom variable. Both conventional as well as Lamarckian GAs are used, along with elitism. (Lamarckian GAs employ a local search, with replacement on a small fraction of the population within each generation. In the Baldwinian approach, unlike in Lamarckian, the original population is not updated by the solution found in the local search.)

A Generic Evolutionary Method for Molecular Docking (GEMDOCK) [130] has been developed for flexible-ligand docking. The potential energy function, involving numerous atomic interactions, is often computationally too expensive to implement using evolutionary strategies. Hence rapid recognition of potential ligands is emphasized using a robust, simpler scoring function, encountering fewer local minima. Discrete and continuous search techniques are combined with local search to speed up convergence. The energy function encompasses electrostatic, steric and hydrogen bonding potentials of the molecules. A new rotamer-based mutation operator helps reduce the search space of ligand structure conformations. GEMDOCK is an automatic system that generates all related docking variables, like atom formal charge, atom type and the ligand binding site of a protein. A major problem in GOLD, *viz.*, its sensitiveness to docking hydrophobic ligands, is reduced in GEMDOCK [130]. However, its empirical scoring function is yet to incorporate important functional group interactions between ligands and proteins as in GOLD.

In a slightly different approach, the prediction of the conserved or displaced status of water molecules in the binding site, upon ligand binding, was made [131] by using a k-nearest-neighbors classifier. GAs determine the optimal feature-weight values for the classifier. Fitness is based on the percentage of correct predictions made.

The side-chain packing problem deals with the prediction of side-chain conformations. This is a crucial aspect of protein folding, since it determines feasible backbone conformations. GAs have been used in prediction of side-chain packing [132] to search for low-energy hydrophobic core sequences and structures, using a custom Rotamer library as input. Each core position is allocated a set of bits in the chromosome, to encode a specific residue type and a set of torsional angles as specified in the library.

Evolutionary programming has been employed for *faster* finding of deep minima in the energy landscape of protein folding [133]. One folding step of the protein molecule involves (i) calculation of molecular motion of the structure, i.e., rotation around one bond, and (ii) computation of free energy of the new conformation, which is discarded if it increases after the molecular motion. This process is simultaneously repeated to simulate a large set of folding operations, each using a different expanded starting structure for the protein. The program then determines those simulations yielding structures with the lowest free energies. It uses a lattice model of proteins to speed up the simulation, allowing only bond angle changes (0°, $\pm 45^o$, $\pm 90^o$) between adjacent amino acid residues along one or two of the three planes.

Mutations of the program were created by using different types or magnitudes of molecular motions, or different positions of bonds around which rotations are performed, or different sequences of these motions. Positive mutants, i.e., those which performed better than the original program, were used for further mutations. Negative program mutants, i.e., those which did not find a deeper energy minimum within a certain period of time, were discarded. It is observed that only 20 evolution steps yielded a more than 10-fold increase in speed of finding deep minima in the energy landscape of two 64-residue proteins.

Hybridization

Hybrid approaches to applications related to protein secondary structure also exist in literature. A knowledge-based approach was employed to extract inference rules about a biological problem that were then used to configure ANNs [134]. Integration with GAs was attempted to generate an optimal ANN topology [135], and its performance on secondary structure prediction was found to be comparable to that of Qian and Sejnowski [107].

6.7.3 Microarray

Each DNA array contains the measures of the level of expression of many genes. Various distances and/or correlations can be computed from pairwise comparison of these patterns. Let $gene_j(e_{j1}, \ldots, e_{jn})$ denote the expression pattern for the jth gene for $i = 1, \ldots, n$ samples. The *Euclidean distance* between the jth and kth genes, computed as

$$d_{j,k} = \sqrt{\sum_i (e_{ji} - e_{ki})^2}, \tag{6.36}$$

is suitable when the objective is to cluster genes displaying similar levels of expression. There exists considerable literature on the applications of different soft computing paradigms in the area of gene expression data.

FS

Fuzzy c-means [138] is a well-known fuzzy partitive algorithm employed for clustering overlapping data. Use of fuzzy clustering enables genes to simultaneously belong to multiple groups, thereby revealing distinctive features of their function and regulation. Fuzzy c-means algorithm has been applied to cluster microarray data [139]. The value of the fuzzifier m is appropriately tuned for gene selection, based on resultant distribution of distances between genes. The selected genes exhibit tight association to the clusters.

Many proteins serve different functions depending on the demands of the organism, such that a corresponding set of genes is often coexpressed with multiple, distinct groups of genes under different conditions. This type of conditional coregulation of genes is modeled using a heuristically modified version of fuzzy c-means clustering [140], to identify overlapping partitions of genes based on the response of yeast cells to environmental changes.

The temporal order and varying length of sampling intervals are some of the important factors for clustering time-series microarray data into biologically meaningful partitions. However, the shortness and unequal sampling of gene expression time-series data limits the use of conventional modeling in these cases. The fuzzy short time-series algorithm [141] clusters profiles based on the similarity of their relative change in expression level and the corresponding temporal information. Here the short time-series distance measure is incorporated in the fuzzy c-means framework. The performance, on the transcriptional data of sporulation in budding yeast, is evaluated in terms of Dunn's clustering validity index [136].

An interesting image processing application was developed at the preprocessing stage, for the fuzzy filtering of noise from cDNA microarray color images in the two-channel Red-Green space [142]. This sort of reduction in noise impairment facilitated subsequent analysis of the cDNA images. The two-component adaptive vector filter integrates concepts from fuzzy sets, nonlinear filtering, multidimensional scaling and robust order statistics. Robust noise removal is achieved by tuning a membership function, which utilizes distance criteria based on a novel color-ratio model, on cDNA vectorial inputs at each image location.

ANN

The two major mining tasks, modeled here, are clustering and classification. While unsupervised learning is self-organized, supervised learning helps incorporate known biological functions of genes into the knowledge discovery process of gene expression pattern analysis for gene discovery and prediction.

Kohonen's SOM has been applied to the clustering of gene expression data [143, 144, 145]. It generates a robust and accurate clustering of large and noisy data, while providing effective visualization. SOMs require the winning neuron or node in the gene expression space (along with its neighbors) to be rotated in the direction of a selected gene expression profile (pattern). However, the predefinition of a two-dimensional topology of nodes can often be a problem considering its *biological relevance*.

SOTA has also been applied to gene expression clustering [146]. As in SOMs the gene expression profiles are sequentially and iteratively presented at the terminal nodes, and the mapping of the node that is closest (along with its neighboring nodes) is appropriately updated. Upon convergence the node containing the most variable (measured in terms of distance) population of expression profiles is split into sister nodes, causing a growth of a binary tree structure. Unlike conventional hierarchical clustering, SOTA is linear in complexity to the number of profiles. The number of clusters need not be known in advance as in c-means clustering. The algorithm starts from the node having the most heterogeneous population of associated input gene profiles. A statistical procedure is followed for terminating the growing of the tree, thereby eliminating the need for an arbitrary choice of cutting level as in hierarchical models. However, no validation is provided to establish the biological relevance.

A binary tree-structured vector quantization [147] uses (i) self-organizing map (SOM) for visualization, and (ii) partitive c-means clustering for grouping the similar component planes of SOMs and organizing them. Results are provided on cDNA microarray lung cancer data.

Classification of acute leukemia, having highly similar appearance in gene expression data, has been made by combining a pair of classifiers trained with mutually exclusive features [148]. Gene expression profiles were constructed from 72 patients having acute lymphoblastic leukemia (ALL) or acute myeloid leukemia (AML), each constituting one sample of the DNA microarray§. Each pattern consists of 7129 gene expressions. A neural network combines the outputs of the multiple classifiers. Feature selection with nonoverlapping correlation (such as Pearson and Spearman correlation coefficients) encourages the classifier ensemble to learn different aspects of the training data in a wide solution space. The recognition accuracy and generalization capacity are reported to be higher than those involving SOM, decision tree, k-nearest neighbors classifier and support vector machine.

An autoassociative neural network has been used for *simultaneous* pattern identification, feature extraction and classification of gene expression data [149]. The network output approximates a reconstructed version of the input vector. Backpropagation is used to adjust the connection weights. The analysis of the network structure and strength of connections allows the (i) identification of specific phenotype markers, (ii) extraction of peculiar associations among genes and physiological states and (iii) assignment to multiple classes, like different pathological conditions or tissue samples. Results are demonstrated on Leukemia and Colon cancer¶ datasets.

Bayesian regularized neural network has been employed [150] to classify multiple gene expression temporal patterns, with sequential time points under different experimental conditions. The Bayesian setting, along with the regularization, helps overcome experimental as well as biological noise or uncertainty. A feedforward architecture is used, with the input neurons corresponding to the number of time points or experimental conditions in the microarray experiment. Results are provided on the yeast data.

§ http://www.genome.wi.mit.edu/MPR
¶ http://microarray.princeton.edu/oncology

The identification of gene subsets for classifying two-class disease samples has been modeled as a multi-objective evolutionary optimization problem [151], involving minimization of gene subset size to achieve reliable and accurate classification based on their expression levels. NSGA-II, a multi-objective GA, is used for the purpose. This employs elitist selection and an explicit diversity preserving mechanism, and emphasizes the nondominated solutions. Results are provided on three cancer samples, *viz.*, Leukemia, Lymphoma and Colon. An l-bit binary string, where l is the number of selected (filtered) genes in the disease samples, represents a solution. The major difficulties faced in solving the optimization problem include the availability of only a few samples as compared to the number of genes in each sample, and the resultant huge search space of solutions. Moreover many of the genes are redundant to the classification decision, and hence need to be eliminated. The three objectives simultaneously minimized are (i) the gene subset size, (ii) number of misclassifications in training and (iii) number of misclassifications in test samples.

The grouping GA (GGA) [152] is a modified GA, developed to suit the particular structure of grouping problems like clustering. GGA has also been applied to the clustering of microarray data [72]. The clusters of expression profiles are directly encoded in the chromosomes, based on their ordinal numbers, and the fitness function is defined on this set of groupings. The composition of the groups controls the value of the objective function.

GAs have also been used to correctly classify the SRBCT dataset with a selection of 12 genes [153]. There are four classes of tumors, from 88 samples described by 2308 genes. Simulated annealing (SA) [154] is employed to generate a robust clustering of temporal gene expression profiles [155]. An iterative scheme quantitatively evaluates the optimal number of clusters, while simultaneously optimizing the distribution of genes within them. The ith profile is represented by a vector $\{e_{i1}, \ldots, e_{in}\}$, with expression component e_{it} corresponding to the normalized expression level of gene i at time t in the range $[0, 1]$. The distribution of profiles is optimized for c clusters by minimizing the within-cluster distance between them, using

$$E(c) = \frac{1}{c} \sum_{k=1}^{c} \left[\sum_{i \in U_k} \sum_{j \in U_k} d_{i,j} \right], \qquad (6.37)$$

where $d_{i,j}$ is the Euclidean distance [eqn. (6.36)] between profiles belonging to cluster U_k.

Biclustering or simultaneous clustering of both genes and conditions have generated considerable interest over the past few decades, particularly related to the analysis of high-dimensional gene expression data in information retrieval, knowledge discovery and data mining. The objective is to find sub-matrices, i.e., maximal subgroups of genes and subgroups of conditions where the genes exhibit highly correlated activities over a range of conditions, thereby better reflecting the biological reality. GAs are employed [156], by integrating a greedy algorithm as a local search in order to improve the quality of biclustering.

Optimization being done with respect to the mutually conflicting goals of homogeneity and size, they become suitable candidates for multi-objective modeling. A novel multi-objective evolutionary biclustering framework is developed in Ref. [157] by incorporating local search strategies. A new quantitative measure to evaluate the goodness of the biclusters has been presented. The experimental results on benchmark datasets demonstrate better performance as compared to existing algorithms available in literature. Evolutionary segmentation of the yeast genome has also been attempted in literature [158].

RS

A basic issue related to many practical applications of knowledge databases is whether the whole set of attributes in a given information system is always necessary to define a given partition of the universe. Many of the attributes are superfluous, i.e., we can have *'optimal'* subsets of attributes which define the same partition as the whole set of attributes. These subsets are called the *reducts* in rough set theory [159], and correspond to the minimal feature set that are sufficient to represent a decision. Such dimensionality reduction has considerable impact on subsequent decision-making.

Classification rules (in *If–Then* form) have been extracted from microarray data [160], using rough sets with supervised learning. The underlying assumption is that the associated genes are organized in an ontology, involving super- and sub-classes. This biological knowledge is utilized while generating rules in terms of the minimal characteristic features (reducts) of temporal gene expression profiles. A rule is said to *cover* a gene if the gene satisfies the conditional part, expressed as a conjunction of attribute-value pairs. The rules do not discriminate between the super-and sub-classes of the ontology, while retaining as much detail about the predictions without losing precision.

Hybridization

We begin with the *neuro-fuzzy* models. Fuzzy adaptive resonance theory (ART) network [161] has been employed for clustering the time series expression data related to the sporulation of budding yeast [162].

An evolving modular fuzzy neural network, involving dynamic structure growing (and shrinking), adaptive online learning and knowledge discovery in rule form, has been applied to the Leukemia and Colon cancer gene expression data [163]. Feature selection improves classification by reducing irrelevant attributes that do not change their expression between classes. The Pearson correlation coefficient is used to select genes that are highly correlated with the tissue classes. Rule generation provides physicians, on whom the final responsibility for any decision in the course of treatment rests, with a justification regarding how a classifier arrived at a judgement. Fuzzy logic rules, extracted from the trained network, handle the inherent noise in microarray data while offering the knowledge in a human-understandable linguistic form. These rules point to genes (or their combinations) that are strongly associated with specific types of cancer, and may be used for the development of new tests and treatment discoveries.

A dynamic fuzzy neural network, involving self-generation, parameter optimization and rulebase simplification, is used [164] for the classification of cancer data such as Lymphoma[||], small round blue cell tumor (SRBCT)[**], and liver cancer[††]. Initial feature selection is done in terms of t-tests. It is observed that a small number of important genes (5 out of 4026, 8 out of 2308, 24 out of 1648 features, in the three datasets respectively) succeed in attaining 100% classification.

Gastric tumor classification in microarray data is made using rough set based learning [165], implemented with ROSETTA involving genetic algorithms and dynamic reducts [39]. The fitness function incorporates measures involving the classification performance (discernibility) along with the size of the reduct. Thereby precedence is provided to solutions having less number of attributes. A major problem with microarray data being the smaller number of objects with a comparatively larger number of attributes, a preprocessing stage of feature selection based on bootstrapping is made. The dataset consists of 2504 human genes corresponding to the conditional attributes, while the 17 tumor types are clubbed as six different clinical parameters or the decision attributes.

An evolutionary rough c-means clustering algorithm has been applied to microarray gene expression data [167]. Rough sets are used to model the clusters in terms of upper and lower approximations. GAs are used to tune the threshold, and relative importance of upper and lower approximation parameters of the sets. The Davies-Bouldin clustering validity index [136] is used as the fitness function of the GA, that is minimized while arriving at an optimal partitioning. The algorithm performs particularly well over the Colon cancer gene expression data, involving a collection of 62 measurements from colon biopsy samples with 2000 genes (features).

A multi-objective evolutionary-rough feature selection algorithm [168] has been used for classifying microarray gene expression patterns. Since the data typically consist of a large number of redundant features, an initial redundancy reduction of the attributes is done to enable faster convergence. Thereafter

[||] http://llmpp.nih.gov/lymphoma/data/figure1/figure1.cdt
[**] http://research.nhgri.nih.gov/microarray/Supplement/
[††] http://genome-www.stanford.edu/hcc/

rough set theory is employed to generate reducts, which represent the minimal sets of nonredundant features capable of discerning between all objects, in a multi-objective framework.

For a decision table $\mathcal{A}$ with N condition attributes and a single decision attribute d, the problem of finding a d-reduct is equivalent to finding a minimal subset of columns $R(\subseteq \{1, 2, \cdots, N\})$ in the distinction table [cf. Definition 5, eqn. (6.29)], satisfying

$$\forall (k, j) \exists i \in R : b((k, j), i) = 1, \text{whenever } d(x_k) \neq d(x_j).$$

So, in effect, we may consider the distinction table to consist of N columns, and rows corresponding to only those object pairs (x_k, x_j) such that $d(x_k) \neq d(x_j)$. Let us call this shortened distinction table, a d-*distinction table*. Note that, as $\mathcal{A}$ is taken to be consistent, there is no row with all 0 entries in a d-distinction table.

Let the number of objects initially in the two classes be m_1 and m_2 respectively. Then the number of rows in the d-distinction table becomes $(m_1 * m_2) < \frac{m*(m-1)}{2}$, where $m_1 + m_2 = m$. Two *fitness functions* f_1 and f_2 are considered for each individual. We have

$$f_1(\vec{\nu}) = \frac{N - L_{\vec{\nu}}}{N} \qquad (6.38)$$

and

$$f_2(\vec{\nu}) = \frac{C_{\vec{\nu}}}{(m^2 - m)/2}, \qquad (6.39)$$

where $\vec{\nu}$ is the reduct candidate, $L_{\vec{\nu}}$ represents the number of 1's in $\vec{\nu}$, m is the number of objects and $C_{\vec{\nu}}$ indicates the number of object combinations $\vec{\nu}$ can discern between. The fitness function f_1 gives the candidate credit for containing less attributes (fewer 1's), while the function f_2 determines the extent to which the candidate can discern among objects.

The effectiveness of the algorithm is demonstrated on three cancer datasets, *viz.*, Colon, Lymphoma and Leukemia. In case of Leukemia data a two-genes set is selected, whereas the Colon data results in a eight-genes reduct size.

6.7.4 Gene regulatory network

Understanding of regulatory networks is crucial to the understanding of fundamental cellular processes involving growth, development, hormone secretion and cellular communication. Determination of transcriptional factors that control gene expression can offer further insight into the misregulated expressions common in many human diseases. In this section we outline some of the recent literature on the use of soft computing in the area of gene regulatory networks.

ANN

Recurrent neural network has been used to model the dynamics of gene expression [169]. The significance of the regulatory effect of one gene product on the expression of other genes of the system is defined by a weight matrix. Multigenic regulation, involving positive and/or negative feedback, is considered. The process of gene expression is described by a single network, along with a pair of linked networks independently modeling the transcription and translation schemes. Bayesian networks with Bayesian learning were employed [170], in a reverse engineering approach, to infer gene regulatory interactions from simulated gene expression data.

Adaptive Double Self-Organizing Map (ADSOM) [171] provides a clustering strategy for identifying coregulated regions of interest. These are utilized for extracting gene regulatory networks. The model has a flexible topology, and allows simultaneous visualization of clusters. DSOM combines features of SOM with two-dimensional position vectors, to provide a visualization tool for deciding on the required number of clusters. However, its free parameters are difficult to control to guarantee proper convergence. ADSOM updates these free parameters during training, and allows convergence of its position vectors to a fairly consistent number of clusters (provided its initial number of nodes is greater than the expected number of clusters). The effectiveness of AD-SOM in identifying the number of clusters is proven by applying it to publicly available gene expression data from multiple biological systems such as yeast, human and mouse.

EC

Use of GAs for reconstructing genetic networks has been reported in literature [172, 173]. Typically the GA searches for the most likely genetic networks that best fit the data, considering the set of genes to be included in the network along with the strength of their interactions.

Hybridization

Ritchie et al. [174] optimized the back propagation neural network architecture, using genetic programming, in order to improve upon the ability of ANNs to model, identify, characterize and detect nonlinear gene-gene interactions in studies of common human diseases. The performance is reported to be superior in terms of predictive ability and power to detect gene-gene interactions when nonfunctional polymorphisms are present.

Identification of protein-DNA interactions in the promoter region, in terms of DNA motifs that characterize the regulatory factors operating in the transcription of a gene, is important for recognizing genes that participate in a regulation process. This enables determination of their interconnection in a gene regulatory network. A hybrid methodology for this purpose has been developed [175] by combining ANN, fuzzy sets and multi-objective GAs. A time-delayed neural network (TDNN) learns compound binding site motifs from nonspecific DNA sequences by decomposing it into modules corresponding to submotifs. The MCC of eqn. (6.33) is used to discriminate between promoters and nonpromoters. The system can handle multiplicity of RNA polymerase targets and multiple functional binding sites in closely located regulatory regions, along with the associated uncertainty of the motifs.

6.8 Conclusion

Bioinformatics is a new area of science where a combination of statistics, molecular biology and computational methods is used for analyzing and processing biological information like gene, DNA, RNA and proteins. Improper folding of protein structure is responsible for causing many diseases. Therefore, accurate structure prediction of proteins is a major goal of study. With the availability of huge volume of high-dimensional data, there exists a lot of possibilities for the emergent field of biological data mining. Hybrid approaches, combining powerful algorithms and interactive visualization tools with the strengths of fast processors, hold promise for enhanced performance in the near future.

Soft computing, involving paradigms like FS, ANNs, EC and RS, has been used for analyzing the different protein sequences, structures and folds, microarrays as well as regulatory networks. Since this permits approximate, good solutions, instead of the high precision, globally optimum solution, it allows one to arrive at a low cost goal faster.

We have provided, in this chapter, a detailed review on the role of soft computing paradigms and their hybridizations in different aspects of Bioinformatics, mainly involving pattern recognition and data mining tasks. It is categorized based on the domain of operation, the function modeled, and the tool used. The major tasks covered include classification, clustering, feature selection and rule mining. Gene regulatory networks, a relatively new area of study, have also been surveyed.

Fuzzy sets, which constitute the oldest component of soft computing, are suitable for handling the issues related to understandability of patterns, incomplete or noisy data and human interaction, and can provide approximate solutions faster. The characteristics of adaptivity and learning help ANNs to minimize error and self-organize in data-rich environments. The low precision, approximate reasoning of fuzzy sets allows faster convergence. Exhaustive enumeration and evaluation of all gene combinations being NP-hard, EC uses intelligent, goal-directed search while optimizing a fitness function determined by the knowledge about the environment. Rough sets allow dimensionality reduction for high-dimensional data, and are found suitable in mining microarray gene expressions. Different types of hybridizations incorporate the generic and application specific merits of the constituent paradigms.

Knowledge about the domain is often found to be useful in improving performance of a system. For example, the incorporation of the alignment profile generated by Psi-BLAST was found to be advantageous in protein secondary structure determination by ANNs. This is evident from the comparative study projected in Table 6.1. Similarly, the use of prior knowledge about the secondary structure at the input enhances the performance for tertiary structure determination.

Metabolism is the chemical engine that drives a living process. By means of utilization of a vast repertoire of enzymatic reactions and transport processes, organisms process and convert thousands of organic compounds into various biomolecules necessary to support their existence. The cells as well as the organism direct the distribution and processing of metabolites throughout an extensive map of pathways. While we seek to develop strategies to effectively eliminate metabolic pathways due to microorganisms through antibiotics in order to curb bacterial infection, we also strive to enhance the performance of certain other pathways or introduce novel routes for the production of biochemicals of commercial interest. The domain of metabolic pathways and gene regulatory networks open up significant challenges for research involving application of soft computing techniques.

6.9 Exercises

1. Distinguish between fuzziness and randomness. Explain how we need fuzzy sets in everyday life.

2. (a) Draw $M_1(x) = \frac{x}{1+5(x-2)^2}$ and $M_2(x) = 2^{-x}$, for x lying in the interval $x = [0, 5]$ of real numbers.

(b) Determine graphically the membership functions $\overline{M}_1$, $\overline{M}_2$, $M_1 \sqcup M_2$, $M_1 \sqcap M_2$.

3. Write a computer program to implement the backpropagation algorithm for training a multilayer perceptron to learn a 4-input 1-output parity problem. (Odd number of ones implies a 'one' output, and vice versa.)

4. (a) What are the different kinds of learning in artificial neural networks?

(b) What do you understand by overlearning and generalization?

5. (a) Create a computer program for a crossover function that selects a crossing site in a pair of binary strings, performs single-point crossover with probability p_c and generates a pair of offsprings.

(b) Insert a mutation function in your program to toggle a bit value at a particular location, with mutation probability p_m.

(c) Study the effect of varying the mutation probability on the performance of GAs.

6. Five strings have fitness function values of 2, 4, 6, 10, 16, respectively. Write a program for roulette wheel selection to find the expected number of copies of each string in the mating pool, given that a constant population size of five is maintained.

7. Distinguish between fuzzy membership and rough approximations.

8. Write a program for protein secondary structure prediction, using a feedforward neural network with a window on the amino acid sequence at its input.

9. Simulate the c-means, fuzzy c-means and rough c-means clustering algorithms on a microarray gene expression dataset of your choice. Comment on their relative merits and demerits.

References

[1] L. A. Zadeh, "Fuzzy logic, neural networks, and soft computing," *Communications of the ACM*, vol. 37, pp. 77–84, 1994.

[2] S. K. Pal and S. Mitra, *Neuro-fuzzy Pattern Recognition: Methods in Soft Computing*. New York: John Wiley, 1999.

[3] S. Mitra and T. Acharya, *Data Mining: Multimedia, Soft Computing, and Bioinformatics*. New York: John Wiley, 2003.

[4] S. Mitra and Y. Hayashi, "Bioinformatics with soft computing," *IEEE Transactions on Systems, Man, and Cybernetics, Part C: Applications and Reviews*, vol. 36, pp. 616–635, 2006.

[5] L. A. Zadeh, "Fuzzy sets," *Information and Control*, vol. 8, pp. 338–353, 1965.

[6] L. A. Zadeh, "The concept of a linguistic variable and its application to approximate reasoning: Part 1, 2, and 3," *Information Sciences*, vol. 8, 8, 9, pp. 199–249, 301–357, 43–80, 1975.

[7] L. A. Zadeh, "Fuzzy sets as a basis for a theory of possibility," *Fuzzy Sets and Systems*, vol. 1, pp. 3–28, 1978.

[8] L. A. Zadeh, "The role of fuzzy logic in the management of uncertainty in expert systems," *Fuzzy Sets and Systems*, vol. 11, pp. 199–227, 1983.

[9] H. J. Zimmermann, *Fuzzy Sets, Decision Making and Expert Systems*. Boston, MA: Kluwer Academic Publishers, 1987.

[10] G. J. Klir and B. Yuan, *Fuzzy Sets and Fuzzy Logic: Theory and Applications*. NJ: Prentice Hall, 1995.

[11] E. H. Mamdani and S. Assilian, "An experiment in linguistic synthesis with a fuzzy logic controller," *International Journal of Man Machine Studies*, vol. 7, pp. 1–13, 1975.

[12] T. Takagi and M. Sugeno, "Fuzzy identification of systems and its application to modeling and control," *IEEE Transactions on Systems, Man, and Cybernetics*, vol. 15, pp. 116–132, 1985.

[13] J. J. Buckley and T. Feuring, *Fuzzy and Neural: Interactions and Applications.* Studies in Fuzziness and Soft Computing, Heidelberg: Physica-Verlag, 1999.

[14] S. Haykin, *Neural Networks: A Comprehensive Foundation.* New York: Macmillan College Publishing Co. Inc., 1994.

[15] J. Hertz, A. Krogh, and R. G. Palmer, *Introduction to the Theory of Neural Computation.* Reading, MA: Addison-Wesley, 1994.

[16] D. E. Rumelhart and J. L. McClelland, eds., *Parallel Distributed Processing: Explorations in the Microstructures of Cognition*, vol. 1. Cambridge, MA: MIT Press, 1986.

[17] T. Kohonen, *Self-Organization and Associative Memory.* Berlin: Springer-Verlag, 1989.

[18] J. M. Zurada, *Introduction to Artificial Neural Systems.* New York: West Publishing Company, 1992.

[19] D. R. Hush and B. G. Horne, "Progress in supervised neural networks," *IEEE Signal Processing Magazine*, pp. 8–39, January 1993.

[20] D. O. Hebb, *The Organization of Behaviour.* New York: Wiley, 1949.

[21] W. S. McCulloch and W. Pitts, "A logical calculus of the idea immanent in nervous activity," *Bulletin of Mathematical Biophysics*, vol. 5, pp. 115–133, 1943.

[22] A. B. Tickle, R. Andrews, M. Golea, and J. Diederich, "The truth will come to light: Directions and challenges in extracting the knowledge embedded within trained artificial neural networks," *IEEE Transactions on Neural Networks*, vol. 9, pp. 1057–1068, 1998.

[23] S. Mitra and Y. Hayashi, "Neuro-fuzzy rule generation: Survey in soft computing framework," *IEEE Transactions on Neural Networks*, vol. 11, pp. 748–768, 2000.

[24] L. M. Fu, "Knowledge-based connectionism for revising domain theories," *IEEE Transactions on Systems, Man, and Cybernetics*, vol. 23, pp. 173–182, 1993.

[25] G. G. Towell and J. W. Shavlik, "Knowledge-based artificial neural networks," *Artificial Intelligence*, vol. 70, pp. 119–165, 1994.

[26] F. Rosenblatt, "The perceptron: A probabilistic model for information storage and organization in the brain," *Psychological Review*, vol. 65, pp. 386–408, 1958.

[27] F. Rosenblatt, *Principles of Neurodynamics, Perceptrons and the Theory of Brain Mechanisms.* Washington: Spartan Books, 1961.

[28] M. Minsky and S. Papert, *Perceptrons: An Introduction to Computational Geometry.* Cambridge, MA: MIT Press, 1969.

[29] J. Moody and C. J. Darken, "Fast learning in networks of locally-tuned processing units," *Neural Computation*, vol. 1, pp. 281–294, 1989.

[30] D. R. Hush, B. Horne, and J. M. Salas, "Error surfaces for multilayer perceptrons," *IEEE Transactions on Systems, Man, and Cybernetics*, vol. 22, pp. 1152–1161, 1993.

[31] D. E. Goldberg, *Genetic Algorithms in Search, Optimization and Machine Learning.* Reading, MA: Addison-Wesley, 1989.

[32] Z. Michalewicz, *Genetic Algorithms + Data Structures = Evolutionary Programs.* Berlin: Springer-Verlag, 1994.

[33] D. B. Fogel, L. J. Fogel, and V. W. Porto, "Evolving neural networks," *Biological Cybernetics*, vol. 63, pp. 487–493, 1990.

[34] K. Deb, *Multi-Objective Optimization using Evolutionary Algorithms*. London: John Wiley, 2001.

[35] K. Deb, S. Agarwal, A. Pratap, and T. Meyarivan, "A fast and elitist multi-objective genetic algorithm : NSGA-II," *IEEE Transactions on Evolutionary Computation*, vol. 6, pp. 182–197, 2002.

[36] J. R. Koza, *Genetic Programming: On the Programming of Computers by Means of Natural Selection*. Cambridge, MA: MIT Press, 1992.

[37] R. Slowiński, ed., *Intelligent Decision Support, Handbook of Applications and Advances of the Rough Sets Theory*. Dordrecht: Kluwer Academic, 1992.

[38] A. Skowron and C. Rauszer, "The discernibility matrices and functions in information systems," in *Intelligent Decision Support, Handbook of Applications and Advances of the Rough Sets Theory* (R. Slowiński, ed.), pp. 331–362, Dordrecht: Kluwer Academic, 1992.

[39] J. Wroblewski, "Finding minimal reducts using genetic algorithms," Tech. Rep. 16/95, Warsaw Institute of Technology - Institute of Computer Science, Poland, 1995.

[40] C. T. Lin and C. S. George Lee, *Neural Fuzzy Systems - A Neuro-Fuzzy Synergism to Intelligent Systems*. New Jersey: Prentice Hall, 1996.

[41] W. Pedrycz, *Computational Intelligence: An Introduction*. Boca Raton, FL: CRC Press, 1998.

[42] Y. Hayashi and J. J. Buckley, "Approximations between fuzzy expert systems and neural networks," *International Journal of Approximate Reasoning*, vol. 10, pp. 63–73, 1994.

[43] J. S. R. Jang and C. T. Sun, "Functional equivalence between radial basis function networks and fuzzy inference systems," *IEEE Transactions on Neural Networks*, vol. 4, pp. 156–159, 1993.

[44] S. C. Lee and E. T. Lee, "Fuzzy neural networks," *Mathematical Biosciences*, vol. 23, pp. 151–177, 1975.

[45] S. K. Pal and S. Mitra, "Multi-layer perceptron, fuzzy sets and classification," *IEEE Transactions on Neural Networks*, vol. 3, pp. 683–697, 1992.

[46] J. K. Keller and D. J. Hunt, "Incorporating fuzzy membership functions into the perceptron algorithm," *IEEE Transactions on Pattern Analysis and Machine Intelligence*, vol. 7, pp. 693–699, 1985.

[47] J. M. Keller, R. R. Yager, and H. Tahani, "Neural network implementation of fuzzy logic," *Fuzzy Sets and Systems*, vol. 45, pp. 1–12, 1992.

[48] J. S. R. Jang, C. T. Sun, and E. Mizutani, *Neuro-Fuzzy and Soft Computing*. Englewood Cliffs, NJ: Prentice Hall, 1997.

[49] S. Mitra, "Fuzzy MLP based expert system for medical diagnosis," *Fuzzy Sets and Systems*, vol. 65, pp. 285–296, 1994.

[50] D. Nauck, F. Klawonn, and R. Kruse, *Foundations of Neuro-Fuzzy Systems*. Chichester, England: John Wiley & Sons, 1997.

[51] J. M. Keller, R. Krishnapuram, and F. C. -H. Rhee, "Evidence aggregation networks for fuzzy logic inference," *IEEE Transactions on Neural Networks*, vol. 3, pp. 761–769, 1992.

[52] S. Mitra and S. K. Pal, "Logical operation based fuzzy MLP for classification and rule generation," *Neural Networks*, vol. 7, pp. 353–373, 1994.

[53] G. A. Carpenter, S. Grossberg, N. Markuzon, J. H. Reynolds, and D. B. Rosen, "Fuzzy ARTMAP: a neural network architecture for incremental supervised learning of analog multidimensional maps," *IEEE Transactions on Neural Net-

works, vol. 3, pp. 698–713, 1992.

[54] A. Ghosh, N. R. Pal, and S. K. Pal, "Self-organization for object extraction using multilayer neural network and fuzziness measures," *IEEE Transactions on Fuzzy Systems*, vol. 1, pp. 54–68, 1993.

[55] W. Pedrycz, ed., *Fuzzy Evolutionary Computation*. Boston: Kluwer Academic, 1997.

[56] X. Yao, "A review of evolutionary artificial neural networks," *International Journal of Intelligent Systems*, vol. 8, pp. 539–567, 1993.

[57] H. Muhlenbein, "Limitations of multi-layer perceptron networks - step towards genetic neural networks," *Parallel Computing*, vol. 14, pp. 249–260, 1990.

[58] S. Bornholdt and D. Graudenz, "General asymmetric neural networks and structure design by genetic algorithms," *Neural Networks*, vol. 5, pp. 327–334, 1992.

[59] V. Maniezzo, "Genetic evolution of the topology and weight distribution of neural networks," *IEEE Transactions on Neural Networks*, vol. 5, pp. 39–53, 1994.

[60] B. A. Whitehead and T. D. Choate, "Cooperative-Competitive genetic evolution of radial basis function centers and widths for time series prediction," *IEEE Transactions on Neural Networks*, vol. 7, pp. 869–880, 1996.

[61] S. K. Pal and D. Bhandari, "Genetic algorithms with fuzzy fitness function for object extraction using cellular neural networks," *Fuzzy Sets and Systems*, vol. 65, pp. 129–139, 1994.

[62] H. Ishibuchi, M. Nii, and T. Murata, "Linguistic rule extraction from neural networks and genetic-algorithm-based rule selection," in *Proceedings of IEEE International Conference on Neural Networks*, (Houston, USA), pp. 2390–2395, 1997.

[63] R. J. Machado and A. F. da Rocha, "Evolutive fuzzy neural networks," in *Proceedings of 1st IEEE International Conference on Fuzzy Systems* (San Diego, USA), pp. 493–500, 1992.

[64] M. Banerjee and S. K. Pal, "Roughness of a fuzzy set," *Information Sciences (Informatics & Computer Science)*, vol. 93, pp. 235–246, 1996.

[65] S. K. Pal and A. Skowron, eds., *Rough Fuzzy Hybridization: A New Trend in Decision Making*. Singapore: Springer-Verlag, 1999.

[66] S. Mitra, M. Banerjee, and S. K. Pal, "Rough knowledge-based network, fuzziness and classification," *Neural Computing and Applications*, vol. 7, pp. 17–25, 1998.

[67] M. Banerjee, S. Mitra, and S. K. Pal, "Rough fuzzy MLP: Knowledge encoding and classification," *IEEE Transactions on Neural Networks*, vol. 9, pp. 1203–1216, 1998.

[68] S. Mitra, P. Mitra, and S. K. Pal, "Evolutionary modular design of rough knowledge-based network using fuzzy attributes," *Neurocomputing*, vol. 36, pp. 45–66, 2001.

[69] S. K. Pal, S. Mitra, and P. Mitra, "Rough Fuzzy MLP: Modular evolution, rule generation and evaluation," *IEEE Transactions on Knowledge and Data Engineering*, vol. 15, pp. 14–25, 2003.

[70] C. H. Wu and J. W. McLarty, *Neural Networks and Genome Informatics*, vol. 1 of *Methods in Computational Biology and Biochemistry*. Amsterdam: Elsevier, 2000.

[71] L. P. Wang and X. Fu, *Data Mining with Computational Intelligence*. Berlin: Springer, 2005.

[72] G. Fogel and D. Corne, eds., *Evolutionary Computation in Bioinformatics*. San Francisco: Morgan Kaufmann, 2002.

[73] J. Casasnovas and F. Rosselló, "Averaging fuzzy biopolymers," *Fuzzy Sets and Systems*, vol. 152, pp. 139–158, 2005.

[74] L. Pickert, I. Reuter, F. Klawonn, and E. Wingender, "Transcription regulatory region analysis using signal detection and fuzzy clustering," *Bioinformatics*, vol. 14, pp. 244–251, 1998.

[75] R. Farber, A. Lapedes, and K. Sirotkin, "Determination of eukaryotic protein coding regions using neural networks and information theory," *Journal of Molecular Biology*, vol. 226, pp. 471–479, 1992.

[76] G. Schneider, S. Rohlk, and P. Wrede, "Analysis of cleavage-site patterns in protein precursor sequences with a perceptron-type neural network," *Biochem Biophys Res Commun*, vol. 194, pp. 951–959, 1993.

[77] E. C. Uberbacher, Y. Xu, and R. J. Mural, "Discovering and understanding genes in human DNA sequence using GRAIL," *Methods Enzymol*, vol. 266, pp. 259–281, 1996.

[78] S. Brunak, J. Engelbrecht, and S. Knudsen, "Prediction of human mRNA donor and acceptor sites from the DNA sequence," *Journal of Molecular Biology*, vol. 220, pp. 49–65, 1991.

[79] N. I. Larsen, J. Engelbrecht, and S. Brunak, "Analysis of eukaryotic promoter sequences reveals a systematically occurring CT-signal," *Nucleic Acids Res*, vol. 23, pp. 1223–1230, 1995.

[80] A. G. Pedersen and H. Nielsen, "Neural network prediction of translation initiation sites in eukaryotes: Perspectives for EST and genome analysis," *ISMB*, vol. 5, pp. 226–233, 1997.

[81] C. H. Wu, M. Berry, S. Shivakumar, and J. McLarty, "Neural networks for full-scale protein sequence classification: Sequence encoding with singular value decomposition," *Machine Learning*, vol. 21, pp. 177–193, 1995.

[82] A. Browne, B. D. Hudson, D. C. Whitley, M. G. Ford, and P. Picton, "Biological data mining with neural networks: Implementation and application of a flexible decision tree extraction algorithm to genomic problem domains," *Neurocomputing*, vol. 57, pp. 275–293, 2004.

[83] R. Setiono, "Extracting rules from neural networks by pruning and hidden-unit splitting," *Neural Computation*, vol. 9, pp. 205–225, 1997.

[84] J. Hanke and J. G. Reich, "Kohonen map as a visualization tool for the analysis of protein sequences: Multiple alignments, domains and segments of secondary structures," *Comput Applic Biosci*, vol. 6, pp. 447–454, 1996.

[85] Y. D. Cai, H. Yu, and K. C. Chou, "Artificial neural network method for predicting HIV protease cleavage sites in protein," *J. Protein Chem*, vol. 17, pp. 607–615, 1998.

[86] Y. D. Cai, H. Yu, and K. C. Chou, "Prediction of beta-turns," *J. Protein Chem*, vol. 17, pp. 363–376, 1998.

[87] J. Schuchhardt, G. Schneider, J. Reichelt, D. Schomberg, and P. Wrede, "Local structural motifs of protein backbones are classified by self-organizing neural networks," *Protein Eng*, vol. 9, pp. 833–842, 1996.

[88] P. Arrigo, F. Giuliano, F. Scalia, A. Rapallo, and G. Damiani, "Identification of a new motif on nucleic acid sequence data using Kohonen's self organizing map," *Comput Appl Biosci*, vol. 7, pp. 353–357, 1991.

[89] E. A. Ferran, B. Pflugfelder, and P. Ferrara, "Self-organized neural maps of human protein sequences," *Protein Sci*, vol. 3, pp. 507–521, 1994.

[90] H. C. Wang, J. Dopazo, L. G. de la Fraga, Y. P. Zhu, and J. M. Carazo, "Self-organizing tree-growing network for the classification of protein sequences," *Protein Sci*, vol. 7, pp. 2613–2622, 1998.

[91] H. C. Wang, J. Dopazo, and J. M. Carazo, "Self-organizing tree-growing network for classifying amino acids," *Bioinformatics*, vol. 14, pp. 376–377, 1998.

[92] J. Dopazo and J. M. Carazo, "Phylogenetic reconstruction using an unsupervised growing neural network that adopts the topology of a phylogenetic tree," *Journal of Molecular Evolution*, vol. 44, pp. 226–233, 1997.

[93] R. Thomson, T. C. Hodgman, Z. R. Yang, and A. K. Doyle, "Characterizing proteolytic cleavage site activity using bio-basis function neural networks," *Bioinformatics*, vol. 19, pp. 1741–1747, 2003.

[94] C. LeBlanc, C. R. Katholi, T. R. Unnasch, and S. I. Hruska, "DNA sequence analysis using hierarchical ART-based classification networks," *ISMB*, vol. 2, pp. 253–260, 1994.

[95] C. H. Wu, H. L. Chen, and S. C. Chen, "Counter-propagation neural networks for molecular sequence classification: Supervised LVQ and dynamic node allocation," *Applied Intelligence*, vol. 7, pp. 27–38, 1997.

[96] E. E. Snyder and G. D. Stormo, "Identification of protein coding regions in genomic DNA," *Journal of Molecular Biology*, vol. 248, pp. 1–18, 1995.

[97] D. Wang and G. B. Huang, "Protein sequence classification using extreme learning machine," in *Proceedings of the International Joint Conference on Neural Networks (IJCNN'05)*, (Montreal, Canada), August 2005.

[98] C. Notredame and D. G. Higgins, "SAGA: Sequence alignment by genetic algorithm," *Nucleic Acids Research*, vol. 24, pp. 1515–1524, 1996.

[99] C. Notredame, L. Holm, and D. G. Higgins, "COFFEE: An objective function for multiple sequence alignments," *Bioinformatics*, vol. 14, pp. 407–422, 1998.

[100] P. O. Lewis, "A genetic algorithm for maximum-likelihood phylogeny inference using nucleotide sequence data," *Molecular Biology and Evolution*, vol. 15, pp. 277–283, 1998.

[101] M. J. Brauer, M. T. Holder, L. A. Dries, D. J. Zwickl, P. O. Lewis, and D. M. Hillis, "Genetic algorithms and parallel processing in maximum-likelihood phylogeny inference," *Molecular Biology and Evolution*, vol. 19, pp. 1717–1726, 2002.

[102] D. Howard and K. Benson, "Evolutionary computation method for pattern recognition of cis-acting sites," *BioSystems*, vol. 72, pp. 19–27, 2003.

[103] B. C. H. Chang and S. K. Halgamuge, "Protein motif extraction with neuro-fuzzy optimization," *Bioinformatics*, vol. 18, pp. 1084–1090, 2002.

[104] H. Berman, J. Westbrook, Z. Feng, G. Gilliland, T. Bhat, H. Weissing, I. Shindyalov, and P. Bourne, "The Protein Data Bank," *Nucleic Acids Research*, vol. 28, pp. 235–242, 2000.

[105] D. Pelta, N. Krasnogor, C. Bousono-Calzon, J. L. Verdegay, J. Hirst, and E. Burke, "A fuzzy sets based generalization of contact maps for the overlap of protein structures," *Fuzzy Sets and Systems*, vol. 152, pp. 103–123, 2005.

[106] P. Chou and G. Fasmann, "Prediction of the secondary structure of proteins from their amino acid sequence," *Advances in Enzymology*, vol. 47, pp. 45–148, 1978.

[107] N. Qian and T. Sejnowski, "Predicting the secondary structure of globular proteins using neural network models," *Journal of Molecular Biology*, vol. 202,

pp. 865–884, 1988.

[108] B. Rost and C. Sander, "Prediction of protein secondary structure at better than 70% accuracy," *Journal of Molecular Biology*, vol. 232, pp. 584–599, 1993.

[109] B. Rost, "PHD: Predicting one-dimensional protein structure by profile-based neural networks," *Methods in Enzymology*, vol. 266, pp. 525–539, 1996.

[110] Z. Sun, X. Rao, L. Peng, and D. Xu, "Prediction of protein supersecondary structures based on the artificial neural network method," *Protein Eng*, vol. 10, pp. 763–769, 1997.

[111] H. Bohr, J. Bohr, S. Brunak, R. M. J. Cotterill, and H. Fredholm, "A novel approach to prediction of the 3-dimensional structures of protein backbones by neural networks," *FEBS Letters*, vol. 261, pp. 43–46, 1990.

[112] G. L. Wilcox, M. O. Poliac, and M. N. Liebman, "Neural network analysis of protein tertiary structure," *Tetrahedron Comp Meth*, vol. 3, pp. 191–211, 1991.

[113] O. Lund, K. Frimand, J. Gorodkin, H. Bohr, J. Bohr, J. Hansen, and S. Brunak, "Protein distance constraints predicted by neural networks and probability distance functions," *Protein Eng*, vol. 10, pp. 1241–1248, 1997.

[114] M. Milik, A. Kolinski, and J. Skolnick, "Neural network system for the evaluation of side-chain packing in protein structures," *Protein Eng*, vol. 8, pp. 225–236, 1995.

[115] J. M. Chandonia and M. Karplus, "Neural networks for secondary structure and structural class predictions," *Protein Sci*, vol. 4, pp. 275–285, 1995.

[116] I. Dubchak, I. Muchnik, S. R. Holbrook, and S. H. Kim, "Prediction of protein folding class using global description of amino acid sequence," *Proceedings of National Academy of Sciences USA*, vol. 92, pp. 8700–8704, 1995.

[117] A. Gutteridge, G. J. Bartlett, and J. M. Thornton, "Using a neural network and spatial clustering to predict the location of active sites in enzymes," *Journal of Molecular Biology*, vol. 330, pp. 719–734, 2003.

[118] R. Casadio, M. Compiani, P. Fariselli, and F. Vivarelli, "Predicting free energy contributions to the conformational stability of folded proteins," *Intelligent Systems for Molecular Biology*, vol. 3, pp. 81–88, 1995.

[119] S. K. Riis and A. Krogh, "Improving prediction of protein secondary structure using structured neural networks and multiple sequence alignments," *Journal of Computational Biology*, vol. 3, pp. 163–183, 1996.

[120] D. T. Jones, "Protein secondary structure prediction based on position-specific scoring matrices," *Journal of Molecular Biology*, vol. 292, pp. 195–202, 1999.

[121] H. Kaur and G. P. S. Raghava, "Role of evolutionary information in prediction of aromatic-backbone NH interactions in proteins," *FEBS Letters*, vol. 564, pp. 47–57, 2004.

[122] G. Pollastri, D. Przybylski, B. Rost, and P. Baldi, "Improving the prediction of protein secondary structure in three and eight classes using recurrent neural networks and profiles," *Proteins: Structure, Function, and Genetics*, vol. 47, pp. 228–235, 2002.

[123] P. Baldi, S. Brunak, P. Frasconi, G. Pollastri, and G. Soda, "Exploiting the past and the future in protein secondary structure prediction," *Bioinformatics*, vol. 15, pp. 937–946, 1999.

[124] J. D. Szustakowski and Z. Weng, "Protein structure alignment using a genetic algorithm," *Proteins*, vol. 38, pp. 428–440, 2000.

[125] S. Schulze-Kremer, "Genetic algorithms for protein tertiary structure prediction," in *Parallel Problem Solving from Nature II* (R. Männer and B. Manderick,

eds.), pp. 391–400, Amsterdam: North Holland, 1992.

[126] R. König and T. Dandekar, "Improving genetic algorithms for protein folding simulations by systematic crossover," *BioSystems*, vol. 50, pp. 17–25, 1999.

[127] J. Pedersen and J. Moult, "Protein folding simulations with genetic algorithms and a detailed molecular description," *Journal of Molecular Biology*, vol. 269, pp. 240–259, 1997.

[128] G. Jones, P. Willett, R. C. Glen, A. R. Leach, and R. Taylor, "Development and validation of a genetic algorithm for flexible docking," *Journal of Molecular Biology*, vol. 267, pp. 727–748, 1997.

[129] G. Morris, D. Goodsell, R. Halliday, R. Huey, W. Hart, R. Belew, and A. Olson, "Automated docking using Lamarckian genetic algorithm and an empirical binding free energy function," *Journal of Computational Chemistry*, vol. 19, pp. 1639–1662, 1998.

[130] J. M. Yang and C. C. Chen, "GEMDOCK: A generic evolutionary method for molecular docking," *Proteins: Structure, Function, and Bioinformatics*, vol. 55, pp. 288–304, 2004.

[131] M. Raymer, P. Sanschagrin, W. Punch, S. Venkataraman, E. Goodman, and L. Kuhn, "Predicting conserved water-mediated and polar ligand interactions in proteins using a k-nearest neighbors genetic algorithm," *Journal of Molecular Biology*, vol. 265, pp. 445–464, 1997.

[132] J. Desjarlais and T. Handel, "De novo design of the hydrophobic cores of proteins," *Protein Sci*, vol. 4, pp. 2006–2018, 1995.

[133] B. Noelting, D. Juelich, W. Vonau, and K. Andert, "Evolutionary computer programming of protein folding and structure predictions," *Journal of Theoretical Biology*, 2004.

[134] R. Maclin and J. W. Shavlik, "Using knowledge-based neural network to improve algorithms: Refining Chou-Fasman algorithm for protein folding," *Machine Learning*, vol. 11, pp. 195–215, 1993.

[135] F. Vivarelli, G. Giusti, M. Villani, R. Campanini, P. Fariselli, M. Compiani, and R. Casadio, "LGANN: A parallel system combining a local genetic algorithm and neural networks for the prediction of secondary structure of proteins," *Comput Appl Biosci*, vol. 11, pp. 253–260, 1995.

[136] A. K. Jain and R. C. Dubes, *Algorithms for Clustering Data*. NJ: Prentice Hall, 1988.

[137] K. Y. Yeung, C. Fraley, A. Murua, A. E. Raftery, and W. L. Ruzzo, "Model-based clustering and data transformations for gene expression data," *Bioinformatics*, vol. 17, pp. 977–987, 2001.

[138] J. C. Bezdek, *Pattern Recognition with Fuzzy Objective Function Algorithms*. New York: Plenum Press, 1981.

[139] D. Dembele and P. Kastner, "Fuzzy c-means method for clustering microarray data," *Bioinformatics*, vol. 19, pp. 973–980, 2003.

[140] A. P. Gasch and M. B. Eisen, "Exploring the conditional coregulation of yeast gene expression through fuzzy k-means clustering," *Genome Biology*, vol. 3, pp. research0059.1–0059.22, 2002.

[141] C. S. Möller-Levet, F. Klawonn, K. H. Cho, H. Yin, and O. Wolkenhauer, "Clustering of unevenly sampled gene expression time-series data," *Fuzzy Sets and Systems*, vol. 152, pp. 49–66, 2005.

[142] R. Lukac, K. N. Plataniotis, B. Smolka, and A. N. Venetsanopoulos, "cDNA microarray image processing fuzzy vector filtering framework," *Fuzzy Sets and*

Systems, vol. 152, pp. 17–35, 2005.

[143] K. Torkkola, R. M. Gardner, T. Kaysser-Kranich, and C. Ma, "Self-organizing maps in mining gene expression data," *Information Sciences*, vol. 139, pp. 79–96, 2001.

[144] P. Tamayo, D. Slonim, J. Mesirov, Q. Zhu, S. Kitareewan, E. Smitrovsky, E. S. Lander, and T. R. Golub, "Interpreting patterns of gene expression with self-organizing maps: Methods and applications to hematopoietic differentiation," *Proceedings of National Academy of Sciences USA*, vol. 96, pp. 2907–2912, 1999.

[145] P. Törönen, M. Kolehmainen, G. Wong, and E. Castrén, "Analysis of gene expression data using self-organizing maps," *FEBS Letters*, vol. 451, pp. 142–146, 1999.

[146] J. Herrero, A. Valencia, and J. Dopazo, "A hierarchical unsupervised growing neural network for clustering gene expression patterns," *Bioinformatics*, vol. 17, pp. 126–136, 2001.

[147] M. Sultan, D. A. Wigle, C. A. Cumbaa, M. Maziarz, J. Glasgow, M. S. Tsao, and I. Jurisica, "Binary tree-structured vector quantization approach to clustering and visualizing microarray data," *Bioinformatics*, vol. 18 Suppl. 1, pp. S111–S119, 2002.

[148] S. B. Cho and J. Ryu, "Classifying gene expression data of cancer using classifier ensemble with mutually exclusive features," *Proceedings of the IEEE*, vol. 90, pp. 1744–1753, 2002.

[149] S. Bicciato, M. Pandin, G. Didonè, and C. Di Bello, "Pattern identification and classification in gene expression data using an autoassociative neural network model," *Biotechnology and Bioengineering*, vol. 81, pp. 594–606, 2003.

[150] A. Kelemen and Y. Liang, "Bayesian regularized neural network for multiple gene expression pattern classification," in *Proceedings of the International Joint Conference on Neural Networks*, pp. 1:654–1:659, July 2003.

[151] K. Deb and A. Raji Reddy, "Reliable classification of two-class cancer data using evolutionary algorithms," *BioSystems*, vol. 72, pp. 111–129, 2003.

[152] E. Falkenauer, *Genetic Algorithms and Grouping Problems*. Chichester: John Wiley, 1998.

[153] J. M. Deutsch, "Evolutionary algorithms for finding optimal gene sets in microarray prediction," *Bioinformatics*, vol. 19, pp. 45–52, 2003.

[154] S. Kirkpatrick, C. D. Gelatt Jr., and M. P. Vecchi, "Optimization by simulated annealing," *Science*, vol. 220, pp. 671–680, 1983.

[155] A. V. Lukashin and R. Fuchs, "Analysis of temporal gene expression profiles: Clustering by simulated annealing and determining the optimal number of clusters," *Bioinformatics*, vol. 17, pp. 405–414, 2001.

[156] S. Bleuler, A. Prelić, and E. Zitzler, "An EA framework for biclustering of gene expression data," in *Proceedings of Congress on Evolutionary Computation*, pp. 166–173, 2004.

[157] S. Mitra and H. Banka, "Multi-objective evolutionary biclustering of gene expression data," *Pattern Recognition*, 2006.

[158] D. Mateos, J. C. Riquelme, and J. S. Aguilar-Ruiz, "Evolutionary segmentation of yeast genome," in *Proceedings of the ACM Symposium on Applied Computing*, pp. 1026–1027, March 2004.

[159] Z. Pawlak, *Rough Sets, Theoretical Aspects of Reasoning about Data*. Dordrecht: Kluwer Academic, 1991.

[160] H. Midelfart, A. Lægreid, and J. Komorowski, *Classification of gene expression*

data in an ontology, vol. 2199 of *Lecture Notes in Computer Science*, pp. 186–194. Berlin: Springer-Verlag, 2001.

[161] G. A. Carpenter, S. Grossberg, and D. B. Rosen, "Fuzzy ART: fast stable learning and categorization of analog patterns by an adaptive resonance system," *Neural Networks*, vol. 4, pp. 759–771, 1991.

[162] S. Tomida, T. Hanai, H. Honda, and T. Kobayashi, "Analysis of expression profile using fuzzy adaptive resonance theory," *Bioinformatics*, vol. 18, pp. 1073–1083, 2002.

[163] M. E. Futschik, A. Reeve, and N. Kasabov, "Evolving connectionist systems for knowledge discovery from gene expression data of cancer tissue," *Artificial Intelligence in Medicine*, vol. 28, pp. 165–189, 2003.

[164] F. Chu, W. Xie, and L. Wang, "Gene selection and cancer classification using a fuzzy neural network," in *Proceedings of 2004 Annual Meeting of the North American Fuzzy Information Processing Society (NAFIPS 2004)*, vol. 2, pp. 555–559, 2004.

[165] H. Midelfart, J. Komorowski, K. Nørsett, F. Yadetie, A. K. Sandvik, and A. Lægreid, "Learning rough set classifiers from gene expressions and clinical data," *Fundamenta Informaticae*, vol. 53, pp. 155–183, 2002.

[166] P. Lingras and C. West, "Interval set clustering of Web users with rough k-means," Technical Report No. 2002-002, Dept. of Mathematics and Computer Science, St. Mary's University, Halifax, Canada, 2002.

[167] S. Mitra, "An evolutionary rough partitive clustering," *Pattern Recognition Letters*, vol. 25, pp. 1439–1449, 2004.

[168] M. Banerjee, S. Mitra, and H. Banka, "Evolutionary-rough feature selection in gene expression data," *IEEE Transactions on Systems, Man, and Cybernetics, Part C: Applications and Reviews*, 2007.

[169] J. Vohradsky, "Neural network model of gene expression," *FASEB Journal*, vol. 15, pp. 846–854, 2001.

[170] D. Husmeier, "Sensitivity and specificity of inferring genetic regulatory interactions from microarray experiments with dynamic bayesian networks," *Bioinformatics*, vol. 19, pp. 2271–2282, 2003.

[171] H. Ressom, D. Wang, and P. Natarajan, "Clustering gene expression data using adaptive double self-organizing map," *Physiol. Genomics*, vol. 14, pp. 35–46, 2003.

[172] S. Kikuchi, D. Tominaga, M. Arita, K. Takahashi, and M. Tomita, "Dynamic modeling of genetic networks using genetic algorithm and S-system," *Bioinformatics*, vol. 19, pp. 643–650, 2003.

[173] M. Xiong, J. Li, and X. Fang, "Identification of genetic networks," *Genetics*, vol. 166, pp. 1037–1052, 2004.

[174] M. D. Ritchie, B. C. White, J. S. Parker, L. Hahn, and J. H. Moore, "Optimization of neural network architecture using genetic programming improves detection and modeling of gene-gene interactions in studies of human diseases," *BMC Bioinformatics*, vol. 4, pp. 28–36, 2003.

[175] V. Cotik, R. Romero Zaliz, and I. Zwir, "A hybrid promoter analysis methodology for prokaryotic genomes," *Fuzzy Sets and Systems*, vol. 152, pp. 83–102, 2005.

CHAPTER 7

Connections between Machine Learning and Bioinformatics

In the previous chapters, we have described a number of machine learning techniques relevant to bioinformatics and mentioned particular applications. In this chapter, we describe connections between bioinformatics and machine learning from a different perspective—we describe several broad problem areas within bioinformatics and methods for solving problems in these areas. In some cases, especially for sequence analysis, we consider algorithms that were developed independent of the machine learning community. Nevertheless, we describe how these relate to similar ideas in machine learning. In other cases, we find direct applications of standard machine learning algorithms or specializations of them to particular contexts. We focus on three problem areas: DNA and amino acid sequence analysis, gene expression analysis and network inference.

7.1 Sequence Analysis

Perhaps the area most brought to mind by the term bioinformatics is sequence analysis, including problems such as *sequence alignment*, *gene-finding*, identifying intron-exon boundaries, protein structure prediction and identifying transcription factor binding sites. Techniques for solving these problems date back to the 1970's, and many were developed from ideas of string processing in computer science. However, there are strong mathematical connections between standard sequence analysis algorithms and ideas in machine learning and statistics. Methods for probabilistic modeling of discrete data are especially prevalent, including Hidden Markov Models (HMMs), as emphasized by Durbin et al. [21]. We discuss some of these connections in this section, focussing on two paradigmatic problems: sequence alignment and the identification of transcription factor binding sites. Protein structure prediction is discussed at greater length in Chapters 6 and 8. For material on gene- and exon-finding see, for example, [15, 14, 48].

(A)   TGTTGTGTTGAGGTCACGTGCTATTAGCGCTGG
       TGTATCTACTCACAGCAGATAGGCTCTGC
       TGTGTCGGACTCATTGGGGCCGCCGG
       TTGTCGGAGTTACCTGAAGTAGCACGG

(B) TGTTGTGTTGAGGTCACGTGCTATTAGCGCTGG   (C) TGTTGTGTTGAGGTCACGTGCTATTAGCGCTGG
     | | | | | | | | | | | | | | | | | | | | | | | | | | | | | | | | | | | | | | | | | | | | | | | | | | | | | | | | | | | |
     T-T-GT-CGGAG-TTACCTGA-AGTAGCAC-GG   TTGTCGGAGTTACCTGA-AGTAGCACGG

(D) TGTTGTGTTGAGGTCACGTGCTATTAGCGCTGG   (E) TGTTGTGTTGAGGTCACGTGCTATTAGCGCTGG
     | | | | | | | | | | | | | | | | | | | | | | | | | | | | | TGTATCTACTCACAGCAGATAGGCTCTGC
     TGTA-TCT--AC-TCACAGCAGATAGGCTCTGC | | | |
     | | | | | | | | | | | | | | | | | | | | | | | | | | | | | TGTGTCGGACTCATTGGGGCCGCCGG
     TGT-GTC-GGAC-TCAT-TGGGGCC-GC-C-GG | | |
     | | | | | | | | | | | | | | | | | | | | | | | | | | | | | TTGTCGGAGTTACCTGAAGTAGCACGG
     T-T-GTC-GGA-GTTACCTGAAGT-AGCAC-GG

Figure 7.1 *Different types of sequence alignment. (A) Four DNA strings generated by subjecting a common ancestor string (not shown) to 10 random point mutations, insertions and delections. (B) A global pairwise alignment of the first two strings. Characters connected by a vertical line are said to be* aligned. *(C) A local pairwise alignment of the first two strings. (D) A global multiple alignment of all four strings. (E) A local multiple alignment of all four strings. All alignments shown are optimal with respect to the scoring function $S(a,b)$ defined by three rules: $S(a,b) = +5$ if $a = b$; $S(a,b) = -5$ for pairs (A,G), (G,A), (C,T) and (T,C); $S(a,b) = -7$ for all other pairs. The gap penalty function is linear, $\gamma(i) = -10i$.*

7.1.1 Pairwise sequence alignment

We discuss four variants of the sequence alignment problem, depending on whether a *global* or *local* alignment is sought, and depending on whether a single pair of sequences or multiple sequences (three or more) are to be aligned. Figure 7.1 gives examples of these four different types of alignment. Intuitively, the goal of alignment is to put into correspondence parts of the strings that are "similar." Why is this relevant to biology? Imagine two species that are evolutionarily related, so that at some time in the past there was a single species that was their common ancestor. Consider a gene in that ancestor that is passed on to each of the descendent species. As generations pass, each species' copy of that gene may change. Common ways in which DNA sequences changes over time are *point mutations*, in which single nucleotides are replace by different ones, *insertions*, in which one or more nucleotides are added to the middle of the sequence, and *deletions*, in which one or more nucleotides disappear from the middle of the sequence. Such changes are especially likely to occur in parts of the sequence that are not functional, such as introns, parts of the regulatory region between transcription factor binding sites or even, for protein-coding genes, bases in the coding region whose alteration does not change the amino acid sequence of the protein. By aligning the DNA for the two genes, we identify those parts that have not been changed by evolution or that have changed less than other parts. These are good candidates for functional sequence, which helps us identifying the meaningful parts of the gene. This reasoning assumes that changes to functional sequence are on average deleterious to an organism's fitness, which appears to be generally true. Such sequence is said to be under *negative selection*. There are, however, a few known cases of *positive selection*, in which DNA sequences appear to be evolving even faster than nonfunctional sequence, whose accumulated mutations over time are said to follow *neutral drift.*

In the *global pairwise sequence alignment* problem we are given two sequences $x = x_1x_2 \ldots x_n$ and $y = y_1y_2 \ldots y_m$ from a finite alphabet Σ. In DNA sequence alignment, the alphabet would be

$$\Sigma_{DNA} = \{A, C, G, T\} \, ,$$

whereas in amino acid sequence alignment (that is, peptide or protein) alignment, the alphabet would be the standard 1-letter codes for amino acids

$$\Sigma_{AA} = \{A, R, N, D, C, E, Q, G, H, I, L, K, M, F, P, S, T, W, Y, V\} \, .$$

Formally, an alignment is a pair of sequences $x' = x'_1x'_2 \ldots x'_p$ and $y' = y'_1y'_2 \ldots y'_p$ that can be obtained by inserting zero or more gap characters, '−', between each letter of x and y respectively, as well as at the start and end of either sequence. Figure 7.1B shows an example. Observe that two letters can be aligned without being identical. Figure 7.1B, for example, includes a T from the first string aligned with a C from the second string, a T aligned with a G, a C aligned with a T, and so on. A *gap* in either x' or y' is a sequence of one or more consecutive gap characters. Observe that while x and y may be of different lengths, x' and y' must be of the same length.

Alignments are scored based on two factors: a scoring or substitution matrix $S : \Sigma \times \Sigma \to \Re$ and a gap penalty function $\gamma : \{1, 2, 3, \ldots\} \to \Re$. $S(a, b)$ gives a score, which may be positive or negative, for aligning a character a from sequence x with a character b from sequence y. This component of the alignment's score is summed over all characters that are not in correspondence with a gap:

$$\sum_{i=1}^{p} \begin{cases} S(x'_i, y'_i) & \text{if } x'_i \in \Sigma \text{ and } y'_i \in \Sigma \\ 0 & \text{otherwise} \end{cases} \tag{7.1}$$

The gap penalty function gives a negative score for each gap, depending on its length. The overall score of an alignment is then the sum of its S-component (Equation 7.1) and the penalty for each gap appearing in either x' or y'.

A *local pairwise alignment* of x and y is a (global) alignment of a consecutive subsequence $x'' = x_i \ldots x_j$ of x with a consecutive subsequence $y'' = y_k \ldots y_l$ of y. Figure 7.1C shows an example. Local alignments are scored the same as global alignments, except that the score is computed only over the portions aligned.

In this formulation, the scoring matrix S and the gap penalty function can be arbitrary. In practice, these scores often reflect biological beliefs about the relative frequency of different types of evolutionary events. Suppose, for example, that the two sequences to be aligned are separated by 1 million years of evolution. For $a, b \in \Sigma$, $S(a, b)$ could be chosen to represent the log probability that a common ancestral position evolves to the character a in one lineage and b in the other lineage. If there were no gaps in an alignment, x' and y', then the score of the alignment would simply be the sum of scores for aligned letters. This corresponds to the log probability that the entire string x and string y evolved from the same common ancestral sequence over the course of the intervening 1 million years. Alternatively, the score $S(a, b)$ could represent the log probability of evolving a and b from a common ancestor minus the log probability of seeing an a and a b in unrelated positions. The score of the alignment would then be a log-odds ratio for the hypothesis of common ancestry versus the hypothesis of unrelatedness. The presence of gaps in the alignment makes interpreting the score of an alignment more complicated. For aligning amino acid sequences, the PAM and more recent BLOSUM substitution matrices are standard choices. There is less agreement on DNA subtitution matrices and how to estimate them. See Gu and Li [24] for discussion.

Optimal pairwise sequence alignments can be computed by simple dynamic programming algorithms—the Needleman-Wunsch algorithm for global alignments [38] and the Smith-Waterman algorithm for local alignments [52]. For simplicity, we restrict attention to the case that the gap penalty function is linear: $\gamma(i) = -ci$ for some $c > 0$. An affine gap penalty, $\gamma(i) = -c_0 i - c_1$ for $c_0, c_1 > 0$, is more common in practice and can be handled with a simple extension. The general Needleman-Wunsch and Smith-Waterman algorithms allow for arbitrary concave gap penalty functions.

The Needleman-Wunsch algorithm computes the scores of optimal alignments of every prefix of x with every prefix of y. Let $H_{i,j}$ be the score of the optimal alignment of $x_1 x_2 \ldots x_i$ with $y_1 y_2 \ldots y_j$. We include the special cases of $H_{0,0}$, meaning an alignment between two empty sequences, $H_{i,0}$ denoting an alignment of the first i characters of x with an empty sequence, and $H_{0,j}$ denoting an alignment of an empty sequence with the first j characters of y. The scores of the special cases are trivially computed as

$$H_{0,0} = 0 \qquad H_{i,0} = -ci \qquad H_{0,j} = -cj$$

For $i, j \geq 1$, the optimal alignment of $x_1 x_2 \ldots x_i$ with $y_1 y_2 \ldots y_j$ either ends with x_i aligned to y_j, x_i aligned with a gap character, or y_j aligned with a gap character. By maximizing over these three possibilities, the score for the optimal alignment of those prefixes can be written recursively as

$$H_{i,j} = \max(H_{i-1,j-1} + S(x_i, y_j), \ H_{i-1,j} + c, \ H_{i,j-1} + c) \, .$$

This recursive formula can be used in a straightforward dynamic program to compute all the $H_{i,j}$. By additionally keeping track of which of the three possibilities resulted in the maximum value for each $H_{i,j}$, the optimal global alignment can be reconstructed by starting from $H_{n,m}$ and working backwards.

This algorithm was used to compute the global pairwise alignment shown in Figure 7.1B. The algorithm is efficient, requiring only $O(mn)$ space and time.

The Smith-Waterman algorithm solves the local alignment problem similarly. Let $H_{i,j}$ denote the optimal global alignment of any suffixes of the strings $x_1 \ldots x_i$ and $y_1 \ldots y_j$, allowing for the empty suffixes. That is, $H_{i,j}$ is the score of a global alignment of $x_{i'} \ldots x_i$ and $y_{j'} \ldots y_j$ for some $1 \leq i' \leq i+1$ and $1 \leq j' \leq j+1$. The cases $i' = i+1$ or $j' = j+1$ are interpreted as all of $x_1 \ldots x_i$ or all of $y_1 \ldots y_j$ being discarded. An optimal local alignment of x and y is found as an optimal global alignment of suffixes of $x_1 \ldots x_i$ and $y_1 \ldots y_j$, for some i and j. The $H_{i,j}$ are initialized as

$$H_{0,0} = H_{i,0} = H_{0,j} = 0 \ .$$

The recurrence for computing the $H_{i,j}$ is slightly modified from the Needleman-Wunsch algorithm, to allow for the fact that the best alignment of suffixes of $x_1 \ldots x_i$ with $y_1 \ldots y_j$ may be simply empty sequences, which have score zero.

$$H_{i,j} = \max(H_{i-1,j-1} + S(x_i, y_j), \ H_{i-1,j} + c, \ H_{i,j-1} + c, 0) \ .$$

By keeping track of which of the four branches resulted in the maximum value for each $H_{i,j}$ we can reconstruct the optimal local alignment. However, we do not start at $H_{n,m}$. Rather, we find the i and j for which $H_{i,j}$ is maximal and work backwards from there. Intuitively, then, the dynamic programming determines the best prefixes of x and y to discard, and beginning reconstruction at the maximal $H_{i,j}$ determines the best suffixes of x and y to discard. The time and space complexity of the algorithm is $O(mn)$, just as for global pairwise sequence alignment.

7.1.2 Multiple sequence alignment

In the *global multiple sequence alignment* problem, we are given $m \geq 3$ sequences to align: $x_1 = x_{11}x_{12} \ldots x_{1n_1}$, $x_2 = x_{21}x_{22} \ldots x_{2n_2}$, $\ldots$, $x_m = x_{m1}x_{m2} \ldots x_{mn_m}$. The alignment is a set of sequences $x'_1, x'_2, \ldots x'_{n'}$ all of the same length n', comprising the sequences $x_1, x_2, \ldots x_m$ with zero or more gap characters inserted between each character and at the start and end of each sequence, as shown in Figure 7.1D. A *local multiple sequence alignment* of the same sequences is a global alignment of substrings of each sequence—that is, a global alignment after a prefix and suffix of each sequence is discarded (see Figure 7.1E). Alignments are usually scored by the sum-of-pairs scheme, in which the score of a multiple alignment is simply the sum of the scores of all pairwise alignments implied by the multiple alignment.

The Needleman-Wunsch and Smith-Waterman dynamic programming algorithms can be generalized to compute optimal multiple sequence alignments. For the case of global alignments, we use an m-dimensional table, with $H_{i_1 i_2 \ldots i_m}$ denoting the score of the best global alignment of the prefixes $x_{11} \ldots x_{1i_1}$,

$x_{21} \ldots x_{2i_2}, \ldots, x_{m1} \ldots x_{mi_m}$. Initialization is similar to the pairwise case. The recurrence for computing $H_{i_1 i_2 \ldots i_m}$ is a maximum over all $O(2^m)$ possibilities for the inclusion or noninclusion of the final character of each suffix in the final column of the alignment (the alternative being a gap character and the necessity of including the last character in some previous column). We omit spelling out the recurrence. The algorithm takes $O(\Pi_{i=1}^m n_i)$ space and $O(2^m \Pi_{i=1}^m n_i)$ time, as does the similar, generalized Smith-Waterman algorithm for local multiple alignment. These methods were used to compute the alignments shown in Figure 7.1D,E. However, the computations become prohibitive when there are too many sequences to align or when they are too long. Indeed, solving the multiple sequence alignment problem optimally has been shown NP-compete [59], and so heuristic solutions are employed in practice.

Currently, the most common heuristic approach is embodied by the CLUSTAL family of algorithms, [26, 17]. In the CLUSTAL approach, pairwise alignments are first used to create a distance matrix between the sequences, from which a phylogenetic tree is estimated. The multiple alignment is then built from the bottom up by a series of pairwise alignments, according to the structure of the tree. In essence, at each internal node of the (binary) phylogenetic tree, an alignment is made between the sequences on the left branch and the sequences on the right branch. The sequences on each branch, however, are represented compactly by a consensus sequence or a profile, so that the alignment can be done in an efficient pairwise manner. There are different methods for aligning consensus sequences or profiles, and recent versions of CLUSTAL improve the quality of the alignments by attending to a number of biologically relevant details such as the lengths of branches in the tree, position-dependent gap penalties and weights that specify the importance of each input sequence [57].

7.1.3 Identification of transcription factor binding sites

Recall that transcriptional regulation is a primary means by which gene expression is regulated. A transcription factor affects transcription when it binds the DNA at a transcription factor binding site (TFBS), a stretch of DNA usually between 6 and 20 nucleotides long. Each transcription factor binds to different TFBSs, and different TFBSs for the same transcription factor generally do not have identical nucleotide sequences. Finding these TFBSs in the DNA is a major challenge both experimentally and computationally, and is a key step in understanding the functioning of transcriptional regulatory networks.

Computational methods for representing and identifying the characteristic patterns of bases to which a given transcription factor binds began to be developed in the 1970's. (For reviews see [54, 13].) There are two main variants of the problem. In one, we are given a set of nucleotide sequences of known binding sites for a given transcription factor and are asked to create a model which can then be used to predict, or identify, other binding sites for the factor. In the second variant, we are given a set of nucleotide sequences which contain, or are believed to contain, binding sites for a given transcription factor, but we are not told exactly where those binding sites are. In the latter case, we may also be given examples sequences that are believed to not contain any binding sites for the transcription factor. Here we focus on the first variant, deferring discussion of the second variant and other extensions to Section 7.3.2.

Once a set of TFBSs for a particular transcription factor is experimentally determined, the pattern of nucleotides to which the factor binds can be statistically characterized. This is usually done by a *position weight matrix* (PWM), also known as a *position-specific scoring matrix* (PSSM). A PWM has 4 rows, corresponding to the nucleotides A, C, G, T and as many columns as the length of the experimentally determined TFBSs. In the simplest approach, each entry $M_{i,j}$ is set to the empirical frequency of nucleotide i in position j among the experimentally determined TFBSs. This models the distribution of the experimental TFBSs under the assumption that the positions are statistically independent, and employs the maximum likelihood estimate for the probabilities of different nucleotides within each position. This model can be viewed as a trivial HMM in which the hidden state deterministically transitions through a series of states $s_1, s_2, \ldots, s_n$, where n is the number of columns of the matrix and $M_{i,j}$ is the probability of observing nucleotide i when the state of the chain is j. The probability that the transcription factor would bind to any particular stretch of DNA $s_1 s_2 \ldots s_n \in \{A, C, G, T\}$, or more properly, the probability that the TFBS model would generate such a sequence, can be computed as $\Pi_{j=1}^{n} M_{s_j,j}$. In a variant of the PWM idea, the matrix entries are the logarithms of the empirical probabilities, $M'_{i,j} = \log M_{i,j}$, so that the log probability of a transcription factor binding to a stretch of the DNA is $\sum_{j=1}^{n} M'_{s_j,j}$. Another variant considers $M''_{i,j} = \log(M_{i,j}/p_i) = \log M_{i,j} - \log p_i$, where p_i denotes the probability of nucleotide i under some background model. Then the summation $\sum_{j=1}^{n} M''_{s_j,j}$ can be interpreted as a log-odds ratio that the DNA sequence is generated from the TFBS model versus the background model. (Alternatively, the p_i may be interpreted as relating to binding energies. See Stormo [54] for a more detailed discussion of these and related models.) Any of these models can be used to score candidate length-n stretches of DNA in an effort to identify previously unknown TFBSs for the transcription factor.

One potential problem is that when there are relatively few experimentally determined TFBSs, some entries of the matrix M may be zero. (M' or M'' would have $-\infty$ entries.) If $M_{i,j} = 0$, then the model predicts that the transcription factor would never bind to a DNA sequence with nucleotide i in position j. While this may be correct in some cases, the general observation is that transcription factor binding is not so strict, and zeros in the PWM can result in serious false negatives. One solution to this problem is to replace the maximum likelihood parameter estimates with Bayesian parameter estimates, one approach being to assume Dirichlet priors over the $M_{i,j}$, resulting in a nonnegative estimated probability for binding any DNA sequence. Other methods of introducing smoothing or *pseudocounts* to the $M_{i,j}$ based more on biological principles, such as nucleotide mutation rates, have a similar effect and possibly better accuracy. (See Henikoff & Henikoff [25] for one example and for further discussion.) It appears that the assumption that the positions j are statistically independent is not a significant limitation. Attempts at fitting models that allow for dependencies between positions, such as digram models [55] or artificial neural networks [27], resulted in little or no improvement in prediction performance.

On a genome scale, the identification of TFBSs based solely on PWM scores, whether smoothed (e.g., by pseudocounts) or not, suffers from too high a false positive rate. If one scores every length-n segment of the DNA of an organism and selects only those segments with score higher than the score of the lowest-scoring experimentally determined TFBS, many thousands or millions of candidate TFBSs can easily result [54]. Most of these sites are not actually bound by the factor, probably for reasons related to cofactors or chromatin that are not readily apparent from the DNA sequence itself.

Any of several observations can dramatically reduce the false positive rate for TFBS identification. First, functional TFBSs tend to be near a gene, in particular, near to the transcription initiation site. Second, TFBSs tend to occur in modules—stretches of DNA that include multiple binding sites for the same or a variety of transcription factors [19, 7]. Thus, our confidence in a particular identification can be increased if other candidate TFBSs are nearby. Third, comparative genomics approaches, which look for sequence fragments conserved in the promoter regions of orthologous genes across species, can help to identify functional TFBSs [9]. Various schemes based on these observations have been proposed, ranging from simple filtering criterion to more sophisticated statistical models. While PWMs for representing the distribution of nucleotides at binding sites remain the state of the art, the best methods for combining PWM scores with other types of evidence—and exactly what those other types of evidence should be—remain a topic of active research.

7.2 Analysis of High-Throughput Gene Expression Data

Problems related to sequence analysis and their solutions grew slowly in complexity and sophistication over the course of several decades, as more sequence data became available and as our scientific understanding of that data grew. By contrast, the production of quantitative gene expression data was revolutionized in the mid-1990s by the invention of high-throughput expression assays, especially RNA expression microarrays and SAGE. These technologies offer great promise for understanding system-wide phenomena as well as hunting down genes associated with different cellular response or diseases.

However, the invention of these technologies also created an urgent demand for techniques for handling the many machine learning and statistical challenges posed by the data. First, there is significant "noise" in the data, interpreted broadly to include natural variation in mRNA levels and unintentional variations in equipment, collection and measurement protocols and execution. A second challenge is the large number of variables measured (mRNA expression values for thousands or tens of thousands of genes) compared to the number of samples (typically in the tens or low hundreds). Statistical significance is a major and recurring concern in many analyses of gene expression data. Third, when constructing models or predictions based on so many different variables, it can be difficult to extract human-useful knowledge or understanding. Fourth, though the number of variables measured by these technologies is truly astounding, they do not come close to measuring the abundances of all biologically relevant quantities, such as proteins, metabolites, etc. Finally, samples average expression levels over populations of cells, let alone different compartments of cells, potentially obscuring much important information. While there are technological solutions to some of these problems (e.g., microfluidic approaches for measuring cell-specific expression), a wide variety of machine learning methods have been successfully employed to address the challenges in analyzing this data. Indeed, the area remains one of the most actively researched, with hundreds, if not thousands, of papers published each year. We begin our discussion with techniques for handling noise in one-color microarrays specifically, as they are the most commonly used of the high-throughput expression measurement technologies. We then proceed to methods suitable for addressing the other problems mentioned above, regardless of the particular technology generating the data.

7.2.1 Gene expression normalization

There are numerous sources of variability in one-color microarray data (and in all other high-throughput technologies as well). Variability in chip manufacture, in sample collection, in sample processing, in scanning and in the technicians and machines carrying out the analysis all contribute noise to the data. The true variations in mRNA concentrations between samples, which are the signals of interest, must be estimated while accounting for this noise. The initial stages of microarray data extraction and cleaning have become standardized in recent years. First, low-level image processes techniques are used to extract probe-level intensities for each probe on each chip [53]. (Recall that microarray chips contain a large number of spots, the probes, where RNAs bind and are quantitated by fluorescent imaging.) Next, a series of normalization steps synthesizes these readings to produce an estimated expression level for each gene on each chip. Most normalization methods are based on the supposition that most genes do not have different expression levels in different samples. Thus, the simplest normalization techniques rescale the probe intensities extracted from each chip to have the same mean or the same mean and variance (see [10] for several variants). The most popular normalization routine is RMA [29], an acronym for Robust Multichip Average. RMA has three steps: background "subtraction" and quantile normalization (both applied at the probe-level) and a method to integrate probe values to produce single expression values for each gene. Background subtraction accounts for effects such as nonspecific binding or systematic bias in readings between chips. Naive background subtraction methods literally subtract an estimated background signal level from the probe intensities, but this can lead to negative adjusted probe intensities, which is especially problematic if one later intends to transform expression values to a logarithmic scale. Instead, RMA assumes the measured intensity of probe j for gene n on chip i is $P_{ijn} = s_{ijn} + b_{ijn}$, where s_{ijn} is the true biological signal and b_{ijn} is confounding noise. Assuming that for each array the s_{ijn} are distributed exponentially and the b_{ijn} are distributed normally, the background-adjusted intensities are computed as the expectation $B(P_{ijn}) = E(s_{ijn}|P_{ijn})$. Background subtraction adjusts the probe intensities for each chip separately. The second step, quantile normalization, operates on all chips simultaneously. In quantile normalization, the ith largest probe intensity on each chip is set equal to the mean of the ith largest probe intensities across chips. After this step, the adjusted probe intensities on each chip have the exact same empirical distribution. Finally, RMA takes the logarithm of the background-adjusted quantile-normalized probe intensities, which we denote by Y_{ijn} and obtains a log expression score μ_{in} for each gene n on each chip i using the model

$$Y_{ijn} = \mu_{in} + \alpha_{jn} + \epsilon_{ijn} ,$$

where α_{jn} is a probe-specific effect and ϵ_{ijn} is a mean zero noise term. If there are I arrays, J probes per gene and N genes, there are $I \times J \times N$ such equations. A straightforward least squares approach could be used to solve for the $(I + J) \times N$ unknowns (the μ_{in} and α_{jn}), though RMA uses the more robust median polish technique to avoid problems with outlier probes.

7.2.2 Gene expression analysis

Once the expression data matrix is obtained (we assume rows correspond to genes and columns correspond to samples), subsequent analysis depends on the questions the researcher wants to answer. A common starting point is to plot or visualize the data. Clustering and dimensionality-reduction techniques are useful in this regard—and may be useful as preprocessing for further analysis. It has become almost obligatory for publications involving microarray data to include a heat map of the data matrix with the rows of the matrix organized according to a hierarchical clustering, typically constructed by single- or average-linkage agglomerative clustering. (For an example based on the yeast cell cycle data of Spellman et al. [53], see Figure 7.2A.) A dendrogram

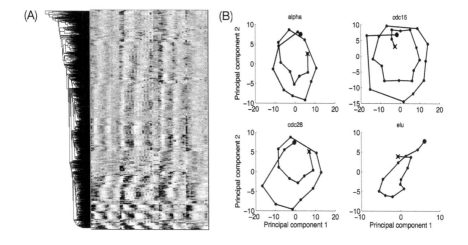

Figure 7.2 *(A) Heatmap and dendrogram for hierarchical clustering of microarray data [53]. Rows correspond to 800 genes estimated to be cell-cycle regulated. Columns correspond to 73 experimental conditions, corresponding to four time series in which the yeast cells progressed through the cell cycle. The clustering is an average-link agglomerative clustering based on Euclidean distance between expression profiles of genes (rows of the data matrix). (B) Plots of each time series projected on to the first two principal components (computed based on all the data). Dots represent samples and lines connect the time series. A large dot marks the start of each time series and an 'X' marks the end of each time series. The title of each plot refers to the means used to synchronize the cells at the same stage of the cell cycle.*

alongside the matrix indicates the structure of the hierarchical clustering, and the branch heights indicate the dissimilarity of the clusters joined. If the samples are not naturally ordered—for example, they correspond to independent samples and not a time series—then the columns are often clustered as well.

Dimensionality reduction can be used to reduce either the number of rows or the number of columns in the data matrix. In the former case, which is the more common of the two, we construct a new coordinate basis for representing different samples. When the basis has two or three dimensions, the samples can be plotted in that space, giving a visual sense of similarity between the samples and the distribution of the data. For example, when the time series comprising the Spellman et al. data [53] are plotted in two dimensions, the series' samples trace out one or two cycles around the origin. More medically relevant applications include, for example, using principal components analysis to visualize similarities between different kinds of stem cells [39], and using self-organizing maps to classify cancerous tumors [19]. Dimensionality reduction and clustering, especially partitional clustering, can also be useful preprocessing steps for solving prediction or network inference problems. By reducing the number of independent variables, one reduces the chance of

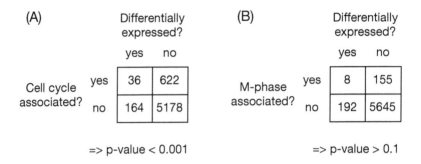

Figure 7.3 *A hypothetical example of testing for gene enrichment. In a yeast (Saccharomyces cerevisiae) microarray experiment, 200 genes are found to be differentially expressed between two experimental conditions, out of 6000 genes total. The experimenter believes that cell cycle effects may explain a statistically significant fraction of the differential expressed genes. (A) Shows a 2-by-2 contingency table in which each gene falls into exactly one category, depending on whether or not it was differentially expressed and whether or not it is known to be associated with the yeast cell cycle (according, for example, to the Gene Ontology[18]). The null hypothesis of no association is rejected with a p-value of less than 0.001 according to Fisher's exact test or a Chi-square test, confirming the experimenter's suspicion. The experimenter further believes that the differentially expressed genes may be associated with the M phase of the cell cycle. (B) This hypothesis is not supported, however, as there is insufficient evidence to reject a null hypothesis of no association to known M phase associated genes (B).*

overfitting and can avoid situations in which the learning problem is under-constrained.

Another approach to understanding microarray data is to test for the *differential expression* of genes between two or more conditions. If there are multiple replicates from each of a pair of conditions, then simple t-tests, applied to each gene, can determine which genes show a statistically significantly different mean expression level. A slightly more sophisticated approach that has been widely adopted is *Significance Analysis of Microarrays* (SAM) [58]. It modifies the t-statistic slightly by including a correction to the sample standard deviation that accounts for greater variability in measurements of genes expression at a low level. Permutation tests can be used to estimate the false discovery rate. Other elaborations are aimed at the situation of very few replicates (e.g., [5, 12]). The ANOVA can be used to detect significant changes in differential expression among three or more conditions [31]. These approaches generate lists of genes which appear to be differentially expressed, providing a starting point for further biological investigation.

When lists of differentially expressed genes are very large—say, hundreds of genes—extracting useful information is challenging. One approach is *gene en-*

richment, where the goal is to identify known *classes* of genes that appear over-represented in the differentially-expressed list. (See Figure 7.3 for an example.) Often, the classes of genes are taken from an ontology, such as the Gene Ontology (GO) database [18], which defines groups of genes from the very general (e.g., genes involved in development) to the specific (e.g., proteins found on the presynaptic membranes of neurons). The microarray data are analyzed to produce a list of genes that are differentially expressed among the conditions. For any particular gene class, a two-by-two contingency table can be constructed, where each gene either is or is not in the class and each gene is or is not on the differentially expressed list. From the contingency table, the statistical significance of an association between the differentially expressed list and the genes in the class can be determined, for example by Fisher's exact test or a chi-square test. When the number of gene classes tested is large, one must be wary of the multiple comparison problem. Further, the tests for each gene class are not independent. The classes in GO, for example, are organized into a hierarchy. If a gene class is disproportionately represented, *enriched*, in the differentially expressed list, then it is more likely that either parents or children of that gene class are also enriched. One way to address these problems is to employ a False Discovery Rate framework, or similar approach to multiple testing, in conjunction with permutation tests to account for correlations in the statistics for different gene classes [62]. See Khatri and Draghici [32] for a recent comparison of different techniques.

7.3 Network Inference

One of the ultimate aims of molecular biology is to identify all the chemical entities involved in life and how they relate or interact. In particular, the problem of *genetic network inference* is to identify regulatory relationships between genes, often mediated by transcription factors, that control so much of cellular function. Techniques in sequence analysis, especially for the discovery of transcription factor binding sites, are one avenue for solving this problem. In this section, we describe a variety of further techniques. First, we consider techniques based on expression data alone. Then we describe techniques that integrate different types of data (e.g., sequence and expression data) to estimate regulatory relationships, as well as mentioning a few approaches for inferring other types of networks, such as protein-protein interaction networks.

7.3.1 Inference from expression data

One of the dreams inspired by microarray data and other high-throughput technologies is that, simply by observing the state of a biological system under different conditions or over time, it should be possible to infer the existence and nature of the interactions between components of the system (e.g., genes or proteins). The *structure* of such a network specifies which genes directly transcriptionally regulate which genes, and can be represented simply by lists, or an interaction matrix, or by a directed graph.The problem of inferring this structure from expression data, or more detailed knowledge of interactions, is also called *reverse engineering* the network or *solving the inverse problem*. Indeed, the goals are similar to ideas of reverse engineering or system identification in the engineering community, though the tools used are more commonly drawn from statistics, machine learning and dynamical modeling.

There are several major caveats to the enterprise, however, which must be kept in mind. First, no technology allows us to observe the full unadulterated state of the system. For example, often mRNA expression is measured but not protein expression, or vice versa. Spatial information, at the subcellular, cellular or tissue levels, is often lost. Further, numerous sources of noise confound the measurements. Second, the chemistry of regulation is so complex that any model we formulate glosses over many details and would never be able to exactly capture the behavior of the system. Often, the approximations made in modeling are gross, and the mismatch between the model formulation and biological reality can lead to erroneous conclusions about both the structure and details of the real system. Finally, it turns out that many genes have strongly correlated expression patterns, which makes it difficult to discriminate among different candidate regulators – an instance of the classic caveat about confusing correlation with causation.

When the data for network inference is a matrix of expression values for a set of genes over a set of conditions, the problem of network inference can be treated as a general problem of testing for association between variables and modeling relationships between them. We discuss four categories of approaches: (1) pairwise tests of assocation, (2) modeling one gene's expression as a function of multiple other genes, (3) joint, or simultaneous, modeling of all the genes with a probabilistic network model and (4) dynamical modeling. These techniques may be applied directly when the number of genes measured is not too large compared to the number of samples. However, most of these approaches make little sense when applied to microarray data or the like, as the problem is vastly underconstrained and the results would not be significant. For such data, it is common to cluster genes and try to model (aggregate) regulatory relationships between clusters of genes, based on prototypical expression patterns for the clusters.

Pairwise tests of assocation: One heuristic for detecting regulatory relationships between genes is to look for genes whose expression is correlated—the reasoning being that one of the genes may be directly activating the other (e.g., [37, 6]). There are a number of problems with this reasoning, and the *co-expression* networks that result are not always viewed as causal explanations of the data. For the moment, however, let us pursue this heuristic. There are many standard tests for a significant association between variables based on a set of observations. For real-valued data, the linear correlation coefficient can

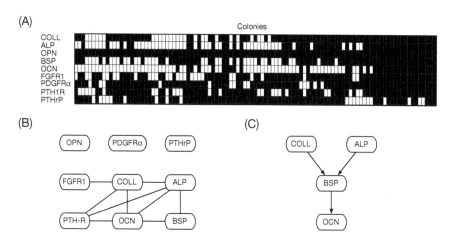

Figure 7.4 *(A) A binary gene expression matrix concerning nine genes (rows) in a set of independent cultures of osteoblasts (patterned after Figure 2 of Madras et al. [34]). Shaded boxes indicate genes that were expressed in each colonies; white boxes indicate genes that were not expressed. (B) Fisher's exact test, a pairwise test of association, with a stringent p-value suggests pairs of genes whose expression is correlated more than would be expected by chance [34]. Several genes have no associations. OPN, for example, becomes expressed before the other genes and is turned on in every colony, so there is no statistical basis for associating it to the other genes. (C) A Bayesian network analysis of four of the genes with the strongest pairwise associations, COLL, ALP, BSP and OCN, suggests a different and more specific relationship among them, with COLL and ALP regulating BSP and BSP regulating OCN [37]. There is insufficient evidence for regulation between COLL and ALP, as suggested by the pairwise tests. More importantly, the COLL-OCN and ALP-OCN correlations in the pairwise model are attributed to indirect effects, through BSP, in the Bayesian network model.*

be used to quantify degree of association, and p-values for the observed degree of association under the null hypothesis of no assocation can be derived in a number of ways. For discretized expression data, tests such as the chi-square, Fisher's exact test or a test based on mutual information are all possibilities. (See Figure 7.4 for an example, and Nagarajan et al. [37] for more detailed analyses.) The first two tests directly produce a p-value for the hypothesis of no association between the variables. The mutual information between two random variables, estimated based on the observed samples, measures strength of association but not significance. Typically, a permutation test is used to assess significance of the mutual information score.

The advantages of pairwise tests are their simplicity and computational efficiency. They can be applied to microarray data, for example, to test for associations between every pair of genes in an organism, even though the number of pairs can easily be in the tens of millions. Only the mutual information test, with its extra burden of the permutations to determine significance, is difficult to apply on such a large scale at this time. Of course, testing many pairs of genes leads to a multiple testing problem, and when applied to microarray data, the results should not be seriously interpreted as reflecting biological reality. Even with very strict p-value thresholds, many false positives are likely to result. However, broad statistical patterns may be of interest (e.g., [39]). One of the drawbacks to the pairwise association approach is that, at least for the tests mentioned above, it does not discriminate which gene regulates which. The tests are symmetric, so if gene pair (x, y) scores highly, so does the pair (y, x). While there are pairs of genes that in reality directly regulate each other, it is far from being the common case. The fact that many genes have highly correlated expression patterns creates several problems for the pairwise association approach. For one thing, if we have a set of highly correlated genes, the pairwise tests would suggest that every pair in the set has a regulatory relationship—a biologically implausible scenario. Further, even if there is a true regulator x of gene y among the set, it would be difficult to identify the correct gene x from among the set of correlated ones. Finally, if there are two genes x and y which together regulate z, but neither x or y are individually strongly correlated with z, the pairwise approach will not find the true regulators. This seems to be a fairly common situation, as regulators may be active only under specific conditions or in certain combinations. The individual correlations are therefore not strong. Indeed, a more common explanation for two genes having correlated expression patterns, called being *coexpressed*, is that the gene share some common regulators—that is, they are *coregulated*. This has motivated another approach to network inference in which genes are first tested for coexpression, and then the promoter regions of coexpressed genes are examined for signatures of common regulators. We discuss this approach in Section 7.3.2.

Predicting each gene's expression based on the expression of other genes: A more general approach than testing for pairwise associations is to solve the network inference problem as a set of regression problems (e.g., [33, 20]). For each gene i, we ask that its expression – column i in the gene expression matrix – be predicted based on the expression of the other genes. This produces a set of N univariate regression problems, each with $N - 1$ input attributes. The regression problem may be solved by any technique one wishes, but often something simple such as linear regression or logistic regression, or naive Bayes. The regulators of gene i are taken to be those genes whose expression values are used by the predictor for gene i's expression. Thus, it is important that the predictor be regularized or somehow sparse, in the sense that it selects a subset of all possible attributes to make its predictions. Otherwise, if we employ a nonsparse predictor and the regression problem is not degenerate, we could easily end up concluding that every gene regulates every other gene. Conversely, if we employ no regularization and the data matrix comes from a microarray experiment or something similar, there are so many combinations of genes that could perfectly predict any given column in the matrix that we would obtain a arbitrary, meaningless set of regulators for each gene. The prediction approach in principle improves upon the pairwise approach in that it allows us to discover combinations of regulators that explain well the behavior of a gene even if each individual factor does not. The prediction approach can be computationally more involved than the pairwise approach, depending on the exact method, but is generally considered tractable.

The prediction formulation of the problem has also been popular for theoretical analysis, where one of the main questions has been how much data are needed to infer network structure and/or parameters. Often, many simplifying assumptions on the form of the data and the inference procedure are made for the sake of the analysis. For example, Liang et al. [33] proposed the REVEAL algorithm for the case that the expression matrix is binary and noiseless, in which, for every gene i, the smallest set of other genes is sought such that the entropy of i's expression conditioned on the entropy of the other genes' expression is zero. The expected number of samples needed to successfully infer the network structure has been studied under a variety of assumptions on the form of the data, including randomly sampled data [2, 40, 30], time series data [41, 40] and knock-out or overexpression data [1, 28]. However, these algorithms assume noiseless data and have not been applied directly in practice.

Joint modeling of the gene expression distribution: Given the oft-cited noisiness of gene expression data and the emphasis on inferring the structure of the network, rather than just parameters, graphical probabilistic models such as Bayeisan networks are a natural formalism for attempting network inference (e.g., [22, 37]). Applying standard Bayesian network structure learning methods to the gene expression matrix results in a candidate network structure. (See Figure 7.4 for an example.) Learned Bayesian networks are a more complete description of the data distribution than obtained by the prediction approach described above. Further, Bayesian network inference methods (meaning computation of probabilities conditioned on observations, in contrast to inference of the network structure) allow principled estimation of unobserved variables and prediction of network state in response to perturbations such as gene knock outs or overexpression. A drawback is that Bayesian network structure learning is a computationally difficult problem, so that for all but the smallest networks, heuristics must be employed to search the space of allowed structures (namely, all acyclic graphs with vertices corresponding to variables). However, variations on the standard structure learning approaches, specifically designed for inferring biological networks, have resulted in both more efficient learning and more accurate learned models (e.g., [50, 43]). A drawback to the Bayesian network formalism is that it requires the structure of the network to be acyclic. This is a poor match for biological networks, in which feedback loops abound. One solution is to focus on dynamical Bayesian networks, which allow feedback loops while maintaining acyclicity, by encoding them as influences from variables at previous time steps on variables at subsequent time steps. We discuss this option further below.

Dynamical modeling of gene expression: Techniques for modeling dynamical systems apply when the expression data comprises one or more time series. Time series analysis methods from statistics, specifically autoregressive modeling, are the easiest to apply. In the form of discrete-time model, an autoregressive-1 model, the expression of all genes at every time point is used to predict the expression of all genes at the next time point. The problem is thus one of multiple regression (or N univariate regression problems), which can be solved in the same manner as described above in the prediction approach; the only difference is in the definition of the inputs and outputs for the predictors. This approach is computationally efficient, like the prediction approach. Because the model can be simulated, it can also be used to make predictions, like the probabilistic modeling approach. However, there is no ambiguity about the direction of regulatory inference (as in a Markov network) and there is no problem with feedback loops (as in a Bayesian network), as the latter can be represented, for example, by allowing the expression of genes x and y at time t to influence the predictions of the expression of genes y and x respectively at the next time point. Of course, far more sophisticated time series models are possible, such as autoregressive or moving-average models with longer histories (so that the prediction for the next time point depends on expression values in the last several time points), or dynamical Bayes networks [36]. One should keep in mind, however, that expression time series are often quite short, 10 time points or fewer and samples may not be spaced evenly in time, whereas many time series modeling techniques are designed for longer series and assume temporally-uniform sampling.

The major alternative to discrete-time models is what are called differential equation models [35, 16]. These models have traditionally been preferred over discrete-time models by the mathematical and computational biology communities, due to their greater realism—expression levels and concentrations in real biological systems change continuously in time, and are not discretely clocked—though the preference is also partly due to influence from physics and chemistry, in which many early mathematical and computational biologists were trained. In a typical ordinary differential equation model, the time derivatives of the expression of all genes at time t, $x(t)$, are given by some parameterized function $\dot{x}(t) = f(x(t), \theta)$. Given an initial condition, $x(0)$, the differential equation can be solved either analytically or numerically, and the parameters θ can be optimized to minimize some notion of error between the model trajectory(-ies) and the observed expression series. Like the time-series models and probabilistic models, differential equation models can be used to make predictions about the effects of perturbations to the system. It is also easy to accommodate temporally-nonuniform time series. A disadvantage is that fitting differential equation models in this way is computationally difficult. The relationships between the error function and model parameters is often highly nonlinear, with many plateaus and local optima. Sophisticated optimization algorithms with long running times, such as simulated annealing, are the norm. However, advances in functional data analysis [45] offer much more computationally tractable means for fitting the parameters of differential equation models. The core idea is to create a smooth or interpolation of the data, $y(t)$ over the whole range of time for which data is observed, and fit the differential equation to minimize the mismatch between $\dot{y}(t)$ and $f(y(t), \theta)$. This is more or less a regression problem, and so can be solved comparatively efficiently—easily hundreds of times faster than the traditional formulation of the fitting problem (see [42] for an application employing similar ideas).

7.3.2 Integrative modeling

There are limitations to what can be deduced from expression data alone, due to factors such as: unobserved variables, noise, inaccuracy in the formulation of the model and even inherent nonidentifiability. Conversely, there are many other sources of information that are useful modeling networks, especially in deducing network structure. Possibilities include: the genome sequence of the organism and related organisms, especially the sequences of promoter regions where transcription factors bind; localization data, such as from chip-chip experiments; chromatin modification data; and, of course, the wealth of knowledge already published in the biological literature. From this situation arose the philosophy of integrative biology—the search for methods that can somehow combine evidence about network structure or function from two or more qualitatively different types of data.

One body of work that exemplifies this idea is the search for evidence of co-regulation among genes that are empirically co-expressed. Recent large-scale studies lend support to the idea that co-expressed genes tend to share regulators (e.g., [3]). However, algorithms for finding potential regulatory elements common to the promoter sequences of a given set of genes stretch back decades. (See [54] for a recent review.) When the set of genes is based on high-throughput expression data, a common approach is to begin by finding a partitional clustering of the genes. Each gene in a cluster thus shares similarities in expression pattern, which may be due to shared regulators. The promoter regions of the genes in the cluster are analyzed for signatures of shared regulation. These nucleotide signatures may be as simple as a consensus sequence [60], a consensus sequence with wildcard or "don't care" characters [46, 11], or position weight matrices (perhaps representing a TFBS) [4, 47, 56]. In one formulation of the problem, we seek maximally specific patterns that are common to all or a specified fraction of the promoters in the gene cluster. Maximally specific may refer to the length of a consensus pattern, for example, or the entropy of a position weight matrix (which should be low). In another formulation of the problem, we seek patterns that are common to the promoters in the gene cluster but absent, mostly absent or unlikely in a set of promoter sequences from genes not in the cluster or in a background sequence model.

Several studies have proposed methods for network inference that incorporate other types of data. No standard algorithm or model has yet emerged, in part because each study aims at integrating different types of data. As such, we simply describe several representative studies. In one trend-setting paper, Segal et al. [49] developed a *probabilistic relational model* (PRM), which can be viewed as a compact representation of a Bayesian network, that combined evidence from promoter sequences, localization data and microarray expression data. Learning the parameters of the PRM implicitly determined transcriptional regulatory relationships between genes. In another paper, Segal et al. [51] combined expression data with binary protein-protein interaction data using a *Markov network*, to identify "pathways." Bernard and Hartemink [8] used dynamic Bayesian networks to model expression time series data in conjunction with localization data. The network modeled the distribution of the time series data, while the localization data was incorporated through the learning scheme, via prior probabilities over structure space. While these studies share their use of probabilistic modeling formalisms to combine different data sources, theirs is not the only approach. Qi et al. [44], for example, approach the problem of estimating protein-protein interaction networks by formulating it as a classification problem. For each pair of proteins, the desired output is binary—whether or not the proteins interact. A classier system (in their case, a combination of random forests and k-nearest neighbor) is trained based on known interactions, and can take any number of different types of evidence as input, including direct experimental evidence and indirect evidence based on expression data, shared functional annotations, sequence data, etc.

7.4 Exercises

1. The Needleman-Wunsch algorithm for global pairwise alignment requires space $O(mn)$ where m and n are the lengths of the strings to be aligned. Suppose we were interested only in the score of the optimal alignment, and not the alignment itself. Describe how to modify the algorithm to use only $O(\min(m, n))$ space.

2. Generalize our description of the Needleman-Wunsch algorithm to allow for an affine gap penalty function.

3. Suppose that a set of binding sites for a transcription factor has been experimental determined, having the sequences: GGATTT, GGATTA, GCATTA, CGAATA, GCAATA, GGTTTA and GCACTA. **(a)** Determine a consensus sequence for these binding sites by identifying the most commonly occurring nucleotide in each position. **(b)** Assume a null hypothesis in which a genome of length 10^9 nucleotides is generated randomly with each character i.i.d. and equally likely. What is the expected number of occurrences of the consensus sequence? (c) Calculate a position weight matrix in which the entries are the log probabilities of each nucleotide (A,C, G or T) occurring in each position.

4. Obtain the Spellman et al. [53] yeast cell cycle data. Compute the principal components of the whole data set, not just those based on the 800 genes estimated to be cell cycle regulated. Plot each time series along the first two principal components. How do your plots compare to those in Figure 7.2B?

5. Generate a random binary data matrix of n rows and 6 columns as follows. Let X_{i1} through X_{i6} represent the entries of row i. Let X_{i1} be 1 with probability p and 0 otherwise. Let $X_{i2} = X_{i1}$ with probability p and the opposite otherwise. Let $X_{i3} = X_{i1}$ OR X_{i2}. Let $X_{i4} = X_{i1}$ AND NOT X_{i2}. Let $X_{i5} = X_{i1}$ XOR X_{i2}. Let X_{i6} be 0 or 1 with equal probability, independent of the other variables. The correct dependencies among the random variable is given by the influence diagram below.

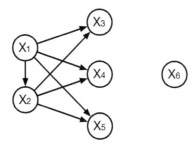

(**a**) Use a pairwise dependency test, such as Fisher's exact test or a mutual information-based test, to detect statistically significant associations between the variables. Which dependencies are correctly identified, which are missed and which are falsely asserted? How do the answers to these questions depend on n and p?

(**b**) Use a Bayesian network structure learning procedure to learn the relationships between the variables. Again, which dependencies are correctly identified, which are missed and which are falsely asserted, and how do the answers to these questions depend on n and p?

References

[1] T. Akutsu, S. Kuhara, O. Maruyama, and S. Miyano. Identification of gene regulatory networks by strategic gene disruptions and gene overexpressions. In

Proceedings of the Ninth ACM-SIAM Symposium on Discrete Algorithms, pages 695–702, 1998.

[2] T. Akutsu, S. Miyano, and S. Kuhara. Identification of genetic networks from a small number of gene expression patterns under the boolean network model. In *Proceedings of the Pacific Symposium on Biocomputing*, pages 17–28, 1999.

[3] D. J. Allocco, I. S. Kohane, and A. J. Butte. Quantifying the relationship between co-expression, co-regulation and gene function. *BMC Bioinformatics*, 5:18, 2007.

[4] T. L. Bailey and C. Elkan. Unsupervised learning of multiple motifs in biopolymers using expectation maximization. *Machine Learning*, 21:51–80, 1995.

[5] P. Baldi and A. D. Long. A bayesian framework for the analysis of microarray expression data: regularized *t*-test and statistical inferences of gene changes. *Bioinformatics*, 17(6):509–519, 2001.

[6] K. Basso, A.A. Margolin, G. Stolovitzky, U. Klein, R. Dalla-Favera, and A. Califano. Reverse engineering of regulatory networks in human B cells. *Nature Genetics*, 37(4):382–390, 2005.

[7] B. P. Berman, Y. Nibu, B. D. Pfeiffer, P. Tomancak, S. E. Celniker, M. Levine, G. M. Rubin, and M. B. Eisen. Exploiting transcription factor binding site clustering to identify cis-regulatory modules involved in pattern formation in the *Drosophila* genome. *Proceedings of the National Academy of Sciences*, 99(2):757–762, 2002.

[8] A. Bernard and A. J. Hartemink. Informative structure priors: Joint learning of dynamic regulatory networks from multiple types of data. In *Pacific Symposium on Biocomputing 10*, pages 459–470, 2005.

[9] M. Blanchette and M. Tompa. Discovery of regulatory elements by a computational method for phylogenetic footprinting. *Genome Research*, 12(5):739–748, 2002.

[10] B. M. Bolstad, R. A. Irizarry, M. Astrand, and T. P. Speed. A comparison of normalization methods for high density oligonucleotide array data based on variance and bias. *Bioinformatics*, 19(2):185–193, 2003.

[11] A. Brazma, I. Jonassen, J. Vilo, and E. Ukkonen. Predicting gene regulatory elements in silico on a genomic scale. *Genome Research*, 8:1202–1215, 1998.

[12] R. Breitling, P. Armengaud, A. Amtmann, and P. Herzyk. Rank products: a simple, yet powerful, new method to detect differentially regulated genes in replicated microarray experiments. *FEBS Letters*, 573(1–3):83–92, 2004.

[13] M. L. Bulyk. Computational prediction of transcription factor binding site locations. *Genome Biology*, 5, 2003.

[14] C. Burge and S. Karlin. Prediction of complete gene structures in human genomic dna. *Journal of Molecular Biology*, 268:78–94, 1997.

[15] C. B. Burge and S. Karlin. Finding the genes in genomic dna. *Current Opinion in Structural Biology*, 8:346–354, 1998.

[16] T. Chen, H. L. He, and G. M. Church. Modeling gene expression with differential equations. In *Proceedings of the Pacific Symposium on Biocomputing*, pages 29–40, 1999.

[17] R. Chenna, H. Sugawara, T. Koike, R. Lopez, T. J. Gibson, D. G. Higgins, and J. D. Thompson. Multiple sequence alignment with the clustal series of programs. *Nucleic Acids Research*, 31:3497–3500, 2003.

[18] The Gene Ontology Consortium. Gene ontology: tool for the unification of biology. *Nature Genetics*, 25:25–29, 2000.

[19] E. M. Crowley, K. Roeder, and M. Bina. A statistical model for locating regu-

latory regions in genomic dna. *Journal of Molecular Biology*, 268(1):8–14, 1997.

[20] P. D'Haeseleer, X. Wen, S. Fuhrman, and R. Somogyi. Linear modeling of mRNA expression levels during CNS development and injury. In *Proceedings of the Pacific Symposium on Biocomputing*, pages 41–52, 1999.

[21] R. Durbin, S. Eddy, A. Krogh, and G. Mitchison. *Biological sequence analysis*. Cambridge University Press, 1998.

[22] N. Friedman, M. Linial, I. Nachman, and D. Pe'er. Using Bayesian Networks to Analyze Expression Data. *Journal of Computational Biology*, 7(3-4):601–620, 2000.

[23] T. R. Golub, D. K. Slonim, P. Tamayo, C. Huard, M. Gaasenbeek, J. P. Mesirov, H. Coller, M. L. Loh, J. R. Downing, M. A. Caligiuri, C. D. Bloomfield, and E. S. Lander. Molecular classification of cancer: Class discovery and class prediction by gene expression monitoring. *Science*, 286:531–537, 1999.

[24] X. Gu and W.-H. Li. Estimation of evolutionary distances under stationary and nonstationary models of nucleotide substitution. *Proceedings of the National Academy of Sciences*, 95(11):5899–5905, 1998.

[25] J. G. Henikoff and S. Henikoff. Using substitution probabilities to improve position-specific scoring matrices. *CABIOS*, 12(2):135–143, 1996.

[26] D. G. Higgins and P. M. Sharp. Clustal: a package for performing multiple sequence alignment on a microcomputer. *Gene*, 73:237–244, 1988.

[27] P. B. Horton and M. Kanehisa. An assessment of neural network and statistical approaches for prediction of *E. coli* promoter sites. *Nucleic Acids Research*, 20:4331–4338, 1992.

[28] T. E. Ideker, V. Thorsson, and R. M. Karp. Discovery of regulatory interactions through perturbation: inference and experimental design. In *Proceedings of the Pacific Symposium on Biocomputing*, pages 302–313, 2000.

[29] R. A. Irizarry, B. Hobbs, F. Collin, Y. D. Beazer-Barclay, K. J. Antonellis, U. Scherf, and T. P. Speed. Exploration, normalization, and summaries of high density oligonucleotide array probe level data. *Biostatistics*, 4(2):249–264, 2003.

[30] W. Just. Reverse engineering discrete dynamical systems from data sets with random input vectors. *Journal of Computational Biology*, 13(8):1435–1456, 2006.

[31] M. K. Kerr, M. Martin, and G. A. Churchill. Analysis of variance for gene expression microarray data. *Journal of Computational Biology*, 7(6):819–837, 2001.

[32] P. Khatri and S. Draghici. Ontological analysis of gene expression data: current tools, limitations and open problems. *Bioinformatics*, 21(18):3587–3595, 2005.

[33] S. Liang, S. Fuhrman, and R. Somogyi. REVEAL, a general reverse-engineering algorithm for inference of genetic network architectures. In *Proceedings of the Pacific Symposium on Biocomputing*, pages 18–29, 1998.

[34] N. Madras, A. L. Gibbs, Y. Zhou, P. W. Zandstra, and J. E. Aubin. Modeling stem cell development by retrospective analysis of gene expression profiles in single progenitor-derived colonies. *Stem Cells*, 20:230–240, 2002.

[35] E. Mjolsness, D. H. Sharp, and J. Reinitz. A connectionist model of development. *Journal of Theoretical Biology*, 152:429–453, 1991.

[36] K. Murphy and S. Mian. Modelling gene expression data using dynamic bayesian networks. 1999.

[37] R. Nagarajan, J. E. Aubin, and C. A. Peterson. Modeling genetic networks from clonal analysis. *Journal of Theoretical Biology*, 230(3):359–373, 2004.

[38] S. Needleman and C. Wunsch. A general method applicable to the search for

similarities in the amino acid sequence of two proteins. *Journal of Molecular Biology*, 48(3):443–53, 1970.

[39] C. Perez-Iratxeta, G. Palidwor, C. J. Porter, N. A. Sanche, M. R. Huska, B. P. Suomela, E. M. Muro, P. M. Krzyzanowski, E. Hughes, P. A. Campbell, M. A. Rudnicki, and M. A. Andrade. Study of stem cell function using microarray experiments. *FEBS Letters*, 579:1795–1801, 2005.

[40] T. J. Perkins, M. Hallett, and L. Glass. Dynamical properties of model gene networks and implications for the inverse problem. *BioSystems*, 84(2):115–123, 2006.

[41] T. J. Perkins, M. T. Hallett, and L. Glass. Inferring models of gene expression dynamics. *Journal of Theoretical Biology*, 230:289–299, 2004.

[42] Theodore J. Perkins, Johannes Jaeger, John Reinitz, and Leon Glass. Reverse engineering the gap gene network of *Drosophila melanogaster*. *PLoS Computational Biology*, 2(5):e51, 2006.

[43] D. Peer, A. Tanay, and A. Regev. MinReg: A scalable algorithm for learning parsimonious regulatory networks in yeast and mammals. *Journal of Machine Learning Research*, 7:167–189, 2006.

[44] Y. Qi, J. Klein-Seetharaman, and Z. Bar-Joseph. Random forest similarity for protein-protein interaction prediction from multiple sources. In *Pacific Symposium on Biocomputing 10*, pages 531–542, 2005.

[45] J. O. Ramsay and B. W. Silverman. *Functional Data Analysis*. Springer-Verlag, 2005.

[46] I. RIgoutsos and A. Floratos. Combinatorial pattern discovery in biological sequences: the teiresias algorithm. *Bioinformatics*, 14(1):55–67, 1998.

[47] F. P. Roth, J. D. Hughes, P. W. Estep, and G. M. Church. Finding dna regulatory motifs within unaligned noncoding sequences clustered by whole-genome mrna quantitation. *Nature Biotechnology*, 16:939–945, 1998.

[48] A. A. Salamov and V. V. Solovyev. Ab initio gene finding in *Drosophila* genomic dna. *Genome Research*, 10:516–522, 2000.

[49] E. Segal, Y. Barash, I. Simon, N. Friedman, and D. Koller. From promoter sequence to expression: A probabilistic framework. In *Proceedings of the Sixth Annual International Conference on Computational Molecular Biology (RECOMB)*, 2002.

[50] E. Segal, M. Shapira, A. Regev, D. Peer, D. Botstein, D. Koller, and N. Friedman. Module networks: identifying regulatory modules and their condition-specific regulators from gene expression data. *Nature Genetics*, 34(2):166–76, 2003.

[51] E. Segal, H. Wang, and D. Koller. Discovering molecular pathways from protein interaction and gene expression data. *Bioinformatics*, 19(1):i264–i272, 2003.

[52] T. F. Smith and M. S. Waterman. Identification of common molecular subsequences. *Journal of Molecular Biology*, 147:195–197, 1981.

[53] P. T. Spellman, G. Sherlock, M. Q. Zhang, V. R. Iyer, K. Anders, M. B. Eisen, P. O Brown, D. Botstein, and B. Futcher. Comprehensive identification of cell cycle-regulated genes of the yeast *Saccharomyces cerevisiae* by microarray hybridization. *Molecular Biology of the Cell*, 9:3273–3297, 1998.

[54] G. D. Stormo. Dna binding sites: representation and discovery. *Bioinformatics*, 16(1):16–23, 2000.

[55] G. D. Stormo, T. D. Schneider, and L. Gold. Quantitative analysis of the relationship between nucleotide sequence and functional activity. *Nucleic acids*

research, 14:6661–6679, 1986.

[56] S. Tavazoie, J. D. Hughes, M. J. Campbell, R. J. Cho, and G. M. Church. Systematic determination of genetic network architecture. *Nature Genetics*, 22:281–285, 1999.

[57] J. D. Thompson, D. G. Higgins, and T. J. Gibson. Clustal w: Improving the sensitivity of progressive multiple sequence alignment through sequence weighting, position-specific gap penalties and weight matrix choice. *Nucleic Acids Research*, 22:4673–4680, 1994.

[58] V. G. Tusher, R. Tibshirani, and G. Chu. Significance analysis of microarrays applied to the ionizing radiation response. *Proceedings of the National Academy of Sciences*, 98:5116–5121, 2001.

[59] L. Wang and T. Jiang. On the complexity of multiple sequence alignment. *Journal of Computational Biology*, 1:337–348, 1994.

[60] T. G. Wolfsberg, A. E. Gabrielian, M. J. Campbell, R. J. Cho, J. L. Spouge, and D. Landsman. Candidate regulatory sequence elements for cell cycle-dependent transcription in *Saccharomyces cerevisia*. *Genome research*, 9(8):775–792, 1999.

[61] Y. H. Yang, M. J. Buckley, S. Dudoit, and T. P. Speed. Comparison of methods for imagine analysis on cdna microarray data. *Journal of Computational and Graphical Statistics*, 11(1):108–136, 2002.

[62] S. Zhong, L. Tian, C. Li, K.-F. Storch, and W. H. Wong. Comparative analysis of gene sets in the gene ontology space under the multiple hypothesis testing framework. In *Proceedings of the 2004 IEEE Computational Systems Bioinformatics Conference (CSB'04)*, pages 425–435, 2004.

CHAPTER 8

Machine Learning in Structural Biology: Interpreting 3D Protein Images

Frank DiMaio, Ameet Soni and Jude Shavlik
University of Wisconsin – Madison

8.1 Introduction

This chapter discusses an important problem that arises in structural biology: given an *electron density map* – a three-dimensional "image" of a protein produced from crystallography – identify the chain of protein atoms contained within the image. Traditionally, a human performs this *interpretation*, perhaps aided by a graphics terminal. However, over the past 15 years, a number of research groups have used machine learning to automate density map interpretation. Early methods had much success, saving thousands of crystallographer-hours, but required extremely high-quality density maps to work. Newer methods aim to automatically interpret poorer and poorer quality maps, using state-of-the-art machine learning and computer vision algorithms.

This chapter begins with a brief introduction to structural biology and x-ray crystallography. This introduction describes in detail the problem of density map interpretation, a background on the algorithms used in automatic interpretation and a high-level overview of automated map interpretation. The chapter also describes four methods in detail, presenting them in chronological order of development. We apply each algorithm to an example density map, illustrating each algorithm's intermediate steps and the resultant interpretation. Each algorithm's section presents pseudocode and program flow diagrams. The chapter concludes with a discussion of the advantages and shortcomings of each method, as well as future research directions.

8.2 Background

Knowledge of a protein's *folding* – that is, the sequence-determined three-dimensional structure – is valuable to biologists. A protein's structure provides great insight into the mechanisms by which a protein acts, and knowing these mechanisms helps increase our basic understanding of the underlying biology. Structural knowledge is increasingly important in disease treatment, and has led to the creation of catalysts with industrial uses. No existing computer algorithm can accurately map sequence to 3D structure; however, several experimental "wet lab" techniques exist for determining macromolecular structure. The most commonly used method, employed for about 80% of structures currently known, is *x-ray crystallography*. This time-consuming and resource-intensive process uses the diffraction pattern of x-rays off a crystallized matrix of protein molecules to produce an electron density map. This electron density map is a three-dimensional "picture" of the electron clouds surrounding each protein atom. Producing a protein structure is then a matter of identifying the location of each of the protein's atoms in this 3D picture.

Density map interpretation – traditionally performed by a crystallographer – is time-consuming and, in noisy or poor-quality maps, often error-prone. Recently, a number of research groups have looked into automatically interpreting electron density maps, using ideas from machine learning and computer vision. These methods have played a significant role in high-throughput structure determination, allowing novel protein structures to quickly be elucidated.

8.2.1 Protein structure

Proteins (also called *polypeptides*) are constructed from a set of building blocks called amino acids. Each of the twenty naturally-occurring amino acids consists of an amino group and a carboxylic acid group on one end, and a variable chain on the other. When forming a protein, adjacent amino groups and carboxylic acid groups condense to form a repeating backbone (or mainchain), while the variable regions become dangling sidechains. The atom at the interface between the sidechain and the backbone is known as the alpha carbon, or C_α for short (see Figure 8.1). The linear list of amino acids composing a protein is often referred to as the protein's primary structure (see Figure 8.2a).

A protein's secondary structure (see Figure 8.2b) refers to commonly occurring three-dimensional structural motifs taken by continuous segments in the protein. There are two such motifs: α-helices, in which the peptide chain folds in a corkscrew, and β-strands, where the chain stretches out linearly. In most proteins, several β-strands run parallel or antiparallel to one another. These regular structural motifs are connected by less-regular structures, called loops (or turns). A protein's secondary structure can be predicted somewhat accurately from its amino-acid sequence [1].

Finally, a protein's three-dimensional conformation – also called its tertiary structure (see Figure 8.2c) – is uniquely determined from its amino acid sequence (with some exceptions). No existing computer algorithm can accurately

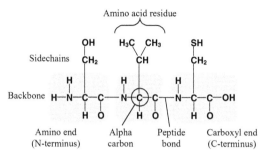

Figure 8.1 *Proteins are constructed by joining chains of amino acids in peptide bonds. A chain of three amino acid residues is illustrated.*

(a) MET–SER–SER–SER–SER–SER–VAL–PRO–ALA–TYR–LEU–GLY–ALA–
LEU–GLY–TYR–MET–ALA–MET–VAL–PHE–ALA–CYS–...

(c)

(b) MET–SER–SER–SER–SER–SER–VAL–PRO–ALA–TYR–LEU–GLY–ALA–

LEU–GLY–TYR–MET–ALA–MET–VAL–PHE–ALA–CYS–...

Figure 8.2 *An illustration of: (a) a protein's* primary structure, *the linear amino-acid sequence of the protein, (b) a protein's* secondary structure, *which describes local structural motifs such as alpha helices and beta sheets, and (c) a protein's* tertiary structure, *the global three-dimensional conformation of the protein.*

map sequence to tertiary structure; instead, we must rely on experimental techniques, primarily x-ray crystallography, to determine tertiary structure.

8.2.2 X-ray crystallography

An overview of protein crystallography appears in Figure 8.3. Given a suitable target for structure determination, a crystallographer must produce and purify this protein in significant quantities. Then, for this particular protein, one must find a very specific setting of conditions (i.e., pH, solvent type, solvent concentration) under which protein crystals will form. Once a satisfactory crystal forms, it is placed in front of an x-ray source. Here, this crystal diffracts a beam of x-rays, producing a pattern of spots on a collector. These spots – also known as *reflections* or *structure factors* – represent the Fourier-transformed

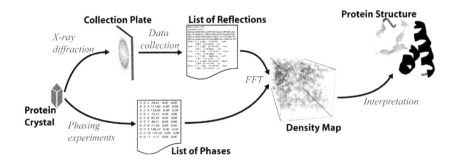

Figure 8.3 *An overview of the crystallographic process.*

electron density map. Unfortunately, the experiment can only measure the intensities of these (complex-valued) structure factors; the phases are lost.

Determining these missing phases is known as the *phase problem* in crystallography, and can be solved to a reasonable approximation using computational or experimental techniques [2]. Only after estimating the phases can one compute the electron-density map (which we will refer to as a "density map" or simply "map" for short).

The electron density map is defined on a 3D lattice of points covering the *unit cell*, the basic repeating unit in the protein crystal. The crystal's unit cell may contain multiple copies of the protein related by crystallographic symmetry, one of the 65 regular ways a protein can pack into the unit cell. Rotation/translation operators relate one region in the unit cell (the asymmetric unit) to all other symmetric copies. Furthermore, the protein may form a multimeric complex (e.g., a dimer, tetramer, etc.) within the asymmetric unit. In all these cases it is up to the crystallographer to isolate and interpret a single copy of the protein.

Figure 8.4 shows a sample fragment from an electron density map, and the corresponding interpretation of that fragment. The amino-acid (primary) sequence of the protein is typically known by the crystallographer before interpreting the map. In a high-quality map, every single (nonhydrogen) atom in the protein can be placed in the map (Figure 8.4b). In a poor-quality map it may only be possible to determine the general locations of each residue. This is known as a C_α trace (Figure 8.4c). This information - though not as comprehensive as an all-atom model - is still valuable to biologists.

The quality of experimental maps as well as the sheer number of atoms in a protein makes interpretation difficult. Certain protein crystals produce a very narrow diffraction pattern, resulting in a poor-quality, "smoothed" density map. This is quantified as the *resolution* of a density map, a numeric value which refers to the minimum distance at which two peaks in the density

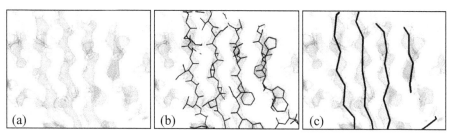

Figure 8.4 *An overview of electron density map interpretation. Given the amino-acid sequence of the protein and a density map (a), the crystallographer's goal is to find the positions of all the protein's atoms (b). Alternatively, a* backbone trace *(c), represents each residue by its C_α location.*

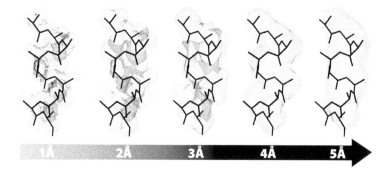

Figure 8.5 *Electron density map quality at various resolutions. The "sticks" show the position of the atoms generating the map.*

map can be resolved. Figure 8.5 shows a simple density map at a variety of resolutions.

Experimentally determined phases are often very inaccurate, and make interpretation difficult as well. As the model is built, these phases are iteratively improved [3], producing a better quality map, which may require resolving large portions of the map. Figure 8.6 illustrates the effect poor phasing has on density-map quality. In addition, noise in diffraction-pattern data collection also introduces errors into the resulting density map.

Finally, the density map produced from x-ray crystallography is not an "image" of a single molecule, but rather an average over an ensemble of all the molecules contained within a single crystal. Flexible regions in the protein are not visible at all, as they are averaged out.

For most proteins, this interpretation is done by an experienced crystallographer, who can, with high-quality data, fit about 100 residues per day. How-

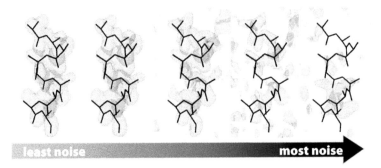

Figure 8.6 *Electron density map quality as noise is added to the computed phases. The "sticks" show the position of the atoms generating the map.*

ever, for poor-quality structures or structures with poor phasing, interpretation can be an order of magnitude slower. Consequently, map interpretation is the second-most time-consuming step in the crystallographic process (after crystal preparation).

A key question for computational methods for interpreting density maps is the following: how are candidate 3D models scored? Crystallographers typically use a model's *R-factor* (for *residual index*) to evaluate the quality of an interpretation. Formally, the *R*-factor is defined, given experimentally determined structure factors F_{obs} and model-determined structure factors F_{obs}, as:

$$R = \frac{\sum ||F_{obs}| - |F_{calc}||}{\sum |F_{obs}|}. \qquad (8.1)$$

The *R*-factor measures how well the proposed 3D structure *explains* the observed electron-density data. Crystallographers usually strive to get *R*-factors under 0.2 (or lower, depending on map resolution), while also building a physically-feasible (i.e., valid bond distances, torsion angles, etc.) protein model without adding too many water molecules. One can always reduce the *R*-factor by placing extra water molecules in the density map; these reductions are a result of *overfitting* the model to the data, and don't correspond to a physically feasible interpretation.

Another commonly used measure is *free R-factor*, or R_{free}. Here, 5-10% of reflections are randomly held out as a test set before refinement. R_{free} is the *R*-factor for these held-aside reflections. Using R_{free} tends to avoid overfitting the protein's atoms to the reflection data.

8.2.3 Algorithmic background

Algorithms for automatically interpreting electron density maps draw heavily from the machine learning and statistics communities. These communities have developed powerful frameworks for modeling uncertainty, reasoning from prior examples and statistically modeling data, all of which have been used by researchers in crystallography. This section briefly describes the statistical and machine learning methods used by the interpretation methods presented in this paper. This section is intended as a basic introduction to these topics. Interested readers should consult Russell and Norvig's text [4] or Mitchell's text [5] for a thorough overview.

Probabilistic models

A *model* here refers to a system that simulates a real-world event or process. Probabilistic models simulate uncertainty by producing different outcomes with different probabilities. In such models, the probabilities associated with certain events are generally not known, and instead have to be estimated from a *training set*, a set of previously solved problem instances. Using *maximum likelihood estimation*, the probability of a particular outcome is estimated as the frequency at which that outcome occurs in the training set.

The *unconditional* or *prior probability* of some outcome A is denoted $P(A)$. Assigning some value to this probability, say $P(A) = 0.3$, means that in the absence of any other information, the best assignment of probability of outcome A is 0.3. As an example, when flipping a (fair) coin, $P(\textit{"heads"}) = 0.5$. In this section, we use "outcome" to mean the setting of some random variable; $P(X = x_i)$ is the probability that variable X takes value x_i. We will use the shorthand $P(x_i)$ to refer to this same value.

The *conditional* or *posterior probability* is used when other, previously unknown, information becomes available. If other information, B, relevant to event A is known, then the best assignment of probability to event A is given by the conditional $P(A|B)$. This reads "the probability of A, *given* B." If more evidence, C, is uncovered, then the best probability assignment is $P(A|B, C)$ (where "," denotes "and").

The *joint probability* of two or more events is the probability of both events occurring, and – for two events A and B – is denoted $P(A, B)$ and is read "the probability of A *and* B." Conditional and joint probabilities are related using the expression:

$$P(A, B) = P(A|B)P(B) = P(B|A)P(A). \qquad (8.2)$$

This relation holds for any events A and B. Two events are *independent* if their joint probability is the same as the product of their unconditional probabilities, $P(A, B) = P(A)P(B)$. If A and B are independent we also have $P(A|B) = P(A)$, that is, knowing B tells us nothing about A.

One computes the *marginal probability* by taking the joint probability and summing out one or more variables. That is, given the joint distribution

$P(A, B, C)$, one computes the marginal distribution of A as:

$$P(A) = \sum_{B} \sum_{C} P(A, B, C). \qquad (8.3)$$

Above, the sums are over all possible outcomes of events B and C. The marginal distribution is important because it provides information about the distribution of some variables (A above) in the full joint distribution, without requiring one to explicitly compute the (possibly intractable) full joint distribution.

Finally, *Bayes' rule* allows one to reverse the direction of a conditional:

$$P(A|B) = \frac{P(B|A)P(A)}{P(B)}. \qquad (8.4)$$

Bayes' rule is useful for computing a conditional $P(A|B)$ when direct estimation (using frequencies from a training set) is difficult, but when $P(B|A)$ can be estimated accurately. Often, one drops the denominator, and instead computes the *relative likelihood* of two outcomes, for example, $P(A = a_1|B)$ versus $P(A = a_2|B)$. If a_1 and a_2 are the only possible outcomes for A, then exact probabilities can be determined by normalization; there is no need to compute the prior $P(B)$.

Case-based reasoning

Broadly defined, *case-based reasoning* (CBR) attempts to solve a new problem by using solutions to similar past problems. Algorithms for case-based reasoning require a database of previously solved problem instances, and some distance function to calculate how "different" two problem instances are. There are two key aspects of CBR systems. First, learning in such systems is *lazy*: the models only generalize to unseen instances when presented with such a new instance. Second, they only use instances "close" to the unseen instance when categorizing it.

The most common CBR algorithm is *k-nearest neighbor* (kNN). In kNN, problem instances are feature vectors, that is, points in some n-dimensional space. The learning algorithm, when queried with a new problem instance $X = \langle x_i, \ldots, x_n \rangle$ for classification or regression, finds the k previously solved problem instances closest to the query in Euclidean space. That is, one chooses the examples minimizing the distance:

$$d(X||Y) = \sqrt{\sum_{i=1}^{n} (x_i - y_i)^2}. \qquad (8.5)$$

Then, the k neighbors "vote" on the query instance's label: usually the majority class label (for classification) or average label (for regression) of the k neighbors is used. One variant of kNN weights each neighbor's vote by its

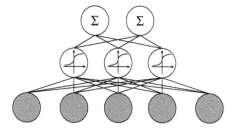

Figure 8.7 *A multilayer, feed-forward neural network. The network consists of an input layer fully connected to a hidden layer of sigmoid units, fully connected to an output layer.*

similarity to the query. Another variant learns weights for each dimension, to be used when computing the distance between two instances.

Neural networks

An *artificial neural network* (ANN) is a nonlinear function estimator used to approximate real or discrete target functions. Inspired by neurons in the brain, ANNs consist of a number of *units* connected in a network. Each unit is connected with multiple *inputs* and a single *output*. A given unit's output is some function of the weighted sum of the inputs. This function is known as the unit's *activation function*. For a *perceptron* – a simple, one-layer network – this function is usually a step function.

More commonly, the network structure has multiple, feed-forward layers, as shown in Figure 8.7. This network consists of a *hidden layer* which fully connects the input and outputs. Learning the weights in the network requires a differentiable activation function; often one uses a sigmoidal function:

$$\sigma(y) = \frac{1}{1 + e^{-y}}. \tag{8.6}$$

Above, y is the weighted sum of inputs, $y = \sum_{i=0}^{N} w_i x_i$.

The *backpropagation* algorithm learns the weights for a multilayer network, given a network structure. Backpropagation uses gradient descent over the network weight space to minimize the squared error between *computed* output values and *desired* output values over some training set. The goal of backpropagation is to find some point in weight space – that is, some setting of all the weights in the network – that (locally) minimizes this squared error.

8.2.4 Approaches to automatic density map interpretation

A number of research groups have investigated automating the interpretation of electron-density maps. This section presents a high-level overview of some of these methods, while the remainder of this chapter takes an in-depth look at four of these methods, describing algorithmically how they have approached this problem.

By far the most commonly used method is ARP/WARP [6, 7, 8]. This atom-based method heuristically places atoms in the map, connects them and refines their positions. To handle poor phasing, ARP/WARP uses an iterative algorithm, consisting of alternating phases in which (a) a model is built from a density map, and (b) the density map's phases are improved using the constructed model. This algorithm is widely used, but has one drawback: fairly high resolution data, about 2.3**A** or better, is needed. Given this high-resolution data, the method is extremely accurate; however, many protein crystals fail to diffract to this extent.

Several approaches represent the density map in an alternate form, in the process making the map more easily interpretable for both manual and automated approaches. One of the earliest such methods, *skeletonization*, was proposed for use in protein crystallography by Greer's BONES algorithm [9]. Skeletonization, similar to the medial-axis transformation in computer vision, gradually thins the density map until it is a narrow ribbon approximately tracing the protein's backbone and sidechains. More recent work by Leherte et al. [10] represents the density map as an acyclic graph: a minimum spanning tree connecting all the critical points (points where the gradient of the density is 0) in the electron density map. This representation accurately approximates the backbone with 3**A** or better data, and separates protein from solvent up to 5**A** resolution.

Cowtan's FFFEAR efficiently locates rigid templates in the density map [11]. It uses *fast Fourier transforms* (FFTs) to quickly match a learned template over all locations in a density map. Cowtan provides evidence showing it locates alpha helices in poorly-phased 8**A** maps. Unfortunately, the technique is limited in that it can only locate large rigid templates (e.g., those corresponding to secondary-structure elements). One must trace loops and map the fit to the sequence manually. However, a number of methods use FFFEAR as a template-matching subroutine in a larger interpretation algorithm.

X-AUTOFIT, part of the QUANTA [12] package, uses a gradient refinement algorithm to place and refine the protein's backbone. Their refinement takes into account the density map as well as bond geometry constraints. They report successful application of the method at resolutions ranging from 0.8 to 3.0**A**.

Terwilliger's RESOLVE contains an automated model-building routine [13, 14] that uses a hierarchical procedure in which helices and strands are located and fitted, then are extended in an iterative fashion, using a library of tripeptides. Finally, RESOLVE applies a greedy fragment-merging routine to overlapping extended fragments. The approach was able to place approximately 50% of the protein chain in a 3.5**A** resolution density map.

Levitt's MAID [15] approaches map interpretation "as a human would," by first finding the major secondary structures, alpha helices and beta sheets, connecting the secondary-structure elements, and mapping this fit to the provided sequence. MAID reports success on density maps at around 2.8**A** resolution.

Ioerger's TEXTAL [16, 17, 18] attempts to interpret poor-resolution (2.2 to 3.0**A**) density maps using ideas from pattern recognition. Ioerger constructs a set of 15 rotation-invariant density features. Using these features at several radii, a subroutine, CAPRA, trains a neural network to identify C_α atoms. TEXTAL then identifies sidechains by looking at the electron density around each putative alpha carbon, efficiently finding the most similar region in a database, and laying down the corresponding sidechain.

Finally, ACMI takes a probabilistic approach to density map interpretation [19]. Residues of the protein are modeled as nodes in a graph, while edges model pairwise structural interactions arising from chemistry. An efficient inference algorithm determines the most probable backbone trace conditioned on these interactions. ACMI finds accurate backbone traces in well-phased 3.0 to 4.0**A** density maps.

The rest of this chapter further describes four of these methods – ARP/WARP, RESOLVE, TEXTAL and ACMI – in detail. Each section presents a method, describing strengths and weaknesses. High-level pseudocode clarifies important subroutines. Throughout the chapter, a small running example is employed to illustrate intermediate steps of the various algorithms. The example uses the density map of protein 1XMT, a 95-amino-acid protein with two symmetric copies in the unit cell.

The running example is not meant as a test of the algorithms (although a comparison appears in [19]), but rather as a real-world illustrative example. The example map is natively at 1.15**A** resolution. Full native resolution is used for ARP/WARP; the map is downsampled to 2.5**A** resolution for RESOLVE, and 3.5**A** resolution for TEXTAL and ACMI. Maps are downsampled by smoothly diminishing the intensities of higher-resolution reflections. The resolution values chosen are – for this particular map – the worst-quality maps for which the respective algorithms produce an accurate trace. The 3.5**A** density map and its crystallographer-determined solution appears in Figure 8.8.

8.3 ARP/WARP

The ARP/WARP (automated refinement procedure) software suite is a crystallographic tool for the interpretation and refinement of electron density maps.

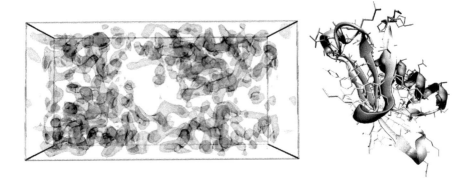

Figure 8.8 *A 3.5**A** resolution electron density map – containing two copies of a 95-amino-acid protein – and its crystallographer-determined solution. This map, at various resolutions, will be used as a running example throughout the section.*

ARP/WARP's WARPNTRACE procedure was the first automatic interpretation tool successfully used for protein models. Today, it remains one of the most used tools in the crystallographic community for 3D protein-image interpretation. ARP/WARP concentrates on the best placement of individual atoms in the map: no attempt is made to identify higher-order constructs like residues, helices or strands. ARP/WARP's "atom-level" method requires high-quality data, however. In general, ARP/WARP requires maps at a resolution of 2.3A or higher to produce an accurate trace.

Figure 8.9 illustrates an overview of WARPNTRACE. WARPNTRACE begins by creating a *free atom model* – a model containing only unconnected atoms – to fill in the density map of the protein. It then connects some of these atoms using a heuristic, creating a *hybrid model*. This hybrid model consists of a partially-connected backbone, together with a set of unconstrained atoms. This hybrid model is refined, producing a map with improved phase estimates. The process iterates using this improved map. At each iteration, WARPNTRACE removes every connection, restarting with a free-atom model.

8.3.1 Free-atom placement

ARP/WARP contains a atom placement method based on ARP, an interpretation method for general molecular models. ARP randomly places unconnected atoms into the density map, producing a *free atom model*, illustrated in Figure 8.10a. To initialize the model, ARP begins with a small set of atoms in the density map. It slowly expands this model by looking for areas above a density threshold, at a bonding distance away from existing atoms. The density threshold is slowly lowered until ARP places the desired number of free atoms.

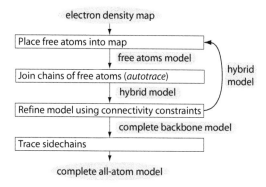

Figure 8.9 *A flowchart of* WARPNTRACE.

For small molecules, ARP's next step is *refining* the free-atom model; that is, iteratively moving atoms to better explain the density map. Free-atom refinement ignores stereochemical information, and moves each atom independently to produce a complete structure. ARP's free-atom refinement, in addition to moving atoms, considers adding or removing atoms from the model. Multiple randomized initial structures are used to improve robustness. Further details are available from Perrakis et al. [7].

However, with molecules as large as proteins, free-atom refinement alone is insufficient to produce an accurate model. Performing free-atom refinement with tens of thousands of atoms leads to overfitting, producing a model that is not physically feasible. For determining protein structures, ARP/WARP makes use of connectivity information in its refinement, using free-atom placement as a starting point. The procedure WARPNTRACE adds connectivity information to the free atom model.

8.3.2 Main-chain tracing

Given a free-atom model of a protein, one can form a crude backbone trace by looking for pairs of free atoms the proper distance apart. WARPNTRACE formalizes this procedure, called *autotracing*, using a heuristic method. The method is outlined in Algorithm 8.1. WARPNTRACE assigns a score – based on density values – to each free atom. The highest scoring atom pairs $3.8 \pm 0.5\text{Å}$ apart become candidate C_α's. The algorithm verifies candidate pairs by overlaying them with a peptide template. If the template matches the map, WARPNTRACE saves the candidate pair.

After computing a list of C_α pairs, WARPNTRACE constructs the backbone using a database of known backbone conformations (taken from solved protein structures). Given a chain of candidate C_α pairs, WARPNTRACE considers all

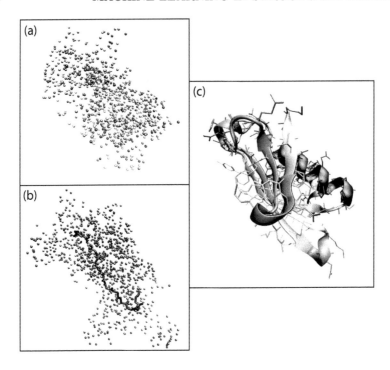

Figure 8.10 *Intermediate steps in* ARP/WARP*'s structure determination: (a) the free atom model, where waters are placed in the map, (b) the hybrid model, after some connections are determined and (c) the final determined structure.*

backbone conformations in the database with matching C_α positions, ordered by length. The longest candidate backbone is then added to the model. The algorithm connects the corresponding free atoms, and removes these atoms from the free-atom pool. The process repeats as long as there remain candidate C_α chains at least 5 residues in length.

Autotracing produces a *hybrid model*, shown in Figure 8.10b. A hybrid model contains a set of connected chains together with a set of free atoms. Autotracing identifies some atom types and connectivity, which enables the use of some stereochemical information in refinement. Added restraints increase the number of observations available, and increase the probability of producing a good model. The tracing is initially very conservative, with many free atoms remaining in the model. Adding too many restraints too early leads to overfitting the model.

Finally, a modified version of ARP refines this hybrid model. ARP uses the refined structure to improve the experimentally determined phases, making the map clearer to interpret. At each iteration of this "autotrace–refine–recompute phases" cycle, WARPNTRACE returns to a free-atom model, by removing pre-

---

Algorithm 8.1: WARPNTRACE's model-building algorithm

Given: electron density map **M**, free-atom model **F**, sequence **seq**
Find: all-atom model
for $i = 1$ to $nIterations$ **do**
    H ← **F**   // *initialize hybrid model*
    CA_pairs ← highest-scoring atom pairs $3.8 \pm 0.5\mathbf{A}$ apart
    foreach $c_i \in$ **CA_pairs do**
        if (c_i does not match backbone template) **then**
            delete c_i from **CA_pairs**
        end
    end

    while a C_α chain of length ≥ 5 remains in **CA_pairs do**
        $bestChain$ ← longest fragment in DB overlapping **CA_pairs**
        remove $bestChain$'s atoms from **CA_pairs**, **H**
        add $bestChain$ to **H**
    end
    H′ ← $refine(\mathbf{H})$
    F ← remove connections from hybrid model **H′**
end
$model$ ← sidechainTrace(**H′**)

---

vious traces. Since the map is better-phased, autotracing produces a more complete model. This, in turn, provides a better refinement, improving the phases further.

This cycle continues for a fixed number of iterations, or until a complete trace is available. Finally, WARPNTRACE adds on side-chains by identifying patterns of free atoms around C_α's. It aligns these free-atom patterns to the sequence to produce a complete model. Figure 8.10c illustrates the complete ARP/WARP-determined trace on the running example.

8.3.3 Discussion

ARP/WARP is the preferred method for automatically interpreting electron density maps, assuming sufficiently high-resolution data are available. Its placement of individual atoms, followed by atom-level refinement, produces an extremely accurate trace with no user action required in 2.3**A** or better density maps. It is widely used by crystallographers to rapidly construct a protein model. Unfortunately, many protein crystals fail to produce maps of sufficient quality, and one must consider alternate methods.

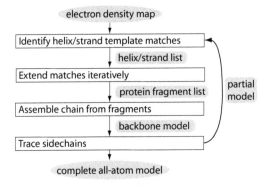

Figure 8.11 *A flowchart of* RESOLVE.

8.4 RESOLVE

While ARP/WARP is extremely accurate with high-resolution data, many protein crystals fail to diffract to a sufficient level for accurate interpretation. In general, ARP/WARP requires individual atoms to be visible in the density map, which happens at about **2.3Å** resolution or better. The next three methods – RESOLVE, TEXTAL and ACMI – all aim to solve maps with > 2.3Å resolution. All three methods take different approaches to the problem; however, all three – in contrast to ARP/WARP – consider higher-level constructs than atoms when building a protein model. This allows accurate interpretation even when individual atoms are not visible.

RESOLVE is a method developed by Terwilliger for automated model-building in poor-quality (around **3Å**) electron density maps [13, 14]. Figure 8.11 outlines the complete hierarchical approach. RESOLVE's method hinges upon the construction of two model secondary structure fragments – a short α-helix and β-strand – for the initial matching. RESOLVE first searches over all of rotation and translation space for these fragments; after placing a small set of overlapping model fragments into the map, the algorithm considers a much larger template set as potential extensions. RESOLVE joins overlapping fragments and, finally, identifies sidechains corresponding to each C_α – conditioned on the input sequence – and places individual atoms into the model.

8.4.1 Secondary structure search

Given an electron density map, RESOLVE begins its interpretation by searching all translations and rotations in the map for a model 6-residue α-helix and a model 4-residue β-strand. RESOLVE constructs these fragments by aligning a collection of helices (or strands) from solved structures; it computes the

Figure 8.12 *The averaged helix (left) and strand (right) fragment used in* RESOLVE*'s initial matching step.*

electron density for each at **3A** resolution, and averages the density across all examples. The "average" models used by RESOLVE are illustrated in Figure 8.12.

Given these model fragments, RESOLVE considers placing them at each position in the map. At each position it considers all possible rotations (at a 30° or 40° discretization) of the fragment, and computes a standardized squared-density difference between the fragment's electron density and the map:

$$t(\vec{x}) = \sum_{\vec{y}} \epsilon_f(\vec{y}) \left(\rho'_f(\vec{y}) - \frac{1}{\sigma_f(\vec{x})} \left[\rho(\vec{y} - \vec{x}) - \bar{\rho}(\vec{x}) \right] \right). \qquad (8.7)$$

Above, $\rho(\vec{x})$ is the map in which we are searching, $\rho'_f(\vec{x})$ is the *standardized* fragment electron density, $\epsilon_f(\vec{x})$ is a masking function that is nonzero only for points near the fragment and $\bar{\rho}(\vec{x})$ and $\sigma_f(\vec{x})$ standardize the map in the masked region ϵ_f centered at $\vec{x}$:

$$\bar{\rho}(\vec{x}) = \frac{\sum_{\vec{y}} \epsilon_f(\vec{y}) \rho(\vec{y} - \vec{x})}{\sum_{\vec{y}} \epsilon_f(\vec{y})}$$

$$\sigma_f^2(\vec{x}) = \frac{\sum_{\vec{y}} \epsilon_f(\vec{y}) \left[\rho(\vec{y} - \vec{x}) - \bar{\rho}(\vec{x}) \right]^2}{\sum_{\vec{y}} \epsilon_f(\vec{y})}. \qquad (8.8)$$

RESOLVE computes the matching function $t(\vec{x})$ quickly over the entire unit cell using FFFEAR's FFT-based convolution [11].

After matching the two model fragments using a coarse rotational step-size, the method generates a list of best-matching translations and orientations of each fragment (shown in Figure 8.13a). Processing these matches in order, RESOLVE refines each fragment's position and rotation to maximize the real-space correlation coefficient (RSCC) between template and map:

$$RSCC(\rho_f, \rho) = \frac{\langle \rho_f \cdot \rho \rangle - \langle \rho_f \rangle \langle \rho \rangle}{\sqrt{\langle \rho_f^2 \rangle - \langle \rho_f \rangle^2} \sqrt{\langle \rho^2 \rangle - \langle \rho \rangle^2}}. \qquad (8.9)$$

Here, $\langle \rho \rangle$ indicates the map mean over a fragment mask. RESOLVE only considers refined matches with an RSCC above some threshold.

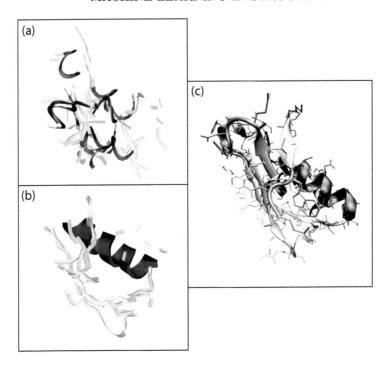

Figure 8.13 *Intermediate steps in* RESOLVE*'s structure determination: (a) locations in the map that match short helical/strand fragments (b) the same fragments after refinement and extension and (c) the final determined structure.*

8.4.2 Iterative fragment extension

At this point, RESOLVE has a set of putative helix and strand locations in the density map. The next phase of the algorithm extends these using a much larger library of fragments, producing a model like that in Figure 8.13b. Specifically, RESOLVE makes use of four such libraries for fragment extension:

(a) 17 α-helices between 6 and 24 amino acids in length

(b) 17 β-strands between 4 and 9 amino acids in length

(c) 9,232 tripeptides containing backbone atoms only *for N-terminus extension*

(d) 4,869 tripeptides containing a partial backbone (the chain $C_\alpha - C - O$ with no terminal N) plus two full residues *for C-terminus extension*

RESOLVE learns these fragment libraries from a set of previously solved protein structures. It constructs the two tripeptide libraries by clustering a larger dataset of tripeptides.

α-helix/β-strand extension

For each potential model's placement in the map, RESOLVE considers extending it using each fragment in either set (a), if the model fragment is a helix, or set (b), if the model fragment is a strand. For each fragment, RESOLVE chooses the longest segment of the fragment such that every atom in the fragment has a density value above some threshold.

To facilitate comparison between these 17 segments of varying length (one for each fragment in the library), each matching segment is given a score $Q = \langle \rho \rangle \sqrt{N}$, with $\langle \rho \rangle$ the mean atom density, and N the number of atoms. The algorithm computes a Z-score:

$$Z = \frac{Q - \langle Q \rangle}{\sigma(Q)}. \tag{8.10}$$

RESOLVE only considers segments with $Z > 0.5$. Notice there may be a large number of overlapping segments in the model at this point.

Loop extension using tripeptide libraries

For each segment in the model, RESOLVE attempts to extend the segment in both the N-terminal and C-terminal direction using the tripeptide library. RESOLVE tests each tripeptide in the library by superimposing the first residue of the tripeptide on the last residue of the current model segment. It then tests the top scoring "first-level" fragments for overlap (steric clashes) with the current model segment. For those with no overlap, a lookahead step considers this first-level extension as a starting point for a second extension. The score for each first-level extension is:

$$score_{\text{first-level}} = \langle \rho_{\text{first-level}} \rangle + \max_{\text{second-level}} \langle \rho_{\text{second-level}} \rangle. \tag{8.11}$$

Above, $\langle \rho_{\text{first-level}} \rangle$ denotes the average map density at the atoms of the first-level extension.

RESOLVE accepts the best first-level extension – taking the lookahead term into account – only if the average density is above some threshold density value. If the density is too low, and the algorithm rejects the best fragment, several "backup procedures" consider additional first-level fragments, or stepping back one step in the model segment. If these backup procedures fail, RESOLVE rejects further extensions.

8.4.3 Chain assembly

Given this set of candidate model segments, RESOLVE's next step is assembling a continuous chain. To do so, it uses an iterative method, illustrated in Algorithm 8.2. The outermost loop repeats until no more candidate segments

---

Algorithm 8.2: RESOLVE's chain-assembly algorithm

Given: electron density map $\mathbf{M}$, set of high scoring fragments $\mathbf{F}$
Find: putative backbone trace $\mathbf{X} = \{\vec{x}_i\}$ including C_β positions
repeat
    $frag_{best} \leftarrow$ top scoring unused segment
    for each $frag_i \in \{\mathbf{F} \backslash frag_{best}\}$ **do**
        if $frag_i$ and $frag_{best}$ overlap at ≥ 2 C_α positions
        and extension does not cause steric clashes **then**
                extend $frag_{best}$ by $frag_i$
        end
    end
until no candidates remain

---

remain. At each iteration, the algorithm chooses the top-scoring candidate segment not overlapping any others. It considers all other segments in the model as extensions: if at least two C_α's between the candidate and extension overlap, then RESOLVE accepts the extension. Finally, the extension becomes the current candidate chain.

8.4.4 Sidechain trace

RESOLVE's final step is, given a set of C_α positions in some density map, to identify the corresponding residue type, and to trace all the sidechain atoms [14]. This sidechain tracing is the first time that RESOLVE makes use of the analyzed protein's sequence. RESOLVE's sidechain tracing uses a probabilistic method, finding the most likely layout conditioned on the input sequence. RESOLVE's sidechain tracing procedure is outlined in Algorithm 8.3.

RESOLVE's sidechain tracing relies on a rotamer library. This library consists of a set of low-energy conformations – or *rotamers* – that characterizes each amino-acid type. RESOLVE builds a rotamer library from the sidechains in 574 protein structures. Clustering produces 503 different low-energy side-chain conformations. For each cluster member, the algorithm computes a density map; each cluster's representative map is the average of its members.

For each C_α, RESOLVE computes a probability distribution of the corresponding residue type. Probability computation begins by first finding the correlation coefficient (see Equation 8.9) between the map and each rotamer. For each rotamer j, the correlation coefficient at the kth C_α is given by cc_{jk}. A Z-score is computed, based on rotamer j's correlation at every other C_α:

$$Z_{jk}^{rot} = \frac{cc_{jk} - \langle cc_j \rangle}{\sigma_j}. \tag{8.12}$$

The algorithm only keeps a *single best-matching rotamer* of each residue type.

Algorithm 8.3: RESOLVE's sidechain-placement algorithm

Given: map $\mathbf{M}$, backbone trace $\mathbf{X} = \{\vec{x}_i\}$ (including C_β's),
      sidechain library $\mathbf{F}$, sequence seq
Find: all-atom protein model
for each sidechain $f_j \in \mathbf{F}$ **do**
    for each C_α $\vec{x}_k \in \mathbf{X}$ **do**
        $cc_{jk} \leftarrow RSCC(\mathbf{M}(\vec{x}_k), f_j))$ // see Equation 8.9
        $Z_{jk} \leftarrow (cc_{jk} - \langle cc_j \rangle)/\sigma_j$
    end
end

// Estimate probabilities P_{ik} that residue type i is at position k
for each residue type i **do**
    $P_{i0} \leftarrow$ a priori distribution of residue type i
    $Z_{ik} \leftarrow \underset{\text{fragment } j \text{ of type } i}{\max} Z_{jk}$
    for each alpha carbon $\vec{x}_k \in \mathbf{X}$ **do**
        $P_{ik} \leftarrow P_{i0} \cdot \dfrac{exp(Z_{ik}^2/2)}{\sum_l P_{l0} \cdot exp(Z_{lk}^2/2)}$
    end
end

// Align trace to sequence, place individual atoms
align seq to chains maximizing product of P_{ik}'s
if (good alignment exists) **then**
    place best-fit sidechain of alignment-determined type at each position
end

That is, for residue type i:

$$Z_{ik}^{res} = \underset{\text{fragment } j \text{ is of type } i}{\max} Z_{jk}^{rot} \tag{8.13}$$

RESOLVE uses a Bayesian approach to compute probability from the Z-score. Amino-acid distributions in the input sequence provide the a priori probability P_{0j} of residue type j. Given a set of correlation coefficients at some position, RESOLVE computes the probability that the residue type is i by taking the product of probabilities that *all other correlation coefficients were generated by chance*. It estimates this probability using the Z-score:

$$P(cc_{ik}) \propto exp(-(Z^r es_{ik})^2/2). \tag{8.14}$$

Substituting and simplifying, the probability of residue type i at position k is:

$$P_{ik} \leftarrow P_{i0} \cdot \frac{exp((Z_{ik}^{res})^2/2)}{\sum_l P_{l0} \cdot exp((Z_{lk}^{res})^2/2)}. \tag{8.15}$$

Finally, given these probabilities, RESOLVE finds the alignment of sequence to structure that maximizes the product of probabilities. The final step is, given an alignment-determined residue type at each position, placing the rotamer of the correct type with the highest correlation coefficient Z-score. RESOLVE's final computed structure on the running example, both backbone and sidechain, is illustrated in Figure 8.13c.

8.4.5 Discussion

Unlike ARP/WARP, RESOLVE uses higher-order constructs than individual atoms in tracing a protein's chain. Searching for helices and strands in the map lets RESOLVE produce accurate traces in poor-quality maps, in which individual atoms are not visible. This method is also widely used by crystallographers. RESOLVE has been successfully used to provide a full or partial interpretation at maps with as poor as 3.5Å resolution. Because the method is based on heuristics, when map quality gets worse, the heuristics fail and the interpretation is poor. Typically, the tripeptide extension is the first heuristic to fail, resulting in RESOLVE traces consisting of disconnected secondary structure elements. In poor maps, as well, RESOLVE is often unable to identify sidechain types. However, RESOLVE is able to successfully use background knowledge from structural biology in order to improve interpretation in poor-quality maps.

8.5 TEXTAL

TEXTAL – another method for density map interpretation – was developed by Ioerger et al. Much like RESOLVE, TEXTAL seeks to expand the limit of interpretable density maps to those with medium to low resolution (2 to 3Å). The heart of TEXTAL lies in its use of computer vision and machine learning techniques to match patterns of density in an unsolved map against a database of known residue patterns. Matching two regions of density, however, is a costly six-dimensional problem. To deal with this, TEXTAL uses of a set of rotationally invariant numerical features to describe regions of density. The electron density around each point in the map is converted to a vector of 19 features sampled at several radii that remain constant under rotations of the sphere. The vector consists of descriptors of density, moments of inertia, statistical variation and geometry. Using rotationally invariant features allows for efficiency in determination – reducing the problem from 6D to 3D – and better generalization of patterns of residues.

TEXTAL's algorithm – outlined in Figure 8.14 – attempts to replicate the process by which crystallographers identify protein structures. The first step is to identify the backbone – the location of each C_α – of the protein. Tracing the backbone is done by CAPRA (C-Alpha Pattern Recognition Algorithm), which uses a neural network to estimate each subregion's distance to the

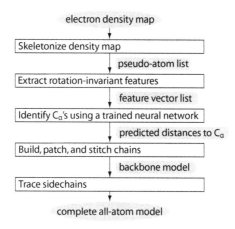

Figure 8.14 *A flowchart of* TEXTAL.

closest C_α given the rotationally invariant features discussed above. CAPRA's putative C_α locations are then sent into the second part of the algorithm, LOOKUP, which identifies the sidechains corresponding to each C_α's. LOOKUP uses these same rotationally invariant features, but instead uses a *nearest neighbor* approach to find a small subset of the database that best matches the region of unknown density. LOOKUP rotationally aligns the best match to the unknown residue, and places the corresponding atoms into the map. Finally, TEXTAL cleans up its trace using a set of post-processing routines.

8.5.1 Feature extraction

The most important component of TEXTAL is its extraction of a set of numerical features from a region of density. These numerical features allow rapid identification of similar regions from different (solved) maps. A key aspect of TEXTAL's feature set is invariance to arbitrary rotations of the region's density. This eliminates the need for an expensive rotational search for each fragment; additionally, a discrete rotational search is likely to underestimate some match scores if the true rotation falls between rotational samples.

TEXTAL uses 76 such numerical features to describe a region of density in a map. These features include 19 rotationally invariant features, sampled at four different radii: 3,4,5 and 6**A**. The use of multiple radii is critical for differentiation between side-chains: large residues often look similar at smaller radii but greatly differ at 6**A**, while small amino acids may have no density in the outer radii and thus are only differentiated at small radii.

The 19 rotation-invariant features fall into four basic classes, shown in Table 8.1. The first class describes statistical properties of these neighborhoods of

Table 8.1 *The rotation invariant features used by* TEXTAL

Class	Description	Quantity
Statistical Features of Density average, standard deviation, skewness, kurtosis		4
Center of Mass distance from center of sphere to center of mass		1
Moments of Inertia magnitude of primary, secondary, tertiary moments; ratios between these moments		6
Spokes/Geometry of Density angles between three "spokes of maximal density" sum of angles, radial densities of each spoke, area of triangle formed by spokes		8

density, treating density values in the neighborhood as a probability distribution. These features include mean, standard deviation, skewness and kurtosis, the last two of which provide descriptions of the lopsidedness and peakedness of the distribution of density values. The second class of features is really just a single feature: the distance from the center of mass to the center of the neighborhood.

A third class of descriptors includes moments of inertia (MOI), which provides six features describing how elliptical is the density distribution. Moments of inertia are calculated as the Eigenvectors of the inertia matrix I:

$$I = \sum_i \left(\rho_i \left| \begin{matrix} y_i^2 + z_i^2 & -x_i y_i & -x_i z_i \\ -x_i y_i & x_i^2 + z_i^2 & -y_i z_i \\ -x_i z_i & -y_i z_i & x_i^2 + y_i^2 \end{matrix} \right| \right).$$

Above, ρ_i is the density at point $\langle x_i, y_i, z_i \rangle$. As a rotation-invariant description, TEXTAL only considers the moments and the ratios between moments, not the axes themselves (the Eigenvectors of the inertia matrix).

The final class of features represent higher-level geometrical descriptors of the region. Three "spokes of maximal density" are extended from the center of the region, spaced $> 75°$ apart. These aim to approximate the direction of the backbone N-terminus, the backbone C-terminus and the sidechain. Rotation-invariant features derived from these spokes include the min, mid, max and sum of the angles, the density sum along each spoke and the area of the triangle formed by connecting the end points of the spokes.

8.5.2 *Backbone tracing*

CAPRA, a subroutine of TEXTAL, produces the initial C_α trace. CAPRA con-

Algorithm 8.4: Textal's capra subroutine for calculating the initial backbone trace

---

Given: electron density map **M**
Find: putative backbone trace $\mathbf{X} = \{\vec{x}_i\}$
$\mathbf{M}' \leftarrow normalize(\mathbf{M})$
pseudoAtoms $\leftarrow skeletonize(\mathbf{M}')$
for $p_i \in$ **pseudoAtoms do**
    $\mathbf{F} \leftarrow$ rotation invariant-features in a neighborhood around p_i
    distance-to-$C_\alpha \leftarrow neuralNetwork(\mathbf{F})$
end
$\mathbf{X} \leftarrow$ construct chain using predicted distances-to-C_α

---

structs a backbone chain using a feed-forward neural network. An overview of the process is illustrated in Algorithm 8.4.

In order to accurately compare maps, capra begins by first normalizing density values in the map, ensuring feature values from different maps are comparable. Next, capra *skeletonizes* the map, creating a trace of *pseudo-atoms* that identifies the medial axis of some density map contour. Figure 8.15a illustrates this skeletonization. This trace is a very crude approximation of the backbone, and may traverse the side-chains or form multiple distinct chains.

A feed-forward neural network – a nonlinear function approximator used for both classification and regression – is trained to learn which pseudo-atoms correspond to actual C_α's. Specifically, the network is trained on a set of previously solved maps to predict the *distance* of each pseudo-atom to the nearest C_α. The rotation invariant features are inputs to the network; a single output node estimates the distance to the closest C_α. A hidden layer of 20 sigmoidal units fully connects input and output layers.

Given a predicted distance-to-C_α for each pseudo-atom, capra uses a greedy trace to find a linear chain linking C_α's together. Further post-modifications have been added to improve performance of capra, such as refining chains and patching missing links. The output for capra, on the sample map at 3.5**A** resolution, is illustrated in Figure 8.15b.

8.5.3 Sidechain placement

After capra returns its predicted backbone trace, textal must next identify the residue type associated with each C_α. This identification is performed by a subroutine lookup. Algorithm 8.5 shows a pseudocode overview of lookup. Essentially, the subroutine compares the density around each C_α to a database of solved maps to identify the residue type. Lookup uses textal's rotation invariant features, and builds a database of feature vectors corresponding to C_α's in solved maps. To determine the residue type of an unknown region of

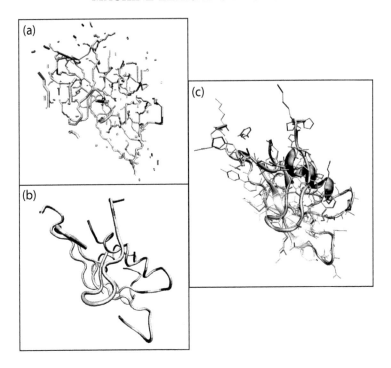

Figure 8.15 TEXTAL's *intermediate steps: (a) the skeletonized density map, which crudely approximates the protein backbone, (b) a backbone trace, which* TEXTAL *builds by determining C_α's from the skeleton points, and (c) the final determined structure.*

density, LOOKUP finds the *nearest neighbors* in the database, using weighted Euclidian distance:

$$D(\rho_1||\rho_2) = \left[\sum_i \lambda_i \cdot \left(F_i(\rho_1) - F_i(\rho_2)\right)^2 \right]^{1/2}. \qquad (8.16)$$

Above, F_i refers to the ith feature in the vector, while ρ_1 and ρ_2 are two regions of density. Feature weights λ_i are optimized to maximize similarity between matching regions and minimize between nonmatching regions, where ground truth is the optimally-aligned *RSCC* (Equation 8.9). TEXTAL sets weights using the SLIDER algorithm [20] which considers features pairwise to determine a good set of weights.

Since information is lost when representing a region as a rotation invariant feature vector, the nearest neighbor in the database does not always correspond to the best-matching region. Therefore, LOOKUP keeps the top k regions – those with the lowest Euclidean distance – and considers these for a more time-consuming *RSCC* computation (see Equation 8.9). Ideally, LOOKUP wants to find the rotation and translation of each of the k regions to maximize

---

Algorithm 8.5: TEXTAL's LOOKUP subroutine for placing sidechains.

Given: electron density map $\mathbf{M}$, backbone trace $\mathbf{X} = \{x_i\}$
Find: all-atom protein model
for $\vec{x}_i \in \mathbf{X}$ **do**
    $\mathbf{F} \leftarrow$ rotation-invariant features in a neighborhood around x_i
    $\mathbf{N} \leftarrow k$ examples in DB minimizing weighted Euclidean distance
    for $\vec{n}_i \in \mathbf{N}$ **do**
        $\vec{n}_i' \leftarrow$ optimal superposition of $\vec{n}_i$ into map at $\vec{x}_i$
        $score_i \leftarrow RSCC(n_i, \mathbf{M}(\vec{x}_i))$ // *see Equation 8.9*
    end
    Choose n_i' maximizing $score_i$
    Add individual atoms of n_i' to model
end

---

this correlation coefficient. It quickly approximates this optimal rotation and translation by aligning the moments of inertia between the template density region ρ_1 and target density region ρ_2. LOOKUP computes the real-space correlation at each alignment, and selects the highest-correlated candidate. Finally, LOOKUP retrieves the translated and rotated coordinated atoms of the top-scoring candidate and places them in the model.

8.5.4 Post-processing routines

Since each residue's atoms are copied from previous examples and glued together in the density map, the model produced by LOOKUP may contain some physically infeasible bond lengths, bond angles or $\phi - \psi$ torsion angles. TEXTAL's final step is improving the model using a few simple post-processing heuristics. First, LOOKUP often reverses the backbone direction of a residue; TEXTAL's post-processing makes sure that all chains are oriented in a consistent direction. Refinement, like that of ARP/WARP, corrects improper bond lengths and bond angles, iteratively moving individual atoms to fit the density map better. Finally, TEXTAL takes into account the target protein's sequence to fix mismatched residues.

TEXTAL makes use of a provided sequence by aligning the map-determined model sequence to the provided input sequence, using a Smith-Waterman dynamic-programming alignment. If there is agreement between the sequences above some threshold, then a second LOOKUP pass corrects residues where the alignment disagrees. In this second pass, LOOKUP is restricted to only consider residues of the type indicated by the sequence alignment. Like RESOLVE, TEXTAL's end result is a complete all-atom protein model, illustrated for our example map in Figure 8.15c.

8.5.5 Discussion

TEXTAL – like RESOLVE – uses higher-order constructs than atoms in order to successfully solve low-quality maps. In practice, TEXTAL works well on maps at around **3A** resolution. TEXTAL's key contribution is the use of rotation-invariant features to recognize patterns in the map. This feature representation allows accurate C_α identification using a neural network; it also does well at classification of amino-acid type. TEXTAL tends to do better than RESOLVE at sidechain identification due to this feature set. One key shortcoming, however, is limiting the initial backbone trace to skeleton points in the density map. In very poor maps, skeletonization is inaccurate; this is in part responsible for TEXTAL's failure in maps worse than **3A**. However, TEXTAL has successfully employed previously solved structures and domain knowledge from structural biology to produce an accurate map interpretation.

8.6 ACMI

ACMI (automatic crystallographic map interpreter) is a recent method developed by DiMaio et al. for tracing protein backbones in poor-quality (3 to **4A** resolution) density maps [19]. ACMI takes a probabilistic approach to electron density map interpretation, finding the most likely layout of the backbone under some likelihood function. This likelihood function takes into account local matches to the density map as well as the global pairwise configuration of residues. The key difference between ACMI and the preceding methods is that ACMI represents each residue not as a single location, but rather as a complete *probability distribution* over the full electron density map. This property – not forcing each residue to a single location – is advantageous as it naturally handles noise in the map, errors in the input sequence and disordered regions in the protein.

Figure 8.16 shows a high-level overview of ACMI. The algorithm is comprised of two main components: a *local matching* component that probabilistically matches individual amino acids to the density map, and a *global constraint* component where the backbone chain is probabilistically refined from the local matches. Global refinement is based on physical laws governing the structure of proteins. ACMI's key is an efficient inference algorithm that determines the most probable *physically feasible* backbone trace in an electron density map.

8.6.1 Local matching

The *local matching* component of ACMI is provided the density map and the protein's amino acid sequence. It computes, for each residue in the protein, a probability distribution $P_i(\vec{w}_i)$ over all locations $\vec{w}_i$ in the unit cell. This

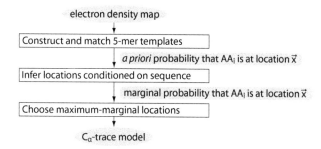

Figure 8.16 *A flowchart of* ACMI.

Algorithm 8.6: ACMI's local-matching algorithm

Given: sequence **seq**, electron density map $\mathbf{M}$
Find: probability distribution $P_i(\vec{w}_i)$ of each residue over map
for each residue $seq_i \in$ **seq do**
    $\mathbf{N} \leftarrow$ instances of 5-mer $\langle seq_{i-2} \ldots seq_{i+2} \rangle$ from PDB
    $\mathbf{C} \leftarrow$ cluster $\mathbf{N}$, extract centroids
    $\lambda_i \leftarrow$ cluster weights
    for each cluster centroid $c_j \in \mathbf{C}$ **do**
        $t(\vec{w}_i) \leftarrow$ Perform 6D search for c_j over density map
        Use a tuning set to convert $t(\vec{w}_i)$ to probabilities $P_{ij}(\vec{w}_i)$
    end
    $P_i(\vec{w}_i) \leftarrow \displaystyle\sum_{\text{clusters } j} \lambda_i P_{ij}(\vec{w}_i)$
end

---

probability distribution reflects the probability that residue i is at position and orientation $\vec{w}_i$ in the unit cell.

ACMI's local match – shown in Algorithm 8.6 – is based upon a 5-mer (that is, a polypeptide sequence five amino-acids in length) search. Using a set of previously solved structures, ACMI first constructs a basis set of structures describing the conformational space of each 5-mer in the protein. ACMI searches for each of these fragments in the map over all translations and rotations. This local search produces – for each residue i – an estimated probability distribution of that residue i's location and orientation in the unit cell, $P_i(\vec{w}_i)$.

Constructing a sequence-specific 5-mer basis set

Given some 5-amino-acid sequence, ACMI uses the Protein Data Bank (PDB), a repository of solved protein structures, to find all the instances of this sequence. If there are fewer than fifty such instances, ACMI considers "near-

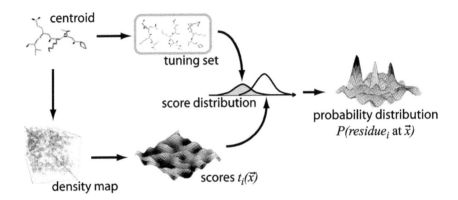

centroid

tuning set

score distribution

probability distribution

$P(residue_i$ at $\vec{x})$

density map   scores $t_i(\vec{x})$

Figure 8.17 *An overview of the 5-mer template matching process. Given a cluster of 5-mers for residue i,* ACMI *performs a 6D search for the fragment in the density map.* ACMI *also matches the fragment to a tuning set of known structures, using Bayes' rule to determine a probability distribution from the match scores* $t(\vec{w})$.

neighbors" (using the PAM-120 score, a measure of amino-acid similarity) until at least fifty distinct conformations are available. It uses these structures to represent the *conformational space* of the given 5-mer.

Searching for all these conformations in the electron density map is wasteful, because many are redundant, so ACMI clusters the structures into a smaller number of distinct conformations, representing each cluster with a single "central" instance (or *centroid*) of that cluster and a numeric weight. ACMI stores noncentroid instances from each cluster as well for tuning. Clustering uses the all-atom root mean squared deviation (RMSd) as a distance metric. ACMI can perform this clustering process in advance for all 3.2×10^6 possible 5-mers.

Searching for 5-mer centroid fragments

Given a protein sequence, ACMI considers the 5-mer sequence centered at each amino-acid. It extracts the fragments which represent the conformational space of that sequence, which it learns from the clustering process, and searches the density map for these fragments. Figure 8.17 illustrates this process graphically. Given these fragments and a resolution limit, ACMI builds an expected density map for each fragment. Then, at each map location, ACMI computes the standardized mean squared electron density difference $t(\vec{x})$ between the map and the fragment. Notice that this $t(\vec{x})$ is the same density map similarity function employed by RESOLVE, shown in Equation 8.7.

ACMI's fragment search is a 6D search: it considers every rotation of a fragment at every location in the map. Using FFFEAR's FFT-based convolution, one can compute this function very efficiently [11]. ACMI then stores – at each position – the best-matching 5-mer fragment and corresponding rotation.

The electron density difference function $t(\vec{x})$ is a good measure of similarity between regions of density; however, it doesn't allow comparison of scores from different templates. ACMI uses each cluster's tuning set in order to convert squared density differences into a probability distribution over the unit cell. To compute probabilities, Bayes' rule is used:

$$P(\text{res. } i \text{ at } \vec{x}_i | t(\vec{x}_i)) = \frac{P(t(\vec{x}_i) | \text{res. } i \text{ is at } \vec{x}_i) \cdot P(\text{res. } i \text{ at } \vec{x}_i)}{P(t(\vec{x}_i))}. \qquad (8.17)$$

ACMI computes terms on the right-hand side: the denominator, $P_i(t(\vec{x}_i))$, is the distribution of match scores over the (unsolved) map. The prior probability, $P_i(\text{res. } i \text{ at } \vec{x}_i)$, is simply a normalization term. ACMI drops this term and simply ensures that probabilities over the map are normalized to sum to the number of copies of the 5-mer in the map. However, the first term in the numerator - the distribution of scores when the 5-mer matches the map - is trickier to compute. ACMI estimates this term using a *tuning set* saved from an earlier step. This tuning set contains instances of a particular 5-mer conformation other than the centroid. Matching the centroid 5-mer structure's density to each tuneset structure's density gives an accurate estimate to this term. As shorthand, we will refer to this probability distribution simply as $P_i(\vec{x}_i)$ for the remainder of this section. Figure 8.18a plots this probability distribution for two residues in the example protein.

8.6.2 Global constraints

Given each residue's independent probability distribution over the unit cell, $P_i(\vec{x}_i)$, ACMI accounts for global constraints (enforced by *physical feasibility*) of a conformation by modeling the protein using a *pairwise Markov field*. A pairwise Markov field defines the joint probability distribution over a set of variables as a product of *potential functions* defined over vertices and edges in an undirected graph (see Figure 8.19).

Markov field model

Formally, the graph $\mathcal{G} = (\mathcal{V}, \mathcal{E})$ consists of a set of nodes $i \in \mathcal{V}$ connected by edges $(i, j) \in \mathcal{E}$. Each node in the graph is associated with a (hidden) random variable $\vec{w}_i \in \mathbf{W}$, and the graph is conditioned on a set of observation variables $\mathbf{M}$. Each vertex has a corresponding *observation potential* $\psi_i(\vec{w}_i, \mathbf{M})$, and each edge is associated with an *edge potential* $\psi_{ij}(\vec{w}_i, \vec{w}_j)$. Then, the probability of a complete trace $\mathbf{W}$ is:

$$P(\mathbf{W}|\mathbf{M}) \propto \prod_{(i,j) \in \mathcal{E}} \psi_{ij}(\vec{w}_i, \vec{w}_j) \times \prod_{i \in \mathcal{V}} \psi_i(\vec{w}_i, \mathbf{M}). \qquad (8.18)$$

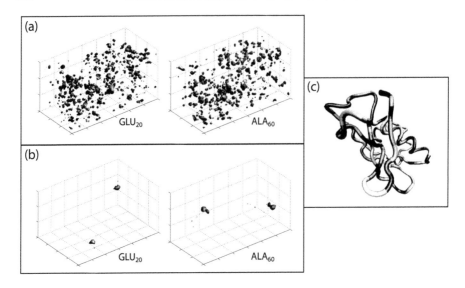

Figure 8.18 *Intermediate steps in* ACMI*'s structure determination: (a) the matching probability distributions* $P_i(\vec{x}_i)$ *on two residues, GLU$_{20}$ and ALA$_{60}$, contoured at* $p = 10^{-4}$, *(b) the marginal probability distributions over the same two residues and (c) the final (backbone-only) model, each residue shaded by likelihood (darker is more likely).*

Probabilistic inference finds the labels $\mathbf{W} = \{\vec{w}_i\}$ maximizing this probability, given some $\mathbf{M}$.

To encode a protein in a Markov field model, ACMI constructs an undirected graph where each node i corresponds to an amino-acid residue in the protein. The label $\vec{w}_i = \langle \vec{x}_i, \vec{q}_i \rangle$ for each amino-acid consists of seven terms: the 3D Cartesian coordinates $\vec{x}_i$ of the residue's alpha Carbon (C_α), and four "orientation" parameters $\vec{q}_i$ (three rotational parameters plus a "bend" angle). The observation potential $\psi_i(w_i, \mathbf{M})$ associated with each residue is the previously computed probability distribution $P_i(\vec{x}_i)$, with – at each position – all the probability mass stored in the orientation of the best match.

Edges in the graph enforce global constraints in a pairwise manner. The graph is fully connected (that is, every pair of residues is connected by an edge); DiMaio breaks the potential functions ψ_{ij} associated with each edge into two types. *Adjacency potentials* ψ_{adj} associated with edges between adjacent residues ensure that these neighboring C_α's maintain the proper 3.8 spacing, and C_α triples maintain the proper bend angle. *Occupancy potentials* ψ_{occ} on all other edges prevent two residues from occupying the same region in three-dimensional space. Thus, the full joint probability of a trace, given a

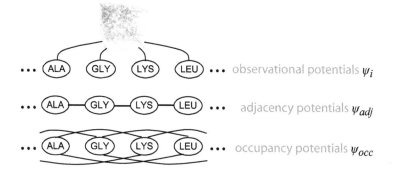

Figure 8.19 ACMI's protein backbone model. The joint probability of a conformation of residues is the product of an observation potential psi$_i$ at each node, (b) an adjacency potential between adjacent residues and (c) an occupancy potential between all pairs of nonadjacent residues.

map, is:

$$P(\mathbf{W}|\mathbf{M}) \propto \prod_{(\vec{w}_i,\vec{w}_j)\in\mathbf{W},|i-j|=1} \psi_{adj}(\vec{w}_i,\vec{w}_j)$$

$$\times \prod_{(\vec{w}_i,\vec{w}_j)\in\mathbf{W},|i-j|>1} \psi_{occ}(\vec{w}_i,\vec{w}_j)$$

$$\times \prod_{\vec{w}_i\in\mathbf{W}} \psi_i(\vec{w}_i,\mathbf{M}). \qquad (8.19)$$

Adjacency potentials

Adjacency potentials, which connect every neighboring pair of residues, are the product of two constraining functions, a distance constraint function and a rotational constraint function:

$$\psi_{adj}(\vec{w}_i,\vec{w}_j) = p_x(||\vec{x}_i - \vec{x}_j||) \cdot p_\theta(\vec{w}_i,\vec{w}_j). \qquad (8.20)$$

In proteins, the C_α - C_α distance is a nearly invariant 3.8**A**. Thus, the potential p_x takes the form of a tight Gaussian ($\sigma = 0.03\mathbf{A}$) around this ideal value, softened a bit for grid effects. ACMI defines the potential p_θ using an alternate parameterization of the angular parameters $\vec{q}_i$.

Specifically, ACMI represents these four degrees of freedom as two pairs of $\theta - \phi$ spherical coordinates: the most likely direction of the forward $(i + 1)$ residue and the backward $(i-1)$ residue. When matching the 5-mer templates into the density map, at each location x_i, ACMI stores four values – θ_b, ϕ_b, θ_f and ϕ_f – indicating the direction of both adjacent C_α in the rotated, best-matching 5-mer.

The angular constraint function p_θ is then – in each direction – a fixed-width Gaussian on a sphere, centered on this preferred orientation, (θ_b, ϕ_b) or (θ_f, ϕ_f).

Occupancy potentials

Occupancy potentials ensure that two residues do not occupy the same location in space. They are defined independently of orientation, and are merely a step function that constrains two (nonadjacent) C_α's be at least 3**A** apart. It is in this structural potential function that ACMI deals with crystallographic symmetry, by ensuring that all crystallographic copies are at least 3**A** apart:

$$\psi_{occ}(\vec{w}_i, \vec{w}_j) = \begin{cases} 1 & \left(\min_{\substack{symmetric \\ transforms\ K}} ||x_i - K(x_j)|| \right) \geq 3.0\mathbf{A} \\ 0 & \text{otherwise.} \end{cases} \tag{8.21}$$

Multiple chains in the asymmetric unit are also handled by ACMI: edges enforcing occupancy constraints fully connect separate chains.

Tracing the backbone

ACMI's backbone trace – shown in Algorithm 8.7 – is equivalent to finding the labels $\mathbf{W} = \{\vec{w}_i\}$ that maximize the probability of Equation 8.18. Since the graph over which the joint probability is defined contains loops, finding an exact solution is infeasible (dynamic programming can solve this in quadratic time for graphs with no loops). ACMI uses belief propagation (BP) to compute an approximation to the marginal probability for each residue i (that is, the full joint probability with all but one variable summed out). ACMI chooses the label for each residue that maximizes this marginal as the final trace.

Belief propagation is an inference algorithm - based on Pearl's polytree algorithm for [21] Bayesian networks - that computes marginal probabilities using a series of local messages. At each iteration, an amino acid computes an estimate of its marginal distribution (i.e., an estimate of the residue's location in the unit cell) as the product of that node's local evidence ψ_i and all incoming messages:

$$\hat{b}_i^n(\vec{w}_i) \propto \psi_i(\vec{w}_i, \mathbf{M}) \times \prod_{j \in \Gamma(i)} m_{j \to i}^n(\vec{w}_i). \tag{8.22}$$

Above, the belief $\hat{b}_i^n$ at iteration n is the best estimate of residue i's marginal distribution,

$$\hat{b}_i^n(\vec{w}_i) \approx \sum_{\vec{w}_0} \cdots \sum_{\vec{w}_{i-1}} \sum_{\vec{w}_{i+1}} \cdots \sum_{\vec{w}_N} P(\mathbf{W}, \mathbf{M}). \tag{8.23}$$

Algorithm 8.7: ACMI's global-constraint algorithm

Given: individual residue probability distributions $P_i(\vec{w}_i)$
Find: *approximate* marginal probabilities $\hat{b}_i(\vec{w}_i)$
$\forall i$ initialize belief $\hat{b}_i^0(\vec{w}_i)$ to $P_i(\vec{w}_i)$
repeat
    for each residue i **do**
        $\hat{b}_i(\vec{w}_i) \leftarrow P_i(\vec{w}_i)$
        for each residue $j \neq i$ **do**
            // *compute incoming message*
            if $|i - j| = 1$ **then**
                $m_{j \to i}^n(\vec{w}_i) \leftarrow \int_{\vec{w}_j} \psi_{adj}(\vec{w}_i, \vec{w}_j) \times \frac{\hat{b}_j}{m_{i \to j}^{n-1}}(\vec{w}_j)d\vec{w}_j$
            else
                $m_{j \to i}^n(\vec{w}_i) \leftarrow \int_{\vec{w}_j} \psi_{occ}(\vec{w}_i, \vec{w}_j) \times \hat{b}_j(\vec{w}_j)d\vec{w}_j$
        end

            // *aggregate messages*
            $\hat{b}_i(\vec{w}_i) \leftarrow \hat{b}_i(\vec{w}_i) \times m_{j \to i}^n(\vec{w}_i)$
        end
    end
until ($\hat{b}_i$'s converge)

Message update rules determine each message:

$$m_{j \to i}^n(\vec{w}_i) \propto \int_{\vec{w}_j} \psi_{ij}(\vec{w}_i, \vec{w}_j) \times \psi_j(\vec{w}_j, \mathbf{M})$$

$$\times \prod_{k \in \Gamma(j) \setminus i} m_{k \to j}^{n-1}(x_j) \, d\vec{w}_j. \qquad (8.24)$$

Computing the message from j to i uses all the messages going into node j *except* the message from node i. When using BP in graphs with loops, such as ACMI's protein model, there are no guarantees of convergence or correctness; however, empirical results show that loopy BP often produces a good approximation to the true marginal [22]. On the example protein, ACMI computes the marginal distributions shown in Figure 8.18b.

To represent belief and messages, ACMI uses a Fourier-series probability density estimate. That is, in the Cartesian coordinates $\vec{x}_i$, marginal distributions are represented as a set of 3-dimensional Fourier coefficients f_k, where, given an upper-frequency limit, (H, K, L):

$$b_i^n(\vec{x}_i) = \sum_{h,k,l=0}^{H,K,L} f_{hkl} \times e^{-2\pi i(\vec{x}_i \cdot \langle h,k,l \rangle)}. \qquad (8.25)$$

In *rotational parameters*, ACMI divides the unit cell into sections, and in each section stores a single orientation $\vec{q}_i$. These orientations correspond to the best-matching 5-mer orientation. More importantly, these stored orientations are not updated by belief propagation: messages are independent of the rotational parameters $\vec{q}_i$.

To make this method tractable, ACMI approximates all the outgoing occupancy messages at a single node:

$$m_{j \to i}^n(\vec{w}_i) \propto \int_{\vec{w}_j} \psi_{occ}(\vec{w}_i, \vec{w}_j) \times \frac{b_j^n(\vec{w}_j) d\vec{w}_j}{m_{i \to j}^{n-1}(\vec{w}_j)}. \qquad (8.26)$$

The approximation drops the denominator above:

$$m_{j \to *}^n(\vec{w}_i) \propto \int_{\vec{w}_j} \psi_{occ}(\vec{w}_i, \vec{w}_j) \times b_j^n(\vec{w}_j) d\vec{w}_j. \qquad (8.27)$$

That is, ACMI computes occupancy messages using all incoming messages to node j *including* the message from node i. All outgoing occupancy messages from node j, then, use the same estimate. ACMI caches the product of all occupancy messages $\prod_i m_{i \to *}$. This reduces the running time of each iteration in a protein with n amino acids from $O(n^2)$ to $O(n)$.

Additionally, ACMI quickly computes all *occupancy* messages using FFTs:

$$\mathcal{F}\left[m_{j \to i}^n(\vec{w}_i)\right] = \mathcal{F}\left[\psi_{occ}(\vec{w}_i, \vec{w}_j)\right] \times \mathcal{F}\left[\left(\prod m_{j \to i}^{n-1}(\vec{w}_i)\right)\right]. \qquad (8.28)$$

This is possible only because the occupancy potential is a function of the *difference* of the labels on the connected nodes, $\psi_{occ}(\vec{w}_i, \vec{w}_j) = f(\|\vec{x}_i - \vec{x}_j\|)$.

Finally, although ACMI computes the complete marginal distribution at each residue, a crystallographer is only interested in a single trace. The backbone trace consists of the locations for each residue that maximize the marginal probability:

$$\vec{w}_i^* = \arg\max_{\vec{w}_i} \hat{b}_i(\vec{w}_i)$$

$$= \arg\max_{\vec{w}_i} \psi_i(\vec{w}_i, \mathbf{M}) \times \prod_{j \in \Gamma(i)} m_{j \to i}^n(\vec{w}_i). \qquad (8.29)$$

ACMI's backbone trace on the example map is illustrated in Figure 8.18c. The backbone trace is shaded by confidence: since ACMI computes a complete probability distribution, it can return not only a putative trace, but also a likelihood associated with each amino acid. This likelihood provides the crystallographer with information about what areas in the trace are likely flawed; ACMI can also produce a high-likelihood partial trace suitable for phase improvement.

8.6.3 Discussion

ACMI's unique approach to density map interpretation allows for an accurate

CONCLUSION 273

trace in extremely poor maps, including those in the 3 to 4**A** resolution range. Unlike the other residue-based methods, ACMI is model-based. That is, it constructs a model *based on the protein's sequence*, then searches the map for that particular sequence. ACMI then returns the most likely interpretation of the map, given the model. This is in contrast to TEXTAL and RESOLVE which search for "some backbone" in the map, then align the sequence to this trace after the fact. This makes ACMI especially good at sidechain identification; even in extremely bad maps, ACMI correctly identifies a significant portion of residues. Unfortunately, ACMI also has several shortcomings. Requiring complete probability distributions for the location of each residue is especially time consuming; this has limited the applicability of the method so far. Additionally, in poor maps, belief propagation fails to converge to a solution, although in these cases a partial trace is usually obtained. By incorporating probabilistic reasoning with structural biology domain knowledge, ACMI has pushed the limit of interpretable maps even further.

8.7 Conclusion

A key step in determining protein structures is interpreting electron density maps. In this area, bioinformatics has played a key role. This chapter describes how four different algorithms have approached the problem of electron density map interpretation:

- The WARPNTRACE procedure in ARP/WARP was the first method developed for automatic density map interpretation. Today, it is still the most widely used method by crystallographers. ARP/WARP uses an "atom-level" technique in which free atoms are first placed into the density map, free atoms are next linked into chains introducing constraints and, finally, the combined model is refined to better explain the map. The method iterates through these three phases, at each iteration using the partial model to improve the map. Because it works at the level of individual atoms, it requires 2.3**A** or better map resolution.

- RESOLVE is a method that searches for higher-level constructs – amino acids and secondary structure elements – than ARP/WARP's atom-level method. Consequently, it produces accurate traces in poor-quality (around 3**A**) electron density maps, unsolvable by ARP/WARP. It, too, is widely used by crystallographers. RESOLVE begins by matching short secondary-structure fragments to the maps, then uses a large fragment library to extend these matches. Finally, overlapping matches are merged in a greedy fashion, and sidechains are traced. Incorporating structural domain knowledge is key to this method's success.

- TEXTAL also accurately interprets medium to low resolution (2 to 3**A**), using residue-level matching. It represents regions of density as a vector of rotation-invariant features. This alternate representation serves several purposes. It is used to learn a classifier to identify C_α's, and it is also used to recognize the residue type of each putative C_α through a database comparison. The classifier is also very well suited to identifying the residue type in a given region, making it more accurate than RESOLVE in this respect. Additionally, TEXTAL's rotation-invariant representation enables fast matching of regions of density. This allows TEXTAL to easily make use of previously solved structures in its map interpretation, providing accurate traces even in poor-quality maps.

- ACMI uses a probabilistic model to trace protein backbones in poor-quality (3 to 4**A** resolution) density maps. It finds the most likely layout of the backbone under a likelihood function which takes into account local matches to the density map as well as the global pairwise configuration of residues. Unlike other methods, ACMI represents each residue not as a single location but as a probability distribution over the entire density map. Also unlike other approaches, it constructs a model based on the protein's sequence, and finds the most likely trace of that particular sequence in the density map. ACMI's probabilistic, model-based approach results in accurate tracing and sidechain identification at poor resolutions.

As each of these methods extended the limit of what resolution maps are automatically interpreted, they brought with them two things: first, a higher-level "basic construct" at which the algorithm searches, and second, better incorporation of structural domain knowledge. These two key aspects are what enables interpretation in maps where individual atoms are not distinguishable, and in the future, what will extend the threshold of interpretation further.

As fewer and fewer fine details are visible in poor-resolution maps, algorithms must use a higher and higher level basic construct – that is, the template for which they search – in order to be robust against blurred details. ARP/WARP has had success when most atoms are visible in the map. If individual atoms are blurred out, the atom-level method will fail. However, a residue-based method like TEXTAL – which will locate amino acids assuming entire residues are not blurred – is robust enough to handle interpretation when atom details are unclear. Similarly, using secondary-structure elements allows even better interpretation. Probabilistic methods like ACMI take this still further: its interpretation considers a single flexible protein-sized element, and is robust enough to handle maps where entire residues are indistinguishable.

Another important feature of these methods is the increasing use of structural domain knowledge. In a way, this makes sense: the crystallographers task is inherently no different than that of the *ab initio* protein folding algorithm. A crystallographer simply has the assistance of a "picture" of the protein (the density map). All four methods use structural domain knowledge, by employing previous solved structures in model building. Primarily, these structures are used in the construction of a set of "representative fragments;" however, ACMI and TEXTAL also use previously solved structures to learn model parameters. Future methods will increasingly rely on such domain knowledge.

In the future, with the rising popularity of high-throughput structure determination, automated map interpretation methods will play a significant role. Improvements in laboratory techniques are producing density maps at a faster rate. In addition, experimental techniques such as cryo-electron microscopy are producing density maps that approach 5-6**A** resolution, and are continually improving. Automatically determining structure from maps of this quality will be increasingly important. To further extend the limit of what maps are interpretable, automated interpretation techniques will need to use more domain knowledge, and consider searching not for a single residue or a single tripeptide, but rather entire proteins, probabilistically. Providing automated systems with increasing amounts of a crystallographer's "expertise" is key to future improvements in these methods.

8.8 Acknowledgments

This work supported by NLM grant 1R01 LM008796-01 and NLM Grant 1T15 LM007359-01. The authors would also like to thank George Phillips and Tom Ioerger for assistance in writing this chapter.

References

[1] B. Rost and C. Sander (1993). Prediction of protein secondary structure at better than 70% accuracy. *J Mol Biol.*

[2] G. Rhodes (2000). *Crystallography Made Crystal Clear.* Academic Press.

[3] J. Abrahams and R. De Graaff (1998). New developments in phase refinement. *Curr Opin Struct Biol.*

[4] S. Russell and P. Norvig (1995). *Artificial Intelligence: A Modern Approach.* Prentice Hall.

[5] T. Mitchell (1997). *Machine Learning.* McGraw-Hill.

[6] V. Lamzin and K. Wilson (1993). Automated refinement of protein models. *Acta Cryst.*

[7] A. Perrakis, T. Sixma, K. Wilson, and V. Lamzin (1997). wARP: Improvement and extension of crystallographic phases by weighted averaging of multiple refined dummy atomic models. *Acta Cryst.*

[8] R. Morris, A. Perrakis, and V. Lamzin (2002). ARP/wARP's model-building algorithms: the main chain. *Acta Cryst.*

[9] J. Greer (1974). Three-dimensional pattern recognition. *J Mol Biol.*

[10] L. Leherte, J. Glasgow, K. Baxter, E. Steeg, and S. Fortier (1997). Analysis of three-dimensional protein images. *JAIR.*

[11] K. Cowtan (2001). Fast Fourier feature recognition. *Acta Cryst.*

[12] T. Oldfield (2001). A number of real-space torsion-angle refinement techniques for proteins, nucleic acids, ligands and solvent. *Acta Cryst.*

[13] T. Terwilliger (2002). Automated main-chain model-building by template-matching and iterative fragment extension. *Acta Cryst.*

[14] T. Terwilliger (2002). Automated side-chain model-building and sequence assignment by template-matching. *Acta Cryst.*

[15] D. Levitt (2001). A new software routine that automates the fitting of protein X-ray crystallographic electron density maps. *Acta Cryst.*

[16] T. Ioerger, T. Holton, J. Christopher, and J. Sacchettini (1999). TEXTAL: A pattern recognition system for interpreting electron density maps. *Proc ISMB.*

[17] T. Ioerger and J. Sacchettini (2002). Automatic modeling of protein backbones in electron density maps via prediction of C-alpha coordinates. *Acta Cryst.*

[18] K. Gopal, T. Romo, E. Mckee, K. Childs, L. Kanbi, R. Pai, J. Smith, J. Sacchettini, and T. Ioerger (2005). TEXTAL: Automated crystallographic protein structure determination. *Proc. IAAI.*

[19] F. DiMaio, J. Shavlik, and G. Phillips (2006). A probabilistic approach to protein backbone tracing in electron density maps. *Proc ISMB.*

[20] K. Gopal, T. Romo, J. Sacchettini, and T. Ioerger (2004). Weighting features to recognize 3D patterns of electron density in X-ray protein crystallography. *Proc CSB.*

[21] J. Pearl (1988). *Probabilistic Reasoning in Intelligent Systems.* Morgan Kaufmann, San Mateo.

[22] K. Murphy, Y. Weiss, and M. Jordan (1999). Loopy belief propagation for approximate inference: An empirical study. *Proc. UAI.*

CHAPTER 9

Soft Computing in Biclustering

Haider Banka and Sushmita Mitra
Indian Statistical Institute, Kolkata, India

9.1 Introduction

Microarray experiments produce gene expression patterns that offer dynamic information about cell function. This is useful while investigating complex interactions within the cell [1, 2]. Microarrays are used in the medical domain to produce molecular profiles of diseased and normal tissues of patients. Such profiles are useful for understanding various diseases, and aid in more accurate diagnosis, prognosis, treatment planning, as well as drug discovery. Gene expression data being typically high-dimensional, it requires appropriate data mining strategies like clustering and biclustering for further analysis [3].

A cluster is a collection of data objects which are similar to one another within the same cluster but dissimilar to the objects in other clusters [4, 5]. The problem is to group N patterns into n_c possible clusters with high *intra-class* similarity and low *inter-class* similarity by optimizing an objective function. Here the genes are typically partitioned into disjoint or overlapped groups according to the similarity of their expression patterns over *all* conditions.

It is often observed that a subset of genes is coregulated and coexpressed under a subset of conditions, but behave almost independently under other conditions. Here the term "conditions" can imply environmental conditions, as well as time points in a gene expression profile. Biclustering refers to the *simultaneous clustering of both genes and conditions* in the process of knowledge discovery about local patterns from microarray data [6]. Since typical biological processes begin and end over a continuous period of time, the genes often need to be identified in the form of biclusters over continuous columns [7, 8].

There exist analogous terminologies for biclustering in the literature [9] like bidimensional clustering, subspace clustering, projected clustering, coclustering, simultaneous clustering, direct clustering, block clustering, two-way clustering, two-mode clustering and two-sided clustering. In fact, the term *subspace clustering* was first proposed in the general data mining domain [10] to find subsets of objects such that they appear as a cluster in a subspace formed by subsets of the features. Hence the subsets of features for various subspace clusters can be different. Two subspace clusters can share some common objects and features, and some objects may not belong to any subspace clusters.

Simultaneous clustering along both dimensions may also be viewed as clustering preceded by dimensionality reduction along columns. But the simultaneity of the process endows it with statistical regularization, thereby generating more interpretable compact clusters with reduced computational complexity. Biclustering can be overlapped or partitional. The two key components in formulating a bicluster are (i) minimizing the discrepancy between the original and the compressed representation [6], measured in terms of the sum of element-wise squared deviation, and (ii) the information-theoretic method [11, 12] which preserves a different set of summary statistics.

The optimization task of finding one or more biclusters by maintaining the two competing constraints, viz., homogeneity and size, is reported to be NP-complete [13]. The high complexity of this problem has motivated researchers to apply various approximation techniques to find near optimal solutions. Some approaches reported in literature include [6, 14, 15] that basically differ in their ways of formulating these conflicting objectives. Soft computing [16], involving fuzzy sets and evolutionary computing, has been successfully explored for faster discovery of biologically relevant overlapped biclusters.

The rest of the chapter is organized as follows. Section 9.2 introduces the basic concepts of the biclustering problem with a state-of-the-art review. The use of multi-objective GA (MOGA) and fuzzy possibilistic sets are investigated in the soft computing framework. These are presented in Sections 9.3 and 9.4. The experimental results and their comparative study, along with some biological validation in terms of gene ontology, are provided in Section 9.5. Finally, Section 9.6 concludes the chapter.

9.2 Biclustering

A bicluster is defined as a pair (g, c), where $g \subseteq \{1, \ldots, m\}$ is a subset of genes and $c \subseteq \{1, \ldots, n\}$ is a subset of conditions/time points. The optimization task [6] is to find the largest bicluster that does not exceed a certain homogeneity constraint. The size (or volume) $f(g, c)$ of a bicluster is defined as the number of cells in the gene expression matrix E (with values e_{ij}) that are covered by it. The homogeneity $\mathcal{G}(g, c)$ is expressed as a mean squared residue score; it represents the variance of the selected genes and conditions. We maximize

$$f(g, c) = |g| \times |c|, \tag{9.1}$$

subject to a low $\mathcal{G}(g,c) \leq \delta$ for $(g,c) \in X$, with $X = 2^{\{1,\ldots,m\}} \times 2^{\{1,\ldots,n\}}$ being the set of all biclusters, where

$$\mathcal{G}(g,c) = \frac{1}{|g| \times |c|} \sum_{i \in g, j \in c} (e_{ij} - e_{ic} - e_{gj} + e_{gc})^2. \tag{9.2}$$

Here

$$e_{ic} = \frac{1}{|c|} \sum_{j \in c} e_{ij}, \tag{9.3}$$

$$e_{gj} = \frac{1}{|g|} \sum_{i \in g} e_{ij} \tag{9.4}$$

are the mean row and column expression values for (g,c) and

$$e_{gc} = \frac{1}{|g| \times |c|} \sum_{i \in g, j \in c} e_{ij} \tag{9.5}$$

is the mean expression value over all cells contained in the bicluster (g,c). A user-defined threshold δ represents the maximum allowable dissimilarity within the bicluster. In other words, the residue quantifies the difference between the actual value of an element e_{ij} and its expected value as predicted from the corresponding row mean, column mean and bicluster mean. A set of genes whose expression levels change in accordance to each other over a set of conditions can thus form a perfect bicluster even if the actual values lie far apart. For a good bicluster, we have $\mathcal{G}(g,c) < \delta$ for some $\delta \geq 0$.

9.2.1 Heuristic and probabilistic approaches

A good survey on biclustering is available in literature [9], with a categorization of the different heuristic approaches made as follows.

- Iterative row and column clustering combination [17]: Apply clustering algorithms to the rows and columns of the data matrix, separately, and then combine the results using some iterative procedure.

- Divide and conquer [18]: Break the problem into smaller sub-problems, solve them recursively, and combine the solutions to solve the original problem.

- Greedy iterative search [6, 19]: Make a locally optimal choice, in the hope that this will lead to a globally good solution.

- Exhaustive biclustering enumeration [20]: The best biclusters are identified, using an exhaustive enumeration of all possible biclusters existent in the data, in exponential time.

- Distribution parameter identification [21]: Identify best-fitting parameters, by minimizing a criterion through an iterative approach.

The pioneering work by Cheng and Church [6] employs a set of heuristic algorithms to find one or more biclusters, based on a uniformity criteria. One bicluster is identified at a time, iteratively. There are iterations of masking null values and discovered biclusters (replacing relevant cells with random numbers), coarse and fine node deletion, node addition and the inclusion of inverted data. The computational complexity for discovering k biclusters is of the order of $O(mn \times (m + n) \times k)$, where m and n are the number of genes and conditions respectively. Here similarity is computed as a measure of the coherence of the genes and conditions in the bicluster. Although the greedy local search methods are by themselves fast, they often yield suboptimal solutions. An improvement over the result in Ref. [6] is visible [22], by slight modification of the node addition algorithm while incorporating a parameter to reduce the blind search space.

Sometimes the masking procedure may result in a phenomenon of random interference, thereby adversely affecting the subsequent discovery of high quality biclusters. In order to circumvent this problem, a two-phase probabilistic algorithm (FLOC) [19] is designed to simultaneously discover a set of possibly overlapping biclusters all at the same time. Initial biclusters (or seeds) are chosen randomly from the original data matrix. Iterative gene and/or condition additions and/or deletions are performed with a goal of achieving the best potential residue reduction. The time complexity of FLOC is lower for p iterations ($p \ll n + m$); i.e., $O((n + m)^2 \times k \times p)$.

The Plaid model [21] tries to capture the approximate uniformity in a submatrix of the gene expression data, while discovering one bicluster at a time in an iterative process with the Expectation-Maximization (EM) algorithm. Gene expression data is considered as a sum of multiple "layers," where each layer may represent the presence of a particular biological process in the bicluster. The input matrix is described as a linear function of variables corresponding to its biclusters, and an iterative maximization process is pursued for estimating the function. It searches for patterns where the genes differ in their expression levels by a constant factor. An extension of the Plaid model is described in Ref. [23].

Another probabilistic model is explored in Ref. [24], using gene regulation for the task of identifying overlapped biological processes and their regulatory mechanism. A key feature is that it allows genes to participate in multiple processes, thus providing a biologically plausible model for the process of gene regulation. Moreover, known genes are grouped to function together while recovering existing regulatory relationships. Novel hypotheses are developed for determining the regulatory role of previously uncharacterized proteins.

Statistical Subspace Clustering (SSC), for both continuous as well as discrete attributes, is described in Ref. [25], using EM algorithm for analyzing high dimensional data. It generates a set of rules, defined over fewer relevant dimensions, while also reducing the number of user supplied input parameters. The algorithm is robust both in terms of classification accuracy and interpretability.

Bipartite graphs are employed in Refs. [20] and [26], with a bicluster being defined as a subset of genes that jointly respond across a subset of conditions. The objective is to identify the maximum-weighted subgraph. Here [20] a gene is considered to be responding under a condition if its expression level changes significantly, over the connecting edge, with respect to its normal level. This involves an exhaustive enumeration, with a restriction on the number of genes that can appear in the bicluster. A simultaneous discovery of all biclusters is made, including overlapped ones.

A partition based co-clustering (also called checkerboard biclustering and spectral biclustering [27]) is used to partition rows and columns in a disjoint manner into separate clusters. A coupled two-way method [17] has been devised to iteratively generate a set of biclusters, at a time, in cancer datasets. It repeatedly performs one-way hierarchical clustering on the rows and columns of the data matrix, while using stable clusters of rows as attributes for column clustering and vice versa. The Euclidean distance is used as the similarity measure, after normalization of the data. However, the clustering results are sensitive to initial setting, and are sometimes difficult to interpret.

Gene ontology information, involving hierarchical functional relationships like "part of" or "overlapping," has been incorporated into the clustering process called Smart Hierarchical Tendency Preserving Algorithm (SHTP) [28]. A fast approximate pattern matching technique has been employed [29] to determine maximum-sized biclusters, with the number of conditions being greater than a specified minimum. The worst case complexity of the procedure is claimed to be $O(m^2n)$. Rich probabilistic models have been used [30] for discovering relations between expressions, regulatory motifs and gene annotations. The outcome is a collection of disjoint biclusters, generated in a supervised manner.

Efficient frequent pattern mining algorithms have been successfully integrated in DBF [31] to deterministically generate a set of good quality biclusters. Here the changing tendency between two conditions is modeled as an item, with the genes corresponding to transactions. A frequent itemset with the supporting genes forms a bicluster. In the second phase, these are iteratively refined by adding more genes and/or conditions.

9.2.2 Information-theoretic approach

Attribute clustering algorithm (ACA) [32] finds optimal meaningful clusters, by maximizing intra-group attribute interdependence. A new interdependence information measure is introduced to capture the interrelationship among attributes. The performance of gene classification, obtained from a pool of top significant genes, is used to evaluate the clusters.

A unified framework for partitional co-clustering, based on matrix approximation, has been developed [11, 12]. Here the approximation error is measured using a large class of loss functions called Bregman divergences, that include squared Euclidean distance and KL-divergence as special cases. Multiple structurally different co-clustering schemes are modeled in terms of a new *minimum Bregman information* principle that generalizes both the *maximum entropy* and *standard least squares*. It leads to a generalization for an entire class of the algorithms based on particular choices of the Bregman divergence and the set of summary statistics to be preserved.

9.2.3 Handling time domain

There exist a number of investigations dealing with time-series data [33, 34]. A polynomial-time algorithm based on the Monte-Carlo method for finding coclusters is described in Ref. [35]. Basic linear algebra and arithmetic tools are used in the time domain [36] to search biclusters having constant values on rows and columns, as well as those with coherent expressions.

A linear time continuous columns algorithm has been designed [37]. This reports all relevant biclusters, extracted in linear time corresponding to the size of the data matrix. The complexity is reduced after discretizing the expression values into *up, down* and *no change* symbols, followed by the generation of a suffix tree along with string processing. The performance is demonstrated on synthetic and real world data.

To overcome some of the existing limitations, an attempt has been made [8] to identify time lagged gene clusters based on the changing tendency of genes over time. The slope of the discretized expression matrix is converted into symbols 0, -1, 1. Finally identification of time lagged subgroups of genes with similar patterns is possible over consecutive time points. A binary reference model Bimax [38] provides an interesting comparative evaluation of biclustering.

9.2.4 Incorporating soft computing

Soft computing tools like fuzzy sets and evolutionary algorithms have also been used for biclustering. A possibilistic fuzzy approach has been investigated [39] to find one bicluster at a time by assigning membership to each gene and condition of the bicluster while minimizing the mean squared residue and maximizing the size. After convergence, the membership is defuzzified. However this methodology is found to be over sensitive to parameter tuning. Section 9.4 presents the algorithm in greater detail.

GAs have been employed incorporating local search strategy for identifying biclusters in gene expression data [14, 15]. Sequential evolutionary biclustering (SEBI) [40] detects high quality overlapped biclusters by introducing the concept of penalty into the fitness functions, in addition to some constrained weightage among the parameters.

An order preserving sub matrix (OPSM) has been developed in Refs. [7, 41, 42, 43], using greedy search heuristic and evolutionary strategies. However, there exist problems related to local optima, consumption of excessive computational resources or imposition of too strict constraints on the clustering criterion.

Simulated annealing (SA) based biclustering [44] is found to provide improved performance over that of Ref. [6], and is also able to escape from local minima. Unlike the optimization techniques like GA, that appreciate only improvements in the chosen fitness functions, SA also allows a probabilistic acceptance of temporary disimprovement in fitness scores. However, the results are often dependent on the temperature schedule.

When there are two or more conflicting characteristics to be optimized, one requires an appropriate formulation of the fitness function in terms of an aggregation of the different criteria involved. In such situations *multi-objective evolutionary algorithm* (MOEA) [45] provide an alternative, more efficient, approach to searching for optimal solutions. It has found successful application in biclustering of microarray gene expression patterns [15]. This is described in greater detail in Section 9.3.

9.3 Multi-Objective Biclustering

Multi-objective evolutionary algorithm is a global search heuristic, primarily used for optimization tasks. In this section we present the general framework and implementation details of multi-objective evolutionary algorithm for biclustering [15]. The aim is to find sub-matrices, where the genes exhibit highly correlated activities over a range of conditions. Since these two objectives are mutually conflicting, they become suitable candidates for multi-objective modeling. A new quantitative measure to evaluate the goodness of the bicluster is also introduced. Local search heuristics are employed, to speed up convergence by refining the chromosomes.

9.3.1 Representation

Each bicluster is represented by a fixed sized binary string called chromosome or individual, with a bit string for genes appended by another bit string for conditions. The chromosome corresponds to a solution for this optimal bicluster generation problem. A bit is set to one if the corresponding gene and/or condition is present in the bicluster, and reset to zero otherwise. Fig. 9.1 depicts such an encoding of genes and conditions in a chromosome.

The initial population is generated randomly. Uniform single-point crossover,

Figure 9.1 *An encoded chromosome representing a bicluster.*

single-bit mutation and crowded tournament selection are employed in the multi-objective framework. Both parent and offspring population, in each generation, are combined to select the best members as the new parent population. Diversity is maintained within the biclusters by using the crowding distance operator.

9.3.2 Multi-objective framework

We observe here that one needs to concentrate on generating maximal sets of genes and conditions while maintaining the "homogeneity" of the biclusters. These two characteristics of biclusters, being conflicting to each other, are well suited for multi-objective modeling. In order to optimize this pair of conflicting requirements, the fitness function f_1 is always maximized while function f_2 is maximized as long as the residue does not exceed the threshold δ. They are formulated as

$$f_1 = \frac{g \times c}{|G| \times |C|}, \tag{9.6}$$

$$f_2 = \begin{cases} \frac{\mathcal{G}(g,c)}{\delta} & \text{if} \quad \mathcal{G}(g,c) \le \delta \\ 0 & \text{otherwise,} \end{cases}$$

where g and c are the number of ones in the genes and conditions within the bicluster, $\mathcal{G}(g,c)$ is its mean squared residue score as defined by eqns. (9.2)-(9.5), δ is the user-defined threshold for the *maximum acceptable* dissimilarity or mean squared residue score of the bicluster and G and C are the total number of genes and conditions of the original gene expression array.

Note that f_1 is maximum for $g = G$ and $c = C$, *i.e.*, when the submatrix (g,c) is equal to the whole input dataset. Now as the size of the bicluster increases, so does the mean squared residue. Thereby f_2 is allowed to increase as long as it does not exceed the homogeneity constraint δ. Beyond this we assign a lower fitness value of zero to f_2, which is ultimately removed during nondominated front selection of MOEA.

9.3.3 Local search

Since the initial biclusters are generated randomly, it may happen that some irrelevant genes and/or conditions get included in spite of their expression values lying far apart in the feature space. An analogous situation may also arise during crossover and mutation in each generation. These genes and conditions, with dissimilar values, need to be eliminated deterministically. Furthermore, for good biclustering, some genes and/or conditions having similar expression values need to be incorporated as well. In such situations, local search strategies [6] can be employed to add or remove multiple genes and/or conditions. It was observed that, in the absence of local search, stand-alone single-objective or multi-objective EAs could not generate satisfactory solutions [14, 15]. The algorithm starts with a given bicluster and an initial gene expression array (G, C). The irrelevant genes or conditions having mean squared residue above (or below) a certain threshold are now selectively eliminated (or added) using the following conditions. A "node" refers to a gene or a condition in the sequel.

1. Multiple nodes deletion.

   (a) Compute e_{ic}, e_{gj}, e_{gc} and $\mathcal{G}(g, c)$ of the bicluster by eqns. (9.2)-(9.5).
   (b) Remove all genes $i \in g$ satisfying

   $$\frac{1}{|c|} \sum_{i \in g} (e_{ij} - e_{ic} - e_{gj} + e_{gc})^2 > \alpha \times \mathcal{G}(g, c). \qquad (9.7)$$

   (c) Recompute e_{ic}, e_{gj}, e_{gc} and $\mathcal{G}(g, c)$.
   (d) Remove all conditions $j \in c$ satisfying

   $$\frac{1}{|g|} \sum_{j \in c} (e_{ij} - e_{ic} - e_{gj} + e_{gc})^2 > \alpha \times \mathcal{G}(g, c). \qquad (9.8)$$

2. Single node deletion, corresponding to a refinement of **Step 1**.

   (a) Recompute e_{ic}, e_{gj}, e_{gc} and $\mathcal{G}(g, c)$ of the modified bicluster by **Step 1**.
   (b) Remove the node with largest mean squared residue (done for both gene and condition), one at a time, until the mean squared residue drops below δ.

3. Multiple nodes addition.

   (a) Recompute e_{ic}, e_{gj}, e_{gc} and $\mathcal{G}(g, c)$ of the modified bicluster of **Step 2**.
   (b) Add all genes $i \notin g$ satisfying

   $$\frac{1}{|c|} \sum_{i \in g} (e_{ij} - e_{ic} - e_{gj} + e_{gc})^2 \leq \mathcal{G}(g, c). \qquad (9.9)$$

   (c) Recompute e_{ic}, e_{gj}, e_{gc} and $\mathcal{G}(g, c)$.
   (d) Add all conditions $j \notin c$ satisfying

   $$\frac{1}{|g|} \sum_{j \in c} (e_{ij} - e_{ic} - e_{gj} + e_{gc})^2 \leq \mathcal{G}(g, c). \qquad (9.10)$$

It is proven that node deletion decreases the mean squared residue score of the bicluster [6]. Here the parameter α determines the rate of node deletion. Usually a higher value of α implies a decrease in multiple node deletion, such that the resulting bicluster size increases. This leads to an increase in execution time, during fine tuning in **Step 2** of the algorithm.

9.3.4 Algorithm

The Nondominated Sorting Genetic Algorithm (NSGA II) in combination with the local search procedure of Section 9.3.3 is used for generating the set of biclusters. The main steps of the proposed algorithm, repeated over a specified number of generations, are outlined as follows.

1. Generate a random population of size P.
2. Delete or add multiple nodes (genes and conditions) from each individual of the population, as discussed in Section 9.3.3.
3. Calculate the multi-objective fitness functions f_1 and f_2, using eqns. (9.6)-(9.7).
4. Rank the population using the dominance criteria of MOEA.
5. Calculate crowding distance.
6. Perform selection using crowding tournament selection.
7. Perform crossover and mutation (as in conventional GA) to generate offspring population of size P.
8. Combine parent and offspring population.
9. Rank the mixed population using dominance criteria and crowding distance, as above.
10. Replace the parent population by the best $|P|$ members of the combined population.

9.3.5 Quantitative evaluation

The bicluster should satisfy two requirements simultaneously. On one hand, the expression levels of each gene within the bicluster should be similar over the range of conditions, *i.e.*, it should have a low mean squared residue score. On the other hand, the bicluster should simultaneously be larger in size. Note that the mean squared residue represents the variance of the selected genes and conditions with respect to the coherence (homogeneity) of the bicluster.

In order to quantify how well the biclusters satisfy these two requirements, we introduce *Coherence Index CI* as a measure of evaluating their goodness. Here CI is defined as the ratio of mean squared residue score to the size of the formed bicluster. Let there be P biclusters of size $|g_k| \times |c_k|$, $\forall k \in P$ in eqn. (9.1), with mean squared residue score $\mathcal{G}_k(g_k, c_k)$ from eqns. (9.2)-(9.5). We define

$$\mathcal{G}_k(g_k, c_k) = \frac{1}{|g_k| \times |c_k|} \sum_{i \in g_k, j \in c_k} (e_{ij} - e_{ic_k} - e_{g_k j} + e_{g_k c_k})^2, \qquad (9.11)$$

$$f_k(g_k, c_k) = |g_k| \times |c_k|, \qquad (9.12)$$

resulting in

$$CI = \min_{k \in P} \frac{\mathcal{G}_k(g_k, c_k)}{f_k(g_k, c_k)}. \qquad (9.13)$$

The kth bicluster for $k \in P$ is considered to be good, if it has minimum CI_k among all $j \in P$ and $j \neq k$. A small mean square residue indicates that the corresponding gene set has consistent value over the samples. Note that an increase in bicluster size also leads to a decrease in the value of CI.

9.4 Fuzzy Possibilistic Biclustering

Often centralized clustering algorithms impose a *probabilistic constraint*, according to which the sum of the membership values of a point in all the clusters must be equal to one. Although this competitive constraint allows the unsupervised learning algorithms to find the barycenter (center of mass) of fuzzy clusters, the obtained evaluations of membership to clusters are not interpretable as a *degree of typicality*. Moreover isolated outliers can sometimes hold high membership values to some clusters, thereby distorting the position of the centroids. The possibilistic approach to clustering [46, 47] assumes that the membership function of a data point in a *fuzzy* set (or cluster) is absolute, *i.e.*, it is an evaluation of a degree of typicality not depending on the membership values of the same point in other clusters. In this section we first describe the basics of possibilistic clustering, before embarking on the fuzzy possibilistic biclustering approach.

9.4.1 Possibilistic clustering

Let $X = \{x_1, \ldots, x_r\}$ be a set of unlabeled data points, $Y = \{y_1, \ldots, y_s\}$ a set of cluster centers (or prototypes) and $U = [u_{pq}]$ the *fuzzy membership matrix*. In the Possibilistic c-means (PCM) algorithm the constraints on the elements of U are relaxed to [46]

$$u_{pq} \in [0, 1] \quad \forall p, q; \qquad (9.14)$$

$$0 < \sum_{q=1}^{r} u_{pq} < r \quad \forall p; \qquad (9.15)$$

$$\bigvee_{p} u_{pq} > 0 \quad \forall q. \qquad (9.16)$$

Roughly speaking, these requirements simply imply that a cluster cannot be empty and each pattern must be assigned to at least one cluster. This turns a standard fuzzy clustering procedure into a mode seeking algorithm.

In Ref. [47] the objective function consists of two terms, the first being the objective function of conventional c-means [48] while the second is a penalty (regularization) term considering the entropy of clusters as well as their overall membership values. It is defined as

$$J_m(U, Y) = \sum_{p=1}^{s} \sum_{q=1}^{r} u_{pq} E_{pq} + \sum_{p=1}^{s} \frac{1}{\beta_p} \sum_{q=1}^{r} (u_{pq} \log u_{pq} - u_{pq}), \qquad (9.17)$$

where $E_{pq} = \|\mathbf{x}_q - \mathbf{y}_p\|^2$ is the squared Euclidean distance, and the parameter β_p (*scale*) depends on the average size of the p-th cluster that must be assigned before the clustering procedure. Thanks to the regularizing term, points with a high degree of typicality have high u_{pq} values while those points not very representative have low u_{pq} values in all the clusters. Note that if we take $\beta_p \to \infty$ $\forall p$ (*i.e.*, the second term of $J_m(U, Y)$ is omitted), we obtain a trivial solution of the minimization of the remaining cost function (*i.e.*, $u_{pq} = 0$ $\forall p, q$), as no probabilistic constraint is assumed.

The pair (U, Y) minimizes J_m, under the constraints (9.14)–(9.16), only if [47]

$$u_{pq} = e^{-E_{pq}/\beta_p} \quad \forall p, q, \qquad (9.18)$$

and

$$\mathbf{y}_p = \frac{\sum_{q=1}^{r} \mathbf{x}_q u_{pq}}{\sum_{q=1}^{r} u_{pq}} \quad \forall p. \qquad (9.19)$$

These two equations may be interpreted as formulae for recalculating the membership functions and the cluster centers [49] following the Picard iteration technique.

A good initialization of centroids needs to be performed before applying PCM (say) by using fuzzy c-means [46, 47]. The PCM works as a refinement algorithm, allowing us to interpret the membership as the cluster typicality degree. Moreover, PCM displays high outlier rejection capability by making their membership very low.

Note that the lack of probabilistic constraints makes the PCM approach equivalent to a set of s independent estimation problems [50]

$$(u_{pq}, \mathbf{y}) = \arg \bigwedge_{u_{pq}, \mathbf{y}} \left[\sum_{q=1}^{r} u_{pq} E_{pq} + \frac{1}{\beta_p} \sum_{q=1}^{r} (u_{pq} \log u_{pq} - u_{pq}) \right] \quad \forall p, \quad (9.20)$$

that can be solved independently one at a time through a Picard iteration of eqns. (9.18)–(9.19).

9.4.2 Possibilistic biclustering

Here we generalize the concept of biclustering in a fuzzy set theoretical approach. For each bicluster we assign two vectors of membership, one for the rows and one other for the columns, denoting them respectively by **a** and **b**. In a crisp set framework row i and column j can either belong to the bicluster ($a_i = 1$ and $b_j = 1$) or not ($a_i = 0$ or $b_j = 0$). An element x_{ij} of X belongs to the bicluster if both $a_i = 1$ and $b_j = 1$, *i.e.*, its membership u_{ij} to the bicluster may be defined as [39]

$$u_{ij} = \text{and}(a_i, b_j), \tag{9.21}$$

with the cardinality of the bicluster expressed as

$$n = \sum_i \sum_j u_{ij} . \tag{9.22}$$

A fuzzy formulation of the problem can help to better model the bicluster and also to improve the optimization process. Now we allow membership u_{ij}, a_i and b_j to lie in the interval $[0, 1]$. The membership u_{ij} of a point to the bicluster is obtained by the average of row and column membership as [39]

$$u_{ij} = \frac{a_i + b_j}{2}. \tag{9.23}$$

The fuzzy cardinality of the bicluster is again defined as the sum of the memberships u_{ij} for all i and j from eqn. (9.22). In the gene expression framework we can now generalize eqns. (9.2)–(9.5)

$$\mathcal{G} = \sum_i \sum_j u_{ij} d_{ij}^2 \tag{9.24}$$

where

$$d_{ij}^2 = \frac{(e_{ij} + e_{gc} - e_{ic} - e_{gj})^2}{n} \tag{9.25}$$

$$e_{ic} = \frac{\sum_j u_{ij} e_{ij}}{\sum_j u_{ij}} \tag{9.26}$$

$$e_{gj} = \frac{\sum_i u_{ij} e_{ij}}{\sum_i u_{ij}} \tag{9.27}$$

$$e_{gc} = \frac{\sum_i \sum_j u_{ij} e_{ij}}{\sum_i \sum_j u_{ij}}. \tag{9.28}$$

The objective is to maximize the bicluster cardinality n while minimizing the residual $\mathcal{G}$ in the fuzzy possibilistic paradigm. Towards this aim we treat one bicluster at a time, with the fuzzy memberships a_i and b_j being interpreted as typicality degrees of gene i and condition j with respect to the bicluster. These requirements are fulfilled by minimizing the functional J_{B} as

$$J_{\mathrm{B}} = \sum_i \sum_j \left(\frac{a_i + b_j}{2} \right) d_{ij}^2 + \lambda \sum_i (a_i \ln(a_i) - a_i) + \mu \sum_j (b_j \ln(b_j) - b_j). \quad (9.29)$$

The parameters λ and μ control the size of the bicluster by penalizing the small values of the memberships. Their values can be estimated by simple statistics over the training set, and then hand-tuned to incorporate possible a priori knowledge while obtaining desired results.

Setting the derivatives of J_{B} with respect to the memberships a_i and b_j to zero, we have

$$\frac{\partial J}{\partial a_i} = \sum_j \frac{d_{ij}^2}{2} + \lambda \ln(a_i) = 0, \quad (9.30)$$

$$\frac{\partial J}{\partial b_j} = \sum_i \frac{d_{ij}^2}{2} + \mu \ln(b_j) = 0. \quad (9.31)$$

These lead to the solutions

$$a_i = \exp \left(-\frac{\sum_j d_{ij}^2}{2\lambda} \right), \quad (9.32)$$

$$b_j = \exp \left(-\frac{\sum_i d_{ij}^2}{2\mu} \right). \quad (9.33)$$

As in the case of standard PCM, the necessary conditions for the minimization of J_{B} together with the definition of d_{ij}^2 [eqn. (9.25)] can be used by the Possibilistic Biclustering (PBC) algorithm to find a numerical solution for the optimization problem (Picard iteration).

PBC Algorithm

1. Initialize the memberships **a** and **b**
2. Compute d_{ij}^2 $\forall i, j$ using eqn. (9.25)
3. Update a_i $\forall i$ using eqn. (9.32)
4. Update b_j $\forall j$ using eqn. (9.33)
5. **If** $\|\mathbf{a}' - \mathbf{a}\| < \varepsilon$ and $\|\mathbf{b}' - \mathbf{b}\| < \varepsilon$ then **stop**
   else jump to **Step 2**

The parameter ε is a threshold controlling the convergence of the algorithm. The memberships' initialization can be made randomly or by using some a priori information about relevant genes and conditions. Moreover the PBC algorithm can also be used as a refinement step for other existing algorithms, using their results for initialization. After convergence the memberships **a** and **b** can be defuzzified with respect to a threshold (say, 0.5) for subsequent comparative analysis.

9.5 Experimental Results

The two soft computing based biclustering algorithms described in Section 9.3 and 9.4.2 were implemented on the benchmark gene expression *Yeast* data. As the problem suggests, the size of an extracted bicluster should be as large as possible while satisfying a homogeneity criterion. The threshold δ was selected as 300 for *Yeast* data in Refs. [6, 51, 31, 19, 14, 29]. There being no definite guidelines available in literature for the choice of δ, and with a view to providing a fair comparison with existing methods, we have often used the same parameter settings for δ and α; *viz.*, $\delta = 300$ with $\alpha = 1.2$. We have also made a detailed study on the variation of these parameters. The crossover and mutation probabilities of 0.75 and 0.03 were selected after several experiments with random seeds. However it was noticed that the crossover and mutation parameters had insignificant effect on the results, as compared to that of δ and α. Due to memory constraints, we restricted the population size to 50. Additionally, we investigated the effect of lower δ values in order to demonstrate the biological relevance of the extracted smaller biclusters.

The *Yeast cell cycle* data* is a collection of 2884 genes (attributes) under 17 conditions (time points), having 34 null entries with -1 indicating the missing values. All entries are integers lying in the range of 0 to 600. The missing values are replaced by random number between 0 to 800, as in [6].

9.5.1 Multi-objective evolutionary

Table 9.1 *Best biclusters for Yeast data after 50 generations with $\delta = 300$*

α	Bicluster size	No. of genes	No. of conditions	Mean squared residue	CI
1.1	6447	921	7	206.77	0.032
1.2	8832	1104	8	249.61	0.028
1.3	9846	1094	9	263.48	0.027
1.4	11754	1306	9	298.54	0.025
1.5	12483	1387	9	299.88	0.024
1.6	12870	1287	10	299.85	0.023
1.7	12970	1297	10	299.87	0.023
1.8	**14828**	**1348**	**11**	**286.27**	**0.019**
1.9	13783	1253	11	299.95	0.022

Table 9.1 summarizes the best biclusters for *Yeast* data after 50 generations,

* http://arep.med.harvard.edu/biclustering

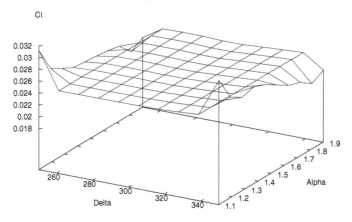

Figure 9.2 *Plot of CI for different choices of α and δ on Yeast data.*

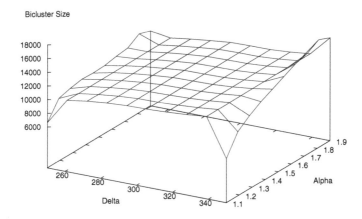

Figure 9.3 *Plot of bicluster size for different choices of α and δ on Yeast data.*

with $\delta = 300$, for different values of α. The population size is chosen to be 50. The largest sized bicluster is found at $\alpha = 1.8$ for each δ, with coherence index CI being minimal and indicating the goodness of the discovered partitions. The minimum value of CI is 0.019 when $\delta = 300$ and $\alpha = 1.8$, with a corresponding bicluster size of 14,828 being the best in the table. As explained earlier, a low mean squared residue indicates a high coherence of the discovered biclusters. It may also include some trivial biclusters containing insignificant fluctuations in their expression values, and are not of interest to our study. Hence δ is used as an upper limit on the allowable dissimilarity among genes and conditions. However, a higher δ is indicative of diminishing homogeneity.

Figs. 9.2 and 9.3 depict the 3D plots of CI and bicluster size, against the variations of parameters δ and α. It is observed that with increasing α and δ, the bicluster size also increases while CI proportionately decreases.

Biological validation

The biological relevance of smaller biclusters for the *Yeast cell-cycle* data, with $\delta=20$, was investigated in terms of the statistically significant Gene Ontology (GO) annotation database[†]. Here genes are assigned to three structured, controlled vocabularies (ontologies) that describe gene products in terms of associated biological processes, components and molecular functions in a species-independent manner.

We have measured the degree of enrichment *i.e.*, *p*-values[‡] using a cumulative hypergeometric distribution, that involves the probability of observing the number of genes from a particular GO category (*viz.*, function, process, component) within each bicluster. The probability p for finding at least k genes, from a particular category within a cluster of size n, is expressed as

$$p = 1 - \sum_{i=0}^{k-1} \frac{\binom{f}{i}\binom{g-f}{n-i}}{\binom{g}{n}}, \tag{9.34}$$

where f is the total number of genes within a category and g is the total number of genes within the genome [52]. The *p*-values are calculated for each functional category in each cluster. Statistical significance is evaluated for the genes in each bicluster by computing *p*-values, that signify how well they match with the different GO categories. Note that a smaller *p*-value, closer to zero, is indicative of a better match.

Table 9.2 shows the significant shared GO terms (or parent of GO terms) used to describe the set of genes (12 and 18) in each small bicluster (generated by tuning λ and μ), for the process, function and component ontologies. Only the three significant common terms with increasing order of *p*-value (*i.e.*, decreasing order of significance) are displayed. For the first cluster, the genes (TRM82, TRM112) are particularly involved in the process of tRNA and RNA methaylation, tRNA modification; while genes (GBP2, AGE1, TOM20, VPS21, PRS4, SSK22, NHP10, SOK1, etc.) are employed in cellular process, intracellular transport, etc. The values within parentheses after each GO term in columns 2–4 of the table, such as (2, 0.00024) in the first row, indicate that out of twelve genes in the first cluster two belong to this process and their statistical significance is provided by a *p*-value of 0.00024. Note that the genes in the cluster share other GO terms also, but with a lower significance (*i.e.*, have higher *p*-value).

From the table we notice that each extracted bicluster is distinct along each category. For example, the most significant processes in the second cluster are cell organization and biogenesis (genes LSM2, RRP7, NHP10, TOM20, ECM9, EMG1, SEC65), rRNA processing (genes LSM2, RRP7, EMG1) and

[†] http://db.yeastgenome.org/cgi-bin/GO/goTermFinder
[‡] The *p*-value of a statistical significance test represents the probability of obtaining values of the test statistic that are equal to or greater in magnitude than the observed test statistic

Table 9.2 *Significant Shared GO Terms (Process, Function, Component) of the selected 12, 18 genes for Yeast data*

No. of Genes	Process	Function	Component
12	tRNA methaylation (2, 0.00024), RNA methaylation (2, 0.00027), biopolymer methaylation (2, 0.0014), tRNA modification (2, 0.0015), cellular process (12, 0.0052), intracellular transport (4, 0.006), establishment of cellular localization (4, 0.0072)	tRNA (guanine) methyltransferase activity (6.07e-05), tRNA methyl-transferase activity (2, 0.00027), RNA methyl transferase activity (2, 0.0006), methyltransferase activity (2, 0.007)	cytosolic small ribosomal subunit (sensu Eukaryota) (2, 0.0046), Eukaryotic 48S initiation complex (2, 0.004), Eukaryotic 43S preinitiation complex (2, 0.0062), small ribosomal subunit (2, 0.010)
18	cell organization and biogenesis (9, 0.0048), rRNA processing (3, 0.008), primary transcript processing (2, 0.0120), membrane lipid biosynthesis (2, 0.0130)	protein transporter activity (2, 0.0067)	nucleus (4, 0.0053), small nucleolar ribonucleo protein complex (2, 0.0089)

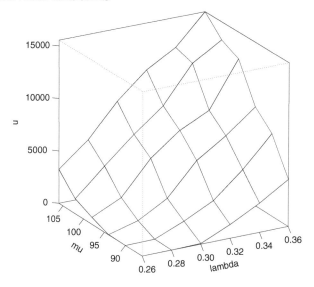

Figure 9.4 *Size of the biclusters vs. parameters* λ *and* μ.

membrane lipid biosynthesis (genes VRA7, FEN1). Looking at the function category of each cluster, we discover that the most significant terms for the first cluster are tRNA (guanine) methyltransferase activity (genes TRM82, TRM112), and for the second cluster it is protein transporter activity (genes TOM20, PSE1). Finally, the extracted biclusters also differ in terms of their cellular component. The genes (RPS22A, RPS16A) of the first bicluster belong to a small ribosomal subunit, those of the second bicluster (LSM2, RRP7, EMG1, RPC19) to the nucleus, and genes (NCL1, PRS5, NTC20, SHM1, GBP2, IES3, PSE1) of the third bicluster to the cell. This indicates that the methodology is capable of finding potentially biologically significant biclusters.

9.5.2 Possibilistic fuzzy

The parameters λ and μ of eqn. (9.29) were varied, considering a threshold of 0.5 for the memberships **a** and **b** [eqns. (9.32) and (9.33)] of 0.5 for the defuzzification. The effect of these parameters on the size of the bicluster, in case of possibilistic fuzzy method, is demonstrated in Fig. 9.4. Results correspond to an average over 20 runs. It is observed that an increase in these parameters leads to a larger bicluster. Thus PBC is observed to be slightly sensitive to initialization of memberships, while being strongly sensitive to variations in parameters λ and μ. The parameter ε depends on the desired precision on the final memberships. Here it was set at 10^{-2}.

Table 9.3 provides a set of biclusters along with the homogeneity $\mathcal{G}$ of eqn.

(9.24). We are able to generate biclusters of a desired size by tuning the parameters λ and μ.

Table 9.3 *Best biclusters for Yeast data, with respect to the parameters λ and μ.*

λ	μ	No. of genes	No. of conditions	Bicluster size	Mean squared residue
0.25	115	448	10	4480	56.07
0.19	200	457	16	7312	67.80
0.30	100	654	8	5232	82.20
0.32	100	840	9	7560	111.63
0.26	150	806	15	12090	130.79
0.31	120	989	13	12857	146.89
0.34	120	1177	13	15301	181.57
0.37	110	1309	13	17017	207.20
0.39	110	1422	13	18486	230.28
0.42	100	1500	13	19500	245.50
0.45	95	1622	12	19464	260.25
0.45	95	1629	13	21177	272.43
0.46	95	1681	13	21853	285.00
0.47	**95**	**1737**	**13**	**22581**	**297.40**
0.48	95	1797	13	23361	310.72

9.5.3 Comparative study

Table 9.4 is a comparative study on biclusters obtained using some of the existing methods in the literature [6, 19, 31, 40], using thresholds of $\delta = 300$, along with the MOEA and PBC algorithms of Sections 9.3 and 9.4.2 ([15, 39]). Yang et al. [19] identified 100 biclusters with an average of 195 genes and 12.8 conditions. For the *Yeast* data, DBF [31] discovered 100 biclusters, with most of them having sizes in the range of 2000 to 3000 and the largest one being of size 4500. Although FLOC [19] generates larger biclusters than DBF, the latter has a lower mean squared residue that is indicative of a higher similarity among genes in the discovered biclusters. Both these methods are regarded as an improvement over the pioneering algorithm of Ref. [6], with respect to mean squared residue and bicluster size. The largest size of a bicluster in Ref. [6] is found to be 4485.

Table 9.4 *Comparative study on Yeast data.*

Method	Average residue	Average bicluster size	Average no. of genes	Average no. of conditions	Largest bicluster size
FLOC [19]	187.54	1825.78	195	12.8	2000
DBF [31]	114.7	1627.2	188	11	4500
Cheng-Church [6]	204.29	1576.98	167	12	4485
GA [15]	52.87	570.86	191.12	5.13	1408
SEBI [40]	205.18	209.92	13.61	15.25	–
MOEA[15]	234.87	10301.71	1095.43	9.29	14,828
PBC [39]	297.40	22581	1737	13	–

The biclusters discovered in Ref. [51] from *Yeast* data are of sizes (124×8), (124×9), (19×8), (19×9), (63×9), (23×9), (20×8) and (20×9), with the two entries within the parentheses corresponding to the numbers of genes and conditions, respectively.

GA has also been used with local search [14], to generate overlapped biclusters. An initial deterministic selection of biclusters, having similar size, is made for a uniform distribution of chromosomes in the population. Thereafter GA is used with minimization of a fitness function, defined as

$$F(g,c) = \begin{cases} \frac{1}{f(g,c)} & \text{if } \mathcal{G}(g,c) \leq \delta \\ \frac{\mathcal{G}(g,c)}{\delta} & \text{otherwise.} \end{cases} \qquad (9.35)$$

The best bicluster generated from *Yeast* data is reported as 12,350, with an average size of 8,600.

The simulated annealing based algorithm [44] is able to find significant biclusters of size 18,460 with $\delta = 300$, but it suffers from the "random interference" problem. The results are also data dependent.

9.6 Conclusions and Discussion

A gene expression data set typically contains thousands of genes. However, biologists often have different requirements on cluster granularity for different subsets of genes. For some purpose, biologists may be particularly interested in some specific subsets of genes and prefer small and tight clusters. While for other genes, people may only need a coarse overview of the data structure. However, most of the existing clustering algorithms only provide a crisp set of clusters and may not be flexible to different requirements for cluster granularity on a single data set. For gene expression data, it would be more appropriate to avoid the direct partition of the data set and instead provide a scalable graphical representation of the data structure, leaving the partition problem to the users.

Biclustering is typically employed in situations involving (say) the (i) participation of a small set of genes in a cellular process of interest, (ii) study of an interesting cellular process that is active only over a subset of conditions, (iii) participation of a single gene in multiple pathways, that may or may not be coactive under all conditions. Robustness of the algorithms is also desirable, due to the complexity of the gene regulation processes as well as to intelligently handle the level of noise inherent in the actual experiments. Again, if a biclustering algorithm could integrate partial knowledge as some clustering constraints, we can expect the partitions to be biologically more meaningful.

Uncovering genetic pathways (or chains of genetic interactions) is equivalent to generating clusters of genes with expression levels that evolve coherently under subsets of conditions, i.e., discovering biclusters where a subset of genes are coexpressed under a subset of conditions. Such pathways can provide clues on (say) genes that contribute towards a disease. This emphasizes the possibilities and challenges posed by biclustering.

However there also exist other application domains, including information retrieval, text mining, collaborative filtering, target marketing, market research, database research and data mining. The tuning and validation of biclustering methods, in comparison to known biological data, is certainly one of the important open issues for future research.

References

[1] "Special Issue on Bioinformatics," *Pattern Recognition*, vol. 39, 2006.

[2] G. P. Shapiro and P. Tamayo, "Microarray data mining: facing the challenges," *ACM SIGKDD Explorations Newsletter*, vol. 5, pp. 1–5, 2003.

[3] S. Mitra and T. Acharya, *Data Mining: Multimedia, Soft Computing, and Bioinformatics*. New York: John Wiley, 2003.

[4] J. T. Tou and R. C. Gonzalez, *Pattern Recognition Principles*. London: Addison-Wesley, 1974.

[5] S. Mitra, H. Banka, and W. Pedrycz, "Rough fuzzy collaborative clustering," *IEEE Transactions on Systems, Man and Cybernetics: Part B*, vol. 36, pp. 795–805, 2006.

[6] Y. Cheng and G. M. Church, "Biclustering of gene expression data," in *Proceedings of the 8th International Conference on Intelligent Systems for Molecular Biology (ISMB)*, pp. 93–103, 2000.

[7] Y. Zhang, H. Zha, and C. H. Chu, "A time-series biclustering algorithm for

revealing co-regulated genes," in *Proceedings of the International Conference on Information Technology: Coding and Computing (ITCC'05)*, 2005.

[8] L. Ji and K. L. Tan, "Identifying time-lagged gene clusters using gene expression data," *Bioinformatics*, vol. 21, pp. 509–516, 2005.

[9] S. C. Madeira and A. L. Oliveira, "Biclustering algorithms for biological data analysis: A survey," *IEEE Transactions on Computational Biology and Bioinformatics*, vol. 1, pp. 24–45, 2004.

[10] R. Agarwal, J. Gehrke, D. Gunopulos, and P. Raghavan, "Automatic subspace clustering of high dimensional data for data mining applications," in *Proceedings of ACM SIGMOD International Conference on Management of Data*, pp. 94–105, 1998.

[11] I. Dhilon, S. Mallela, and D. Modha, "Information-theoretic co-clustering," in *Proceedings of the 9th International Conference on Knowledge Discovery and Data Mining (KDD)*, pp. 89–98, 2003.

[12] A. Banerjee, I. Dhillon, J. Ghosh, S. Merugu, and D. S. Modha, "A generalized maximum entropy approach to Bregman co-clustering and matrix approximation," in *Proceedings of the Tenth ACM SIGKDD International Conference on Knowledge Discovery and Data Mining*, pp. 509–514, ACM Press NY, USA, 2004.

[13] R. Peeters, "The maximum edge biclique problem is NP-Complete," *Discrete Applied Mathematics*, vol. 131, pp. 651–654, 2003.

[14] S. Bleuler, A. Prelić, and E. Zitzler, "An EA framework for biclustering of gene expression data," in *Proceedings of Congress on Evolutionary Computation*, pp. 166–173, 2004.

[15] S. Mitra and H. Banka, "Multi-objective evolutionary biclustering in gene expression data," *Pattern Recognition, 2006*, vol. 39, pp. 2464–2477, 2006.

[16] S. K. Pal and S. Mitra, *Neuro-fuzzy Pattern Recognition: Methods in Soft Computing Paradigm*. John Wiley, 1999.

[17] G. Getz, H. Gal, I. Kela, D. A. Notterman, and E. Domany, "Coupled two-way clustering analysis of breast cancer and colon cancer gene expression data," *Bioinformatics*, vol. 19, pp. 1079–1089, 2003.

[18] J. A. Hartigan, "Direct clustering of a data matrix," *Journal of American Statistical Association (JASA)*, vol. 67, pp. 123–129, 1972.

[19] J. Yang, H. Wang, W. Wang, and P. Yu, "Enhanced biclustering on expression data," in *Proceedings of the Third IEEE Symposium on BioInformatics and Bioengineering (BIBE'03)*, 2003.

[20] A. Tanay, R. Sharan, and R. Shamir, "Discovering statistically significant biclusters in gene expression data," *Bioinformatics*, vol. 18, pp. S136–S144, 2002.

[21] L. Lazzeroni and A. Owen, "Plaid models for gene expression data," *Statistica Sinica*, vol. 12, pp. 61–86, 2002.

[22] H. Qu, L.-P. Wang, Y.-C. Liang, et al., "An improved biclustering algorithm and its application to gene expression spectrum analysis," *Genomics, Proteomics and Bioinformatics*, vol. 3, pp. 189–193, 2005.

[23] H. L. Turner, T. C. Bailey, W. J. Krzanowski, and C. A. Hemingway, "Biclustering models for structured microarray data," *IEEE/ACM Transactions on Computational Biology and Bioinformatics*, vol. 2, pp. 316–329, 2005.

[24] A. Battle, E. Segal, and D. Koller, "Probabilistic discovery of overlapping cellular processes and their regulation," in *Proceedings of the Eighth Annual International Conference on Research in Computational Molecular Biology*, pp. 167–176, ACM Press NY, USA, 2004.

[25] L. Candillier, I. Tellier, F. Torre, and O. Bousquet, "SSC: Statistical subspace clustering," in *4th International Conference on Machine Learning and Data Mining in Pattern Recognition* (P. Perner and A. Imiya, eds.), vol. LNAI 3587, pp. 100–109, 2005.

[26] G. Park and W. Szpankowski, "Analysis of biclusters with applications to gene expression data," in *International Conference on Analysis of Algorithms* (C. Martnez, ed.), Theoretical Computer Science Proceedings AD, pp. 267–274, 2005.

[27] Y. Kluger, R. Basri, J. T. Chang, and M. Gerstein, "Spectral biclustering of microarray data: Coclustering genes and conditions," *Genome Reaserch*, vol. 13, pp. 703–716, 2003.

[28] J. Liu, W. Wang, and J. Yang, "A framework for ontology-driven subspace clustering," in *Proceedings of the Tenth ACM SIGKDD International Conference on Knowledge Discovery and Data Mining (KDD '04)*, pp. 623–628, ACM Press, 2004.

[29] A. H. Tewfik and A. B. Tchagang, "Biclustering of DNA microarray data with early pruning," in *Proceedings of ICASSP 2005*, pp. V773–V776, 2005.

[30] E. Segal, B. Taskar, A. Gasch, N. Friedman, and D. Koller, "Rich probabilistic models for gene expression," *Bioinformatics*, vol. 17, pp. S243–S252, 2001.

[31] Z. Zhang, A. Teo, B. C. Ooi, and K. L. Tan, "Mining deterministic biclusters in gene expression data," in *Proceedings of the Fourth IEEE Symposium on Bioinformatics and Bioengineering (BIBE'04)*, 2004.

[32] W. H. Au, K. C. C. Chan, A. K. C. Wong, and Y. Wang, "Attribute clustering for grouping, selection, and classification of gene expression data," *IEEE/ACM Transactions on Computational Biology and Bioinformatics*, vol. 2, pp. 83–101, 2005.

[33] J. Liu, J. Yang, and W. Wang, "Biclustering in gene expression data by tendency," in *Proceedings of the 2004 Computational Systems Bioinformatics Conference (CSB 2004)*, 2004.

[34] Z. B. Joseph, "Analysing time series gene expression data," *Bioinformatics*, vol. 20, pp. 2493–2503, 2004.

[35] A. A. Melkman and E. Shaham, "Sleeved coclustering," in *Proceedings of the Tenth ACM SIGKDD International Conference on Knowledge Discovery and Data Mining (KDD '04)*, pp. 635–640, ACM Press, 2004.

[36] A. B. Tchagang and A. H. Tewfik, "DNA microarray data analysis: A novel biclustering algorithm approach," *Journal on Applied Signal Processing (EURASIP)*, vol. Article ID 59809, pp. 1–12, 2006.

[37] S. C. Madeira and A. L. Oliveira, "A linear time biclustering algorithm for time series gene expression data," in *WABI 2005, LNBI 3692* (R. Casadio and G. Myers, eds.), pp. 39–52, Berlin: Springer-Verlag, 2005.

[38] A. Prelic, S. Bleuler, P. Zimmermann, A. Wille, P. Buhlman, W. Gruissem, L. Hennig, L. Thiele, and E. Zitzler, "A systematic comparison and evaluation of biclustering methods for gene expression data," *Bioinformatics*, vol. 22, pp. 1122–1129, 2006.

[39] M. Filippone, F. Masulli, S. Rovetta, S. Mitra, and H. Banka, "Possibilistic approach to biclustering: An application to oligonucleotide microarray data analysis," in *Proceedings of the International Conference on Computational Methods in Systems Biology*, LNCS, pp. 312–322, Springer, 2006.

[40] F. Divina and J. S. Aguilar-Ruiz, "Biclustering of expression data with evolu-

tionary computation," *IEEE Transactions on Knowledge and Data Engineering*, vol. 18, pp. 590–602, 2006.

[41] A. Ben-Dor, B. Chor, R. Karp, and Z. Yakini, "Discovering local structure in gene expression data: The order preserving submatrix problem," in *6th International Conference on Computational Biology* (NY, USA), pp. 49–57, ACM Press, 2002.

[42] B. J. Gao, O. L. Griffith, M. Ester, and S. J. M. Jones, "Discovering significant OPSM subspace clusters in massive gene expression data," in *Proceedings of the 12th ACM SIGKDD International Conference on Knowledge Discovery and Data Mining*, pp. 922–928, ACM Press, 2006.

[43] S. Bleuler and E. Zitzler, "Order preserving clustering over multiple time course experiments," in *EvoWorkshops 2005, LNCS 3449* (F. Rothlauf et al., ed.), pp. 33–43, Berlin: Springer-Verlag, 2005.

[44] K. Bryan, P. Cunningham, and N. Bolshakova, "Biclustering of expression data using simulated annealing," in *18th IEEE Symposium on Computer-Based Medical Systems (CSMB 2005)*, pp. 93–103, 2000.

[45] K. Deb, *Multi-Objective Optimization using Evolutionary Algorithms*. London: John Wiley, 2001.

[46] R. Krishnapuram and J. M. Keller, "A possibilistic approach to clustering," *IEEE Transactions on Fuzzy Systems*, vol. 1, pp. 98–110, 1993.

[47] R. Krishnapuram and J. M. Keller, "The possibilistic c-means algorithm: insights and recommendations," *IEEE Transactions on Fuzzy Systems*, vol. 4, pp. 385–393, 1996.

[48] R. O. Duda and P. E. Hart, *Pattern Classification and Scene Analysis*. Wiley, 1973.

[49] F. Masulli and A. Schenone, "A fuzzy clustering based segmentation system as support to diagnosis in medical imaging," *Artificial Intelligence in Medicine*, vol. 16, pp. 129–147, 1999.

[50] O. Nasraoui and R. Krishnapuram, "Crisp interpretations of fuzzy and possibilistic clustering algorithms," *EUFIT 95: 3rd European Congress on Intelligent Techniques and Soft Computing*, vol. 3, pp. 1312–1318, 1995.

[51] H. Cho, I. S. Dhilon, Y. Guan, and S. Sra, "Minimum sum-squared residue co-clustering of gene expression data," in *Proceedings of 4th SIAM International Conference on Data Mining*, 2004.

[52] S. Tavazoie, J. D. Hughes, M. J. Campbell, R. J. Cho, and G. M. Church, "Systematic determination of genetic network architecture," *Nature Genet.*, vol. 22, pp. 281–285, 1999.

CHAPTER 10

Bayesian Machine-Learning Methods for Tumor Classification Using Gene Expression Data

Bani K. Mallick
Texas A & M University

Debashis Ghosh
University of Michigan

Malay Ghosh
University of Florida

10.1 Introduction

Precise classification of tumors is crucial for cancer diagnosis and treatment. The ability to target specific therapies to pathogenetically distinct tumor types is very important for cancer treatment because it maximizes efficacy and minimizes toxicity [19]. Diagnostic pathology has traditionally relied on macro- and microscopic histology and tumor morphology as the basis for tumor classification. The downside of it, however, is the inability to discriminate among tumors with similar histopathologic features, as these features vary in clinical course and in response to treatment. This justifies the growing interest in changing the basis of tumor classification from morphologic to molecular, using microarrays which provide expression measurements for thousands of genes simultaneously [35, 10]. The idea is to carry out classification on the basis of different expression patterns. Several studies using microarrays to profile colon, breast and other tumors have demonstrated the potential power of this idea [2, 21]. Gene expression profiles may offer more information than, and provide an alternative to, morphology-based tumor classification systems. The introduction of microarray technology has raised a variety of statistical issues, ranging from low-level issues such as preprocessing and normalization of the data [44, 53] to differential expression of genes in experimental conditions

[12, 24]. In this chapter, we will be focusing on the problem of classification based on microarray data.

In such analyses, there exists a set of observations that contains vectors of gene expression data as well as the labels of the corresponding tissues. These observations are utilized to fit a classification model which can be applied to predict the tissue types for new samples. Golub et al. [19] employed supervised learning methods and derived discriminant decision rules using the magnitude and the threshold of prediction strength. However, they did not provide the procedure for selecting a cutoff value, which is an essential ingredient for their approach. Heuristic rules for selecting the threshold could be one option, but this would undoubtedly be subjective. Moler et al. [31] proposed a naive Bayes model and Xiong et al. [52] conducted linear discriminant analysis for tumor classification. Brown et al. [7] used a support vector machine to classify genes rather than samples. Dudoit et al. [11] compared several discriminant methods for classification of tumors.

A major problem with microarray data analysis is that the sample size n is small compared to the number of genes p. This is usually called the "large p, small n" problem [50]. Faced with this problem, we need a dimension reduction method to reduce the high-dimensional gene space. Most existing approaches perform a preliminary selection of genes based on some 'filtering' criterion and use only $5 - 10\%$ of the genes for classification. In the approach described here, we can utilize all the genes rather than eliminating most of them based on a crude criterion.

Several authors have proposed such methods for the analysis of gene expression data. Efron et al. [12] have used a nonparametric empirical Bayes approach for gene profiling whereas Ibrahim et al. [24] suggested a parametric mixture model and used a gene selection algorithm along with the L measure statistic for evaluating models. Both groups were mainly interested in the issue of differential expression. West et al. [49] considered a classification approach based on probit regression. Lee et al. [26] used a variable selection approach with mixture priors in a similar setup. In this article we construct Bayesian binary classification models for prediction based on a reproducing kernel Hilbert space (RKHS) [3, 33] approach. The methods are quite general and, in particular, can be used for tumor classification.

Usually RKHS has been used in a decision theoretic framework with no explicit underlying probabilistic model. As a result, it is not possible to assess the uncertainty either of the classifier itself or of the predictions based on it.

Our goal here is to present a full probabilistic model-based approach to RKHS-based classification. First we consider the *logistic classifier* in this framework, and then extend it to *support vector machine* (SVM) classifiers [8, 39]. As with other regularization methods, there are smoothing or regularization parameter(s) which need to be tuned appropriately for efficient classification. One popular approach is to use generalized approximate cross validation (GACV)

[48] to tune the smoothing parameters. In this chapter we take a different approach, namely, developing a hierarchical model where the unknown smoothing parameter will be interpreted as a shrinkage parameter [9]. We will put a prior distribution on it and obtain its posterior distribution via the Bayesian paradigm. The prediction results from this model turn out to be quite competitive to those from classical RKHS results. In this way, we obtain not only the point predictors but also the associated measures of uncertainty. Furthermore, we can extend the model to incorporate multiple smoothing parameters, which will lead to significant improvements in prediction for the examples considered. These models will be compared with respect to their classification errors. One numerical issue here is that the posterior distributions are not available in an explicit form. Hence a Markov chain Monte Carlo (MCMC) based computation approach [18, 15] will be used to generate samples from the posterior. Our models are applied to three publicly available microarray datasets: the leukemia dataset of Golub et al. [19], the hereditary breast cancer data from Hedenfalk et al. [21] and the colon tumor study of Alon et al. [2].

Before proceeding further, we briefly review the current literature on Bayesian learning, and compare our proposal with some existing methods. Tipping [42, 43] and Bishop and Tipping [6] introduced relevance vector machines (RVM's) in place of SVM's. Their objective, like ours, was to obtain entire predictive distributions for future observations rather than just point predictors. In the classification context, they began with a likelihood based on binary data with a logit link function, as in Section 11.4.1 of this chapter. The logits were assumed to have a regression structure with a fairly general basis function including the one considered here. Then a Gaussian distribution was assigned to the vector of regression coefficients (which they call "weights"). Finally the Bayesian procedure amounted to finding the posterior modes of these regression coefficients, or some approximations thereof. Figueiredo [13] took a similar approach, but used the probit instead of the logit link. He also used Jeffreys' prior instead of the usual Gaussian prior for regression coefficients. Zhu and Hastie [54] proposed a frequentist approach using only a subset of the regression vectors. They referred to the resulting procedure as Import Vector Machine (IVM) and used iteratively reweighted least squares (IRLS) as the fitting method.

The present approach, though similar in spirit, is operationally quite different from the above. First, the logits or the probits are not deterministic functions of the basis vectors, but include in addition, a random error to account for any unexplained sources of variation. For classification models with binary data it is well known that conjugate priors do not exist for the regression coefficients and hence the computation becomes much harder. In this chapter, we exploit the idea of this random residual component within the model which enables us to calculate the marginal probabilities for a particular model structure analytically, and we also develop an algorithm based on that. As a result of adopting a Gaussian residual effect, many of the conditional distributions for the model parameters are of the standard form, which greatly aids the computations. Also, rather than estimating the hyperparameters, we assign distributions to them, thus accounting for uncertainty due to estimation of hyperparameters. Finally, a key feature of our method is the treatment of model uncertainty through a prior distribution over the kernel parameter.

RVM introduces sparseness in the model by considering heavy-tailed priors such as double exponential for the regression coefficients [6, 13]. This opportunity exists also for the SVM as considered in this chapter, even though the binary probabilities are then modeled differently. In fact, in our examples, with a Bayesian hierarchical setup the SVM shows more sparseness than the logistic model. Several authors exploited this sparseness property to select significant genes [37, 26]. To do this, we would need to consider a linear structure. Our main emphasis, however, is to obtain predictive distributions for future observations to be used for classification rather than direct estimation of the parameters. So we use a nonlinear model with unknown kernel parameter.

The idea of multiple smoothing parameters used in this chapter has also been addressed elsewhere. In the machine learning literature, this is known as automatic relevance determination (ARD) [29, 32]. An advantage of using multiple parameters is that it enables one to detect the varying influence of different regression coefficients for prediction or classification.

Section 10.2 introduces the RKHS-based classification method. The hierarchical classification model is introduced in Section 10.3. Section 10.4 provides the different likelihoods for the logistic and the support vector machine classification models. Implementation of the Bayesian method is discussed in Section 10.5. Section 10.6 discusses prediction and model choice. Section 10.7 contains the examples. Section 10.8 presents some simulation results. Finally, some concluding remarks are made in Section 10.9.

One may wonder about the possibility of using neural networks as an alternative to the RKHS-based classification. But unlike the RKHS, the neural networks cannot project the prediction problem into the data space and, thus, are not particularly suitable for the analysis of microarray data (where the number of regressors p vastly exceeds n, the sample size). So, once again, preliminary gene selection is needed to use neural networks.

10.2 Classification Using RKHS

For a binary classification problem, we have a training set $\{y_i, \mathbf{x}_i\}, i = 1, \ldots, n$, where y_i is the response variable indicating the class to which the ith observation belongs, and $\mathbf{x}_i$ is the vector of covariates of size p. Our goal is to predict the posterior probability of belonging to one of the classes given a set of new covariates, based on the training data. Usually the response is coded as $y_i = 1$ for class 1 and $y_i = 0$ (or -1) for the other class. We utilize the training data

$\mathbf{y} = (y_1, \cdots, y_n)^T$ and $\mathbf{X}^T = (\mathbf{x}_1, \cdots, \mathbf{x}_n)$ to fit a model $p(\mathbf{y}|\mathbf{x})$ and use it to obtain $P(y_* = 1|\mathbf{y}, \mathbf{x}_*)$ for a future observation y_* with covariate $\mathbf{x}_*$.

For our problem, we have binary responses as $y_i = 1$ indicates that the tumor sample i is from class 1 and $y_i = 0$ (or -1) indicates that it belongs to class 2, for $i = 1, \ldots, n$. Gene expression data on p genes for n tumor samples are summarized in the form of an $n \times p$ matrix; its $(i, j)^{th}$ entry x_{ij} is the measurement of the expression level of the jth gene for the ith sample ($i = 1, \ldots, n; j = 1, \ldots, p$). It is assumed that the expression levels x_{ij} represent rigorously processed data that have undergone image processing as well as within- and between-slide normalization.

To develop a general model for classification, we need to specify a probability model for $p(\mathbf{y}|\mathbf{x})$ where $\mathbf{x}$ is high-dimensional. To simplify the structure, we introduce the latent variables $\mathbf{z} = (z_1, \ldots, z_n)$ and assume that $p(\mathbf{y}|\mathbf{z}) = \prod_{i=1}^{n} p(y_i|z_i)$, i.e., the y_i are conditionally independent given the z_i. In the next stage, the latent variables z_i are modeled as $z_i = f(\mathbf{x}_i) + \epsilon_i$, $i = 1, \ldots, n$, where f is not necessarily a linear function, and ϵ_i's, the random residual effects, are independent and identically distributed $N(0, \sigma^2)$. The use of a residual component is consistent with the belief that there may be unexplained sources of variation in the data. To develop the complete model, we need to specify $p(\mathbf{y}|\mathbf{z})$ and f.

In the machine learning literature, most of the binary classification procedures emerged from a loss function based approach. In the same spirit, we model $p(\mathbf{y}|\mathbf{z})$ on the basis of a loss function $l(\mathbf{y}, \mathbf{z})$, which measures the loss for reporting $\mathbf{z}$ when the truth is $\mathbf{y}$. Mathematically, minimizing this loss function is equivalent to maximizing $-l(\mathbf{y}, \mathbf{z})$, where $\exp[-l(\mathbf{y}, \mathbf{z})]$ is proportional to the likelihood function. This duality between "likelihood" and "loss," particularly viewing the loss as the negative of the log-likelihood, is referred to in the Bayesian literature as a "logarithmic scoring rule" (see for example, Bernardo, p. 688 [4]). Specific choices of the loss functions and the corresponding likelihood functions are discussed in Section 10.4.

To model the high-dimensional function $f(\mathbf{x})$, we will employ a rich function-space setting. Specifically, we will assume that f belongs to an RKHS which has the advantage of assuring that point evaluation is well defined while, at the same time, providing a natural choice for basis function approximation. A Hilbert space H is a collection of functions on a set T with an associated inner product $< g_1, g_2 >$ and norm $||g_1|| = < g_1, g_1 >^{1/2}$ for $g_1, g_2 \in H$. A RKHS H with reproducing kernel K (usually denoted as H_K) is a Hilbert space having an associated function K on $T \times T$ with the properties: (i) $K(\cdot, \mathbf{x}) \in H$, and (ii) $< K(\cdot, \mathbf{x}), g(\cdot) >= g(\mathbf{x})$ for all $\mathbf{x} \in T$ and for every g in H. Here $K(\cdot, \mathbf{x})$ and $g(\cdot)$ are functions defined on T with values at $\mathbf{x}^* \in T$ equal to $K(\mathbf{x}^*, \mathbf{x})$ and $g(\mathbf{x}^*)$ respectively. The reproducing kernel function provides the fundamental building blocks of H as a result of the following lemma from Parzen [33].

Lemma 1: If K is a reproducing kernel for the Hilbert space H, then $\{K(\cdot, \mathbf{x})\}$ span H.

To prove the lemma it suffices to prove that the only function g in H orthogonal to $K(\cdot, \mathbf{x})$ is the zero function; but this is obvious, since by the reproducing property $< g(\cdot), K(\cdot, \mathbf{x}) >= 0$ for every $\mathbf{x} \in T$ implies $g(\mathbf{x}) = 0$ for all $\mathbf{x}$.

A consequence of Lemma 1 is that functions of the form $g_N(\cdot) = \sum_{j=1}^{N} \beta_j K(\cdot, \mathbf{x}_j)$, where $\mathbf{x}_j \in T$ for each $j = 1, \cdots, N$, are dense in H. More precisely, for any $g \in H_K$, there are choices of N and $\beta_1, \cdots, \beta_N$ such that a g_N can be constructed to approximate g to any desired level of accuracy. Thus, the reproducing kernel functions are the natural choice for basis expansion modeling in an RKHS setting.

In the present problem, $\mathbf{x}_1, \ldots, \mathbf{x}_n$ are the observed covariate values, $\mathbf{z}$ the latent responses, and we take the unknown function $f \in H_K$ where choice of K is discussed below. To find optimal f based on $\mathbf{z}, \mathbf{x}$, we minimize $\sum_{i=1}^{n}(z_i - f(\mathbf{x}_i))^2 + ||f||^2$ with respect to f. Arguing as in chapter 1 of Wahba [46], this minimizer must admit the representation

$$f(\cdot) = \beta_0 + \sum_{j=1}^{n} \beta_j K(\cdot, \mathbf{x}_j). \tag{10.1}$$

This reduces the optimization problem to a finite dimension n which is not large for gene expression data. Also, inference about f boils down to inference about $\boldsymbol{\beta} = (\beta_0, \beta_1, \cdots, \beta_n)'$.

With the present Bayesian formulation we need a prior for $\boldsymbol{\beta}$. We will provide a flexible and computationally convenient hierarchical prior for $\boldsymbol{\beta}$ in the next section. In addition we will allow the kernel functions to depend on some unknown parameters to enrich the class of kernels and express them as $K(\cdot, \cdot|\boldsymbol{\theta})$. Hence, K becomes a function of an unknown parameter $\boldsymbol{\theta}$, but this dependence will be implicit through the remainder of the paper for notational simplicity.

Different choices of the reproducing kernel K generate different function spaces. Two common choices are the Gaussian kernel $K(\mathbf{x}_i, \mathbf{x}_j) = \exp\{-||\mathbf{x}_i - \mathbf{x}_j||^2/\theta\}$ and the polynomial kernel $K(\mathbf{x}_i, \mathbf{x}_j) = (\mathbf{x}_i \cdot \mathbf{x}_j + 1)^\theta$. Here $\mathbf{a} \cdot \mathbf{b}$ denotes the inner product of two vectors $\mathbf{a}$ and $\mathbf{b}$. Both these kernels contain a single parameter θ.

10.3 Hierarchical Classification Model

We can construct a hierarchical model for classification as

$$p(y_i|z_i) \propto \exp\{-l(y_i, z_i)\}, \; i = 1, \ldots, n, \tag{10.2}$$

where the $y_1, y_2, \cdots, y_n$ are conditionally independent given $z_1, z_2, \cdots, z_n$ and l is any specific choice of the loss function as explained in the previous section.

We relate z_i to $f(\mathbf{x}_i)$ by $z_i = f(\mathbf{x}_i) + \epsilon_i$, where the ϵ_i are residual random effects. This formulation is analogous to a generalized mixed linear model, but is slightly more general than the latter in that the likelihood is not necessarily restricted to the one-parameter exponential family.

As explained in the previous section, we express f as

$$f(\mathbf{x}_i) = \beta_0 + \sum_{j=1}^{n} \beta_j K(\mathbf{x}_i, \mathbf{x}_j | \boldsymbol{\theta}) \qquad (10.3)$$

where K is a positive definite function of the covariates (inputs) $\mathbf{x}$ and we allow some unknown parameters $\boldsymbol{\theta}$ to enrich the class of kernels.

The random latent variable z_i is thus modeled as

$$z_i = \beta_0 + \sum_{j=1}^{n} \beta_j K(\mathbf{x}_i, \mathbf{x}_j | \boldsymbol{\theta}) + \epsilon_i = \mathbf{K}_i' \boldsymbol{\beta} + \epsilon_i, \qquad (10.4)$$

where the ϵ_i are independent and identically distributed $N(0, \sigma^2)$ variables, and

$$\mathbf{K}_i' = (1, K(\mathbf{x}_i, \mathbf{x}_1 | \boldsymbol{\theta}), \dots, K(\mathbf{x}_i, \mathbf{x}_n | \boldsymbol{\theta})), \ i = 1, \dots, n.$$

To complete the hierarchical model, we need to assign priors to the unknown parameters $\boldsymbol{\beta}$, $\boldsymbol{\theta}$ and σ^2. We assign to $\boldsymbol{\beta}$ the Gaussian prior with mean $\mathbf{0}$ and variance $\sigma^2 \mathbf{D}_*^{-1}$, where $\mathbf{D}_* \equiv \mathrm{Diag}(\lambda_1, \lambda, \cdots, \lambda)$ is a $(n+1) \times (n+1)$ diagonal matrix, λ_1 being fixed at a small value, but λ is unknown. This amounts to a large variance for the intercept term. We will assign a proper uniform prior to $\boldsymbol{\theta}$, an inverse Gamma prior to σ^2 and a Gamma prior to λ. A $\mathrm{Gamma}(\alpha, \xi)$ distribution for a random variable, say U, has probability density function proportional to $\exp(-\xi u) u^{\alpha-1}$, while the reciprocal of U will then be said to have a $\mathrm{IG}(\alpha, \xi)$ distribution. Our model is thus given by

$$p(y_i | z_i) \propto \exp\{-l(y_i, z_i)\};$$

$$z_i | \boldsymbol{\beta}, \boldsymbol{\theta}, \sigma^2 \overset{\mathrm{ind}}{\sim} N_1(z_i | \mathbf{K}_i' \boldsymbol{\beta}, \sigma^2); \qquad (10.5)$$

$$\boldsymbol{\beta}, \sigma^2 \sim N_{n+1}(\boldsymbol{\beta} | \mathbf{0}, \sigma^2 \mathbf{D}_*^{-1}) \mathrm{IG}(\sigma^2 | \gamma_1, \gamma_2), \qquad (10.6)$$

$$\boldsymbol{\theta} \sim \Pi_{q=1}^{p} U(a_{q1}, a_{q2})$$

$$\lambda \sim \mathrm{Gamma}(m, c), \qquad (10.7)$$

where $U(a_{q1}, a_{q2})$ is the uniform probability density function over (a_{q1}, a_{q2}).

This Bayesian model is similar to the RKHS model as the precision parameter or the ridge parameter λ is similar to the smoothing parameter and the exponent of the Gaussian prior for $\boldsymbol{\beta}$ is essentially equivalent to the quadratic penalty function in RKHS context.

We can extend this model using multiple smoothing parameters so that the prior for $(\boldsymbol{\beta}, \sigma^2)$ is

$$\boldsymbol{\beta}, \sigma^2 \sim N_{n+1}(\boldsymbol{\beta} | \mathbf{0}, \sigma^2 \mathbf{D}^{-1}) \mathrm{IG}(\sigma^2 | \gamma_1, \gamma_2), \qquad (10.8)$$

where $\mathbf{D}$ is a diagonal matrix with diagonal elements $\lambda_1, \ldots, \lambda_{n+1}$. Once again λ_1 is fixed at a small value, but all other λ's are unknown. We assign independent Gamma(m, c) priors to them. Let $\boldsymbol{\lambda} = (\lambda_1, \ldots, \lambda_{n+1})'$.

To avoid the problem of specifying the hyperparameters m and c of $\boldsymbol{\lambda}$, we can use Jeffreys' independence prior $p(\boldsymbol{\lambda}) \propto \Pi_{i=1}^{n+1} \lambda_i^{-1}$. This is a limiting form of the gamma prior when both m and c go to 0. Figueiredo [13] observed that this type of prior promoted sparseness, thus reducing the effective number of parameters in the posterior. Sparse models are preferable as they predict accurately using fewer parameters.

10.4 Likelihoods of RKHS Models

We now consider several possible expressions for $l(y_i, z_i)$ in (2).

10.4.1 Logistic classification model

If we code the responses y_i as 0 or 1 according to the classes, then the probability function is $p(y_i|z_i) = p_i^{y_i}(z_i)\{1 - p_i(z_i)\}^{1-y_i}$, where $p_i(z_i) = \exp(z_i)/\{1 + \exp(z_i)\}$. Then the log-likelihood is $\sum_{i=1}^{n} y_i z_i - \sum_{i=1}^{n} \log\{1 + \exp(z_i)\}$. We can use this log-likelihood function and the Bayesian model given in (10.5)-(10.7) or (10.5), (10.7) and (10.8) for prediction purposes. In the probit classification model, the setup is similar, except that $p_i(z_i) = \Phi(z_i)$, where Φ denotes the standard normal distribution function.

10.4.2 The support vector machine model

Here we describe the support vector machine (SVM) classification method. For further details, the reader is referred to Cristiani and Shawe-Taylor [8], Schölkopf and Smola [39] and Herbrich [22]. We code the class labels as $y_i = 1$ or $y_i = -1$. The idea behind support vector machines is to find a linear hyperplane that separates the observations with $y = 1$ from those with $y = -1$ that has the largest minimal distance from any of the training examples. This largest minimal distance is known as the *margin*. In some cases, we can construct such a hyperplane using the original data space. However, this will not work in most cases, so that we map the inputs $\mathbf{x}$ to vectors $\phi(\mathbf{x})$ into some higher-dimensional space, where the two groups are likely to be linearly separable. When linear separability holds, writing $f(\mathbf{x}_i) = \sum_{j=1}^{n} \beta_j \phi_j(\mathbf{x}_i)$, among all hyperplanes, $f(\mathbf{x}_i) + \beta_0 = 0$ which separate the training examples, i.e., satisfy $y_i(f(\mathbf{x}_i) + \beta_0) > 0$ for all i, the SVM has the largest margin. Equivalently one can fix the margin as 1, and minimize $||\boldsymbol{\beta}||^2$ subject to the constraint $y_i(f(\mathbf{x}_i) + \beta_0) \geq 1$ for all i. If the problem is not linearly separable we need to introduce the slack variables $\xi_i > 0$. Then SVMs minimize $||\boldsymbol{\beta}||^2$

among all hyperplanes with margin 1, subject to the constraint $y_i(f(\mathbf{x}_i) + \beta_0) > 1 - \xi_i$ for all $i = 1, \ldots, n$. In other words, we are trying to find the separating hyperplane that maximizes the margin among all classifiers that satisfy the slack variable conditions. As shown by Wahba [47] or Pontil et al. [34], this optimization problem amounts to finding β which minimizes $\frac{1}{2}\|\beta\|^2 + C\sum_{i=1}^{n}\{1 - y_i f(\mathbf{x}_i)\}_+$, where $[a]_+ = a$ if $a > 0$ and is 0 otherwise, $C \geq 0$ is a penalty term and f is defined in (3). The problem can be solved using nonlinear programming methods. Fast algorithms for computing SVM classifiers can be found in Chapter 7 of Cristiani and Shawe-Taylor [8].

In a Bayesian formulation, this optimization problem is equivalent to finding the posterior mode of β, where the likelihood is given by $\exp[-\sum_{i=1}^{n}\{1 - y_i f(\mathbf{x}_i)\}_+]$, while β has the $N(\mathbf{0}, C\mathbf{I}_{n+1})$ prior. However, in our formulation with latent variables $\mathbf{z}$, we begin instead with the density

$$p(\mathbf{y}|\mathbf{z}) \propto \exp\left\{-\sum_{i=1}^{n}[1 - y_i z_i]_+\right\}, \tag{10.9}$$

and assume independent $N(f(\mathbf{x}_i), \sigma^2)$ priors for the z_i. The rest of the prior is the same as that given in (10.6) and (10.7) or (10.7) and (10.8).

If we use the density in (10.9), the normalizing constant may involve $\mathbf{z}$. Following Sollich [40], one may bypass this problem by assuming a distribution for $\mathbf{z}$ such that the normalizing constant cancels out. If the normalized likelihood is

$$p(\mathbf{y}|\mathbf{z}) = \exp\left\{-\sum_{i=1}^{n}[1 - y_i z_i]_+\right\}/c(\mathbf{z}),$$

where $c(\cdot)$ is the normalizing constant, then choosing $p(\mathbf{z}) \propto Q(\mathbf{z})c(\mathbf{z})$, the joint distribution turns out to be

$$p(\mathbf{y}, \mathbf{z}) \propto \exp\left\{-\sum_{i=1}^{n}[1 - y_i z_i]_+\right\}Q(\mathbf{z}), \tag{10.10}$$

as the $c(\cdot)$ cancels from the expression. We will take $Q(\mathbf{z})$ as the product of independent normal pdf's with means $f(\mathbf{x}_i)$ and common variance σ^2. This method will be referred to as the Bayesian support vector machine (BSVM) classification.

The above procedure, analogous to that in Sollich [40], makes the Bayesian analysis somewhat similar to the usual SVM analysis, but the prior on $\mathbf{z}$ seems rather artificial, intended mainly to cancel out a normalizing constant. The other option is to use a Bayesian approach to this problem by evaluating the normalizing constant properly and using it in the likelihood. Then the probability model (cf. Sollich [40]) is

$$p(y_i|z_i) = \begin{cases} \{1 + \exp(-2y_i z_i)\}^{-1} & \text{for } |z_i| \leq 1, \\ [1 + \exp\{-y_i(z_i + \text{sgn}(z_i))\}]^{-1} & \text{otherwise,} \end{cases} \tag{10.11}$$

where $\text{sgn}(u) = 1, 0$ or -1 according as u is greater than, equal or less than 0.

The probability density function given in (10.11) will also be used to perform a Bayesian analysis. The resulting approach will be referred to as *complete SVM* (CSVM) classification and will be compared with BSVM.

10.5 The Bayesian Analysis

For classification problems with binary data and logistic likelihood, conjugate priors do not exist for the regression coefficients. Hence, without the tailored proposal densities needed for the implementation of the Metropolis–Hastings accept-reject algorithm, mixing in the MCMC sampler can be poor as updates are rarely accepted. The construction of good proposals depends on both the model and the data. Introducing latent variables z_i simplifies computation [23], as we will now show.

From the Bayes Theorem,

$$p(\boldsymbol{\beta}, \boldsymbol{\theta}, \mathbf{z}, \sigma^2, \boldsymbol{\lambda}|\mathbf{y}) \propto p(\mathbf{y}|\mathbf{z}, \boldsymbol{\beta}, \boldsymbol{\theta}, \sigma^2, \boldsymbol{\lambda})p(\boldsymbol{\beta}, \mathbf{z}, \boldsymbol{\theta}, \boldsymbol{\lambda}, \sigma^2). \qquad (10.12)$$

This distribution is complex, and implementation of the Bayesian procedure requires MCMC sampling techniques, and in particular, Gibbs sampling [15] and Metropolis–Hastings algorithms [30]. The Gibbs sampler generates posterior samples using conditional densities of the model parameters which we describe below.

First, we note again that conditional on $\mathbf{z}$, all other parameters are independent of $\mathbf{y}$ and furthermore the distributions follow from standard results for the Bayesian linear model. This allows us to adopt conjugate priors for (β, σ^2) and hence perform simulations as well as marginalize over some of the parameter space.

10.5.1 *Conditional distributions*

The prior distributions given in (6) and (7) of Section 2 are conjugate for β and σ^2, whose posterior density conditional on $\mathbf{z}, \boldsymbol{\theta}, \boldsymbol{\lambda}$ is Normal-Inverse-Gamma,

$$p(\boldsymbol{\beta}, \sigma^2|\mathbf{z}, \boldsymbol{\theta}, \boldsymbol{\lambda}) = \text{N}_{n+1}(\boldsymbol{\beta}|\tilde{\mathbf{m}}, \sigma^2\tilde{\mathbf{V}})\text{IG}(\sigma^2|\tilde{\gamma}_1, \tilde{\gamma}_2), \qquad (10.13)$$

where $\tilde{\mathbf{m}} = (\mathbf{K_0}'\mathbf{K_0} + \mathbf{D})^{-1}(\mathbf{K_0}'\mathbf{z})$, $\tilde{\mathbf{V}} = (\mathbf{K_0}'\mathbf{K_0} + \mathbf{D})^{-1}$, $\tilde{\gamma}_1 = \gamma_1 + n/2$ and $\tilde{\gamma}_2 = \gamma_2 + \frac{1}{2}(\mathbf{z}'\mathbf{z} - \tilde{\mathbf{m}}'\tilde{\mathbf{V}}\tilde{\mathbf{m}})$. Here $\mathbf{K_0}' = (\mathbf{K_1}, \cdots, \mathbf{K_n})$, where we recall that $\mathbf{K}_i = [K(\mathbf{x}_i, \mathbf{x}_1), \ldots, K(\mathbf{x}_i, \mathbf{x}_n)]'$.

The conditional distribution for the precision parameter λ_i given the coefficient β_i is Gamma and given by

$$p(\lambda_i|\beta_i) = \text{Gamma}\left(m + \frac{1}{2}, c + \frac{1}{2\sigma^2}\beta_i^2\right), \quad i = 2, \ldots, n+1. \qquad (10.14)$$

Finally, the full conditional density for z_i is

$$p(z_i|z_{-i}, \boldsymbol{\beta}, \sigma^2, \boldsymbol{\theta}, \boldsymbol{\lambda}) \propto \exp\left[-l(y_i, z_i) - \frac{1}{2\sigma^2}\{z_i - \sum_{j=1}^{n} \beta_j K(\mathbf{x}_i, \mathbf{x}_j)\}^2\right].$$

Similarly, the full conditionals are found when $\lambda_2 = \cdots = \lambda_{n+1} = \lambda$ from (10.7) and (10.8).

10.5.2 Posterior sampling of the parameters

We make use of the distributions given in Sections 10.4 and 10.5 through a Gibbs sampler that iterates through the following steps: (i) update $\mathbf{z}$; (ii) update $\mathbf{K}, \boldsymbol{\beta}, \sigma^2$; (iii) update $\boldsymbol{\lambda}$.

For the update to $\mathbf{z}$, we propose to update each z_i in turn conditional on the rest. That is, we update $z_i|\mathbf{z}_{-i}, \mathbf{y}, \mathbf{K}, \sigma^2, \boldsymbol{\beta}$ $(i = 1, \ldots, n)$, where $\mathbf{z}_{-i}$ indicates the $\mathbf{z}$ vector with the ith element removed.

The conditional distribution of z_i does not have an explicit form; we thus resort to the Metropolis–Hastings procedure with a proposal density $T(z_i^*|z_i)$ that generates moves from the current state z_i to a new state z_i^*. The proposed updates are then accepted with probabilities

$$\alpha = \min\left\{1, \frac{p(y_i|z_i^*)p(z_i^*|\mathbf{z}_{-i}, \mathbf{K})T(z_i|z_i^*)}{p(y_i|z_i)p(z_i|\mathbf{z}_{-i}, \mathbf{K})T(z_i^*|z_i)}\right\}; \tag{10.15}$$

otherwise the current state is retained.

We obtain $p(y_i|z_i)$ from (10.9) and

$$p(z_i|\mathbf{z}_{-i}, \mathbf{K}) \propto \exp\{-(z_i - \mathbf{K}_i\boldsymbol{\beta})^2/(2\sigma^2)\}.$$

It is convenient to take the proposal distribution $T(z_i^*|z_i)$ to be a symmetric distribution (say Gaussian) with mean equal to the old value z_i and a prespecified standard deviation.

An update of $\mathbf{K}$ is equivalent to that of $\boldsymbol{\theta}$ and we need a Metropolis–Hastings algorithm to perform it. Now we need the marginal distribution of $\boldsymbol{\theta}$ conditional on $\mathbf{z}$. We can write

$$p(\boldsymbol{\theta}|z) \propto p(z|\boldsymbol{\theta})p(\boldsymbol{\theta}).$$

Let $\boldsymbol{\theta}^*$ denote the proposed change to the parameter. Then we accept this change with acceptance probability

$$\alpha = \min\left\{1, \frac{p(z|\boldsymbol{\theta}^*)}{p(z|\boldsymbol{\theta})}\right\}. \tag{10.16}$$

The ratio of the marginal likelihoods is given by

$$\frac{p(z \mid \boldsymbol{\theta}^*)}{p(z \mid \boldsymbol{\theta})} = \frac{|\tilde{\mathbf{V}}^*|^{1/2}}{|\tilde{\mathbf{V}}|^{1/2}} \left(\frac{\tilde{\gamma}_2}{\tilde{\gamma}_2^*}\right)^{\tilde{\gamma}_1}, \tag{10.17}$$

where $\tilde{\mathbf{V}}^*$ and $\tilde{\gamma}_2^*$ are similar to $\tilde{\mathbf{V}}$ and $\tilde{\gamma}_2$ with $\boldsymbol{\theta}^*$ replacing $\boldsymbol{\theta}$. Updating β, σ^2 and λ is straightforward as they are generated from standard distributions.

10.6 Prediction and Model Choice

For a new sample with gene expression $\mathbf{x}_{new}$, the posterior predictive probability that its tissue type, denoted by y_{new}, is cancerous is

$$p(y_{new}|\mathbf{x}_{new}, \mathbf{y}) = \int p(y_{new} = 1|\mathbf{x}_{new}, \boldsymbol{\phi}, \mathbf{y})p(\boldsymbol{\phi}|\mathbf{y})d\boldsymbol{\phi}, \qquad (10.18)$$

where $\boldsymbol{\phi}$ is the vector of all the model parameters. Assuming conditional independence of the responses, this integral reduces to

$$\int p(y_{new} = 1|\mathbf{x}_{new}, \boldsymbol{\phi})p(\boldsymbol{\phi}|\mathbf{y})d\boldsymbol{\phi}. \qquad (10.19)$$

The associated measure of uncertainty is $p(y_{new} = 1|\mathbf{x}_{new}, \mathbf{y})[1 - p(y_{new} = 1|\mathbf{x}_{new}, \mathbf{y})]$. The integral in (10.19) can be approximated by the Monte Carlo estimate

$$\sum_{i=1}^{M} p(y_{new} = 1|\mathbf{x}_{new}, \boldsymbol{\phi}^{(i)})/M, \qquad (10.20)$$

where $\boldsymbol{\phi}^{(i)}$ $(i = 1\ldots, M)$ are the MCMC posterior samples of the parameter $\boldsymbol{\phi}$.

To select from the different models, we will generally use misclassification error. When a test set is provided, we first obtain the posterior distributions of the parameters (training the model) based on the training data and use them to classify the test samples. For a new observation from the test set, say $y_{i,tst}$, we will obtain the probability $p(y_{i,tst} = 1|\mathbf{y}_{trn}, \mathbf{x}_{tst})$ by using an equation similar to (10.19), and approximate it by its Monte Carlo estimate as in equation (10.20). When this estimated probability exceeds .5, the new observation is classified as 1. Otherwise, it is classified as 0 or -1, depending on whether we use the logistic or the SVM likelihood. Rather than using a fixed classification rule, we can classify y_{tst} conditional on this probability.

If there is no test set available, we use a *hold-one-out* cross-validation approach. We will exploit the technique described in Gelfand [16] to simplify our computation. For the cross-validation predictive density, in general, writing $\mathbf{y}_{-i}$ as the vector of y_j's minus y_i,

$$p(y_i|\mathbf{y}_{-i}) = \frac{p(\mathbf{y})}{p(\mathbf{y}_{-i})} = \left[\int \{p(y_i|\mathbf{y}_{-i}, \boldsymbol{\phi})\}^{-1}p(\boldsymbol{\phi}|\mathbf{y})d\boldsymbol{\phi}\right]^{-1}. \qquad (10.21)$$

Monte-Carlo integration yields

$$\hat{p}(y_i|\mathbf{y}_{-i}) = M/\sum_{j=1}^{M}\left[p(y_i|\mathbf{y}_{-i}, \boldsymbol{\phi}^{(j)})\right]^{-1},$$

where $\phi^{(j)}$, $j = 1, \ldots, M$ are the MCMC posterior samples of the parameter vector ϕ. This simple expression is due to the fact that y_i's are conditionally independent given ϕ_i's. If we wish to make draws from $p(y_i|\mathbf{y}_{-i,trn})$, then we need to use importance sampling [16].

10.7 Some Examples

We now illustrate the methodology by means of several examples. For all examples, six models were fit: (i) logistic regression with a single penalty parameter; (ii) logistic regression with multiple penalty parameters; (iii) Bayesian support vector (BSVM) classification with a single penalty parameter; (iv) Bayesian support vector (BSVM) classification with multiple penalty parameters; (v) complete likelihood Bayesian support vector (CSVM) classification with a single penalty parameter; and (vi) complete likelihood support vector (CSVM) classification with multiple penalty parameters. We have used the SVM matlab toolbox to obtain the classical SVM (SVM*) results (see *http://eewww.eng.ohio-state.edu/~maj/osu_svm/*). The tuning parameters are chosen using an iterative solving method for the quadratic programming formulation of the support vector machines known as the Sequential Minimization Optimization (SMO). We obtained the RVM [42] matlab code from *http://research.microsoft.com/mlp/RVM/relevance.htm*.

Throughout the examples, we selected γ_1 and γ_2 to give a tight inverse gamma prior for σ^2 with mean 0.1. For λ we chose m and c so that the mean of the gamma distribution is small, say 10^{-3}, but with a large variance; a_{q1} and a_{q2}, the prior parameters of θ, are chosen using the $\mathbf{x}$ in such a way that computation of the kernel function does not over- or underflow. We performed the data analysis with both the Gaussian and polynomial kernels $\boldsymbol{K}$ as introduced in Section 10.3, and the results showed very little difference. The results reported here are based on Gaussian kernels.

In all the examples we used a burn-in of 5,000 samples, after which every 100th sample was retained in the next 50,000 samples. The convergence and mixing of the chain were checked using two independent chains and the methods described in Gelman [17].

10.7.1 Benchmark comparisons

We utilize artificially-generated data in two dimensions in order to compare our models with other popular models. In this artificial data both class 1 and 2 were generated from mixtures of two Gaussians by Ripley [35], with the classes overlapping to the extent that the Bayes error is around 8%.

In addition, we analyzed also three well-known benchmark data sets, compared results with several state-of-the art techniques, and present the results in Table

Table 10.1 *Modal classification error rates and 95% credible intervals for bench-mark datasets. Logistic: Logistic regression; BSVM: Bayesian support vector machine; CSVM: Complete likelihood Bayesian support vector machine; RVM: Relevance vector machine; VRVM: Variational relevance vector machine; Jeff: Analysis with Jeffreys prior; SVM*: Classical support vector machine.*

Method	Ripley's	Pima	Crabs
Logistic (single)	13.0(11,17)	21.4 (20.1,24.3)	5 (4,6)
Logistic (multiple)	9.2 (9,12)	19.4 (18.9,21.4)	2 (1,3)
BSVM (single)	12.4(11.1,16.8)	21 (20,23.9)	4 (2,5)
BSVM (multiple)	8.8(8.4,11.6)	18.9 (18.3,20.6)	1 (0,4)
CSVM (single)	12.7(10.8,16.7)	21.3 (19.9,24.1)	4 (2,5)
CSVM (multiple)	9.1(8.9,12)	19.2 (18.9,21.6)	2 (1,4)
RVM	9.3	19.6	2
VRVM	9.2	19.6	N/A
Jeff(Figrd)	9.6	18.5	0
Neural Networks	N/A	22.5	3.0
SVM*	13.2	21.2	4

1. The first two data sets are Pima Indians diabetes and Leptograpsus crabs [35]. The third one is Wisconsin breast cancer data which contains ten basic features to classify two types of cancers, malignant and benign. We split the data randomly into training/testing partitions of sizes 300 and 269, and report average results over ten partitions.

In addition to the methods listed at the beginning of Section 10.7, we have performed analyses with variational relevance vector machines (VRVM) [6], Bayesian neural networks [51] and the analysis using Jeffreys' prior as described in Figueiredo [13]. The results are given in Table 10.1. All our multiple shrinkage models perform nearly as well as the best available alternatives.

10.7.2 Leukemia data

The leukemia dataset was described in Golub et al. [19]. Bone marrow or peripheral blood samples are taken from 72 patients with either myeloid leukemia (AML) or acute lymphoblastic leukemia (ALL). Following the experimental setup of the original paper, the data are split into training and test sets. The former consists of 38 samples, of which 27 are ALL and 11 are AML; the latter consists of 34 samples, 20 ALL and 14 AML. The dataset contains expression levels for 7129 human genes produced by Affymetrix high-density oligonucleotide microarrays.

Table 10.2 *Modal classification error rates and 95% credible intervals for Leukemia data. Logistic: Logistic regression; BSVM: Bayesian support vector machine; CSVM: Complete likelihood Bayesian support vector machine; RVM: Relevance vector machine; SVM*: Classical support vector machine.*

Model	modal misclassification error	error bound
Logistic (single)	4	(2,6)
Logistic (multiple)	2	(1,4)
BSVM (single)	4	(3,7)
BSVM (multiple)	1	(0,3)
CSVM (single)	5	(3,8)
CSVM (multiple)	2	(1,6)
SVM*	4	
RVM	2	

Golub et al. [19] constructed a predictor using their weighted voting scheme on the training samples, and classify correctly on all samples for which a prediction is made, 29 of the 34, declining to predict for the other five. We have provided our results in Table 10.2 with the modal or most frequent number of misclassification errors (the modal values) as well as the error bounds (maximum and minimum number of misclassifications).

Table 10.2 shows that the results produced by the multiple shrinkage models are superior to the single precision models as well as the classical SVM models. Though all the multiple shrinkage models performed well, the best performer among these appears to be the Bayesian support vector machine model.

The use of RKHS leads to a reduction in the dimension of the model, but the dimension can still be as high as the sample size. In the Bayesian hierarchical modeling framework, due to shrinkage priors, we obtain sparsity automatically [42]. The effective number of parameters is the degrees of freedom (DF) of the model, which can be calculated as the trace of $K(K'K + D^{-1})^{-1}K'$ ([20], p. 52). Due to the presence of the unknown parameter θ in the expression of K, this θ induces a posterior distribution for DF (rather than a fixed value). The posterior distributions of DF for all the three multiple shrinkage models were very similar. There is a complaint against classical support vector machines due to lack of sparsity in the classifier but Bayesian version of it can obtain similar sparse classifier through the shrinkage priors.

Table 10.3 *Modal classification error rates and 95% credible intervals for breast cancer data.*

Model	Modal cross-validation error*	Error bound
Logistic (single)	5	(4,8)
Logistic (multiple)	2	(2,4)
BSVM (single)	4	(3,7)
BSVM (multiple)	0	(0,3)
CSVM (single)	5	(3,8)
CSVM (multiple)	2	(1,4)
Feed-forward neural networks	2	
Probabilistic Neural networks (r=0.01)	3	
kNN(k=1)	4	
SVM	4	
Perceptron	5	

∗: Number of Misclassified Samples

10.7.3 Hereditary breast cancer data

Hedenfalk et al. [21] studied gene expression in hereditary and sporadic breast cancers. Studying such cancers will allow physicians to understand the difference between the cancers from mutations in the BRCA1 and the BRCA2 breast cancer genes. In the study, Hedenfalk et al. examined 22 breast tumor samples from 21 breast cancer patients, and all the patients except one were women and fifteen of women had hereditary breast cancer, 7 tumors with BRCA1, 8 tumors with BRCA2. In the analysis of cDNA microarray, 3226 genes were used for each breast tumor sample. We use our methods to classify BRCA1 versus the other (BRCA2 and sporadic). As a test data set is not available, we have used full hold-one-out cross-validation test and use the number of misclassifications to compare different approaches. We present our results in Table 10.3.

We have compared our cross-validation results with other popular classification algorithms including feedforward neural networks [51], k-nearest neighbors (kNN) [14], classical support vector machines (SVM) [45], perceptrons [36] and probabilistic neural networks [41] in Table 10.2. All these methods have used expression values of only 51 genes as used in the paper of Hedenfalk et al. [21]. All the multiple shrinkage models have performed better than any other methods, with SVM performing the best. The average DF for the multiple shrinkage models are: logistic (10), support vector (12), complete support vector (13).

10.7.4 Colon tumor dataset

Using Affymetrix oligonucleotide arrays, expression levels for 40 tumor and 22 normal colon tissues are measured from 6500 human genes. Of these genes,

Table 10.4 *Colon Tumor data*

Model	Modal misclassification error	Error bound
Logistic (single)	6	(5,7)
Logistic (multiple)	4	(3,6)
BSVM (single)	6	(4,7)
BSVM (multiple)	3	(3,6)
CSVM (single)	6	(4,8)
CSVM (multiple)	4	(4,6)
RVM	5	
Classical SVM	7	

the 2000 with the highest minimal intensity across the tissues are selected for classification purposes and these scores are publicly available. Each score represents a gene intensity derived in a process described in Alon et al. [2].

We randomly split the data into two parts: 42 data points are used to train the model and other 20 data points are used as test set. We have provided the misclassification error results of the test set in Table 4.

Again from the results our conclusions remained the same: the multiple shrinkage models perform better than single shrinkage models and classical SVM in terms of misclassification error. The Bayesian SVM performs slightly better than the other two. The posterior mean DF for support vector and complete support vector are 14.45 and 12.26, respectively, which is lower than 21.74 for the logistic model.

The points misclassified by these methods usually lie at the classification boundary, and are thus difficult to identify properly. An advantage of the Bayesian approach is that one of its outputs is the quantification of uncertainty of the misclassified predictions. For demonstration purposes, we plotted the histogram of the posterior misclassification probability of a misclassified test sample using the logistic model. We observed that the majority of the mass of the posterior distribution is concentrated on [0.5,0.6]. This suggests that the sample is hard to classify.

10.7.5 Simulation study

In order to simulate a realistic dataset for comparing the successful multiple shrinkage Bayesian logistic, support vector and complete support vector models, we used the Leukemia data as a prototype. As realistic values of the parameters θ and β we used the posterior means from the original analysis of the data. Then we followed the structure of our models and performed two

Table 10.5 *Simulated data: The average number of misclassifications in the test data of size 34.*

	Generation Model	
Model	Logistic	CSVM
Logistic	2.5	3.8
BSVM	2.7	2.2
CSVM	3.2	2.1

sets of simulations to generate the responses Y, one using the logistic model and the other using the complete support vector model. We replicate each of the simulations 25 times, generating 25 different data sets. Then we analyze these training data sets using logistic, BSVM and CSVM models and obtain the average misclassifications in the test data for the three models. The average misclassifications should be lowest if we use the true model, but we want to see how the other models perform in this situation.

When the data are actually generated from a logistic model, the average number of misclassifications in the test data by using the logistic, BSVM and CSVM models are respectively 2.5, 2.7, 3.2. Similarly, when the data are actually generated from a logistic CSVM model, the average number of misclassifications in the test data by using the logistic, BSVM and CSVM models are respectively 3.8, 2.2, 2.1. Though none of the data is originally generated from the BSVM model (as it has no normalized distribution), in both cases, it is very near the correct (best) model in terms of average misclassification error.

The results are given in Table 10.5.

10.7.6 Analysis with Jeffreys' prior

As discussed in Section 10.3, sparseness can be promoted using Jeffreys' prior [13]. We reanalyzed the two datasets using the multiple shrinkage models and Jeffreys' prior. The modal number of misclassification and average DF (within parentheses) results are presented in Table 10.6. In terms of misclassification, Jeffreys' prior, in general, is doing worse than the Gaussian prior models but it has smaller DF.

In the analysis of leukemia data, for the logistic model, we obtained the posterior summary of the DF for the Gaussian prior and Jeffreys' prior, and clearly observed the improvement of the latter over the former. Similar improvement has been observed in support vector and complete support vector models.

Table 10.6 *Analysis with Jeffreys' prior: Average number of misclassifications and DF (within parentheses) is reported.*

	Data sets	
Model	Leukemia	Breast cancer
Logistic	2 (6.1)	3 (7.2)
BSVM	2 (5.2)	2 (5.9)
CSVM	3 (5.6)	3 (6.4)

10.8 Concluding Remarks

We have proposed a RKHS-based classification method for microarray data. It is shown that these models in a Bayesian hierarchical setup with priors over the shrinkage (smoothing) parameters performed better than other popular classification methods. Also, multiple shrinkage models always appear to be superior to single parameter shrinkage models. With multiple shrinkage parameters, the regular Bayesian SVM model emerges as the winner in all the examples with the complete SVM finishing a close second all the time. However, the complete SVM provides a more formal probabilistic motivation for the use of SVM's, and are more satisfactory from a Bayesian angle.

We may point out also that although SVM's have been very popular in the machine learning community, one problem with their use in practice in a non-Bayesian framework is the inability to quantify prediction error. By using the Bayesian framework, we are able to calculate the uncertainty associated with the predictions. While Sollich [40] also viewed SVMs from a Bayesian perspective, his approach did not include priors for the hyperparameters, and did not also accommodate any potential error in the model specification.

The advantage of the logistic model is that it can be extended easily to multi-categorical responses with a multinomial model. For support vector machines, on the other hand, this extension is not that simple. Usually the multicategory problem is reduced to a series of binary problems by constructing pairwise classifiers or one-versus-rest classifiers. One notable exception is the work of Lee et al. [26], where a formal Bayes rule is constructed for multicategory support vector machines.

One of the advantages of SVM is that its performance does not deteriorate with high input dimension. When the number of parameters exceeds the number of observations, in contrast to other machine learning methods, SVM's do not require an additional projection to the sample space, and then application of a classification algorithm; the dimension reduction is built automatically into SVM methodology. Preliminary selection of highly informative genes can reduce the noise in the data and thus improve the predictive misclassification rate, with tighter bounds.

Use of the probit model with introduction of latent variables [1] rather than a logistic model accelerates the computation significantly. We tried all the examples with the probit model and the results are almost identical to the logistic results.

West [49], in a recent article, has addressed the general regression analysis problem for large p and small n. He achieves dimension reduction through singular value decomposition (SVD) which expresses the n-dimensional linear transform of the original p-dimensional regression vector. This linear transform is a function of the SVD *factor* matrix and the nonnegative singular values. He used a generalized singular g-prior for the vector of transformed regression coefficients. West suggested also the possibility of extending his results by using a Gaussian process prior. The present RKHS formulation is indeed equivalent to consideration of such priors. However, unlike West, we do not need the SVD in carrying out the Bayesian analysis. Our method also allows for modeling an arbitrary (not necessarily linear) function of the mean. In addition, the likelihood need not necessarily belong to the one-parameter exponential family.

10.9 Acknowledgments

The first author's research was partially supported by the National Science Foundation grant DMS-0203215, National Cancer Institute Grant CA-57030. The second author's research was partially supported by MUNN Idea Grant, Prostate SPORE Seed Grant from the University of Michigan and a Grant from the National Science Foundation and National Institute of General Medical Sciences, 1R01GM72007-01. The third author's research was partially supported by NIH Grant R01-85414. The authors are indebted to Randy Eubank for many constructive suggestions that led to a much improved exposition. The authors are grateful to Emanuel Parzen and Raymond Carroll for helpful conversations.

References

[1] Albert, J. and Chib, S. (1993) Bayesian analysis of binary and polychotomous response data. *J. Am. Statist. Ass.*, **88**, 669–679.

[2] Alon, U., Barkai, N., Notterman, D.A., Gish, K., Ybarra, S., Mack, D., and Levine, A.J. (1999) Broad patterns of gene expression revealed by clustering analysis of tumor and normal colon tissues probed by oligonucleotide arrays. *Proc. Nat. Acad. Sci.*, **96**, 6745–6750.

[3] Aronszajn, N. (1950) Theory of reproducing kernels. *Trans. Am. Math. Soc.*, **68**, 337−404.

[4] Bernardo, J.M. (1979) Expected information as expected utility. *Ann. Statist*, **7**, 686−690.

[5] Bishop, C. (1995) *Neural Networks for Pattern Recognition*. Oxford: Clarendon Press.

[6] Bishop, C. and Tipping, M. (2000) Variational relevance vector machines. In *Proceedings of the 16th Conference in Uncertainty and Artificial Intelligence* (eds C. Boutilier and M. Goldszmidt), pp 46−53. San Francisco: Morgan Kaufmann.

[7] Brown, M.P.S., Grundy, W.N., Lin, D., Cristianini, N., Sugnet, C.W., Furey, T.S. and Haussler, D. (2000) Knowledge-based analysis of microarray gene expression data by using support vector machines. *Proc. Nat. Acad. Sci.*, **97**, 262−267.

[8] Cristiani, N. and Shawe-Taylor, J. (2000) *An Introduction to Support Vector Machines*. Cambridge: Cambridge University Press.

[9] Denison, D., Holmes, C., Mallick, B., and Smith, A.F.M. (2002) *Bayesian Methods for Nonlinear Classification and Regression*. London: Wiley.

[10] DeRisi, J.L., Iyer, V.R., and Brown, P.O. (1997) Exploring the metabolic and genetic control of gene expression on a genomic scale. *Science*, **278**, 680−685.

[11] Dudoit, S., Fridlyand, J., and Speed, T.P. (2002) Comparison of discrimination methods for the classification of tumors using gene expression data. *J. Am. Statist. Ass.*, **97**, 77−87.

[12] Efron, B., Tibshirani, R, Storey, J.D. and Tusher, V. (2001) Empirical Bayes Analysis of a Microarray Experiment. *J. Am. Statist. Ass.*, **96(456)**, 1151–1161.

[13] Figueiredo, M. (2002) Adaptive sparseness using Jeffreys prior. In *Neural Information Processing Systems 14* (eds T. Dietterich, S. Becker, and Z. Ghahramani), pp. 697 – 704. Cambridge, MA: MIT Press.

[14] Fix, E. and Hodges, J.L. (1951) Discriminatory analysis-nonparametric discrimination: consistency properties. USAF School of Aviation Medicine, Randolf Field, Texas.

[15] Gelfand, A. and Smith, A. F. M. (1990) Sampling-based approaches to calculating marginal densities. *J. Am. Statist. Ass.*, **85**, 398−409.

[16] Gelfand, A. (1996) Model determination using sampling-based methods. In *Markov Chain Monte Carlo in Practice* (eds W. Gilks, S. Richardson, and D. J. Spiegelhalter), pp. 145–158. London: Chapman and Hall.

[17] Gelman, A. (1996) Inference and monitoring convergence. In *Markov Chain Monte Carlo in Practice* (eds W. Gilks, S. Richardson, and D. J. Spiegelhalter), pp. 131–140. London: Chapman and Hall.

[18] Gilks, W.R., Richardson, S. and Spiegelhalter, D.J. (1996) *Markov Chain Monte Carlo in Practice*, London: Chapman and Hall.

[19] Golub, T.R., Slonim, D., Tamayo, P., Huard, C., Gaasenbeek, M., Mesirov, J., Coller, H., Loh, M., Downing, J., Caligiuri, M., Bloomfield, C., and Lander, E. (1999) Molecular classification of cancer: class discovery and class prediction by gene expression monitoring. *Science*, **286**, 531–537.

[20] Hastie, T.J. and Tibshirani, R.J. (1990) *Generalized Additive Models*. London: Chapman and Hall.

[21] Hedenfalk, I., Duggan, D., Chen, Y., Radmacher, M., Bittner, M., Simon, R., Meltzer, P., Gusterson, B., Esteller, M., Kallioniemi, O. P., Wilfond, B., Borg, A., and Trent, J. (2001). Gene expression profiles in hereditary breast cancer. *New England Journal of Medicine*, **344**, 539−548.

[22] Herbrich, R. (2002) *Learning Kernel Classifiers*. Cambridge: MIT Press.
[23] Holmes, C. and Held, L. (2003) Bayesian auxiliary variable models for binary and polychotomous regression. Technical Report, Imperial College.
[24] Ibrahim, J.G., Chen, M.H. and Gray, R.J. (2002) Bayesian Models for Gene Expression with DNA Microarray Data. *J. Am. Statist. Ass.*, **97(457)**, 88–100.
[25] Kimeldorf, G. and Wahba, G. (1971) Some results on Tchebycheffian spline functions. *J. Math. Anal. Applic.*, **33**, 82-95.
[26] Lee, K.E., Sha, N., Dougherty, E., Vannucci, M., and Mallick, B. (2003) Gene selection: a Bayesian variable selection approach. *Bioinformatics*, **19**, 90–97.
[27] Lin, X., Wahba, G., Xiang, D., Gao, F., Klein, R., and Klein, B. (2000) Smoothing spline ANOVA models for large data sets with Bernoulli observations and the randomized GACV. *Ann. Statist.*, **28**, 1570–1600.
[28] Lin, Y. (2002) Support vector machines and the Bayes rule in classification. *Data Mining and Knowledge Discovery*, **6**, 259–275.
[29] MacKay, D. (1996) Bayesian non-linear modelling for the 1993 energy prediction competition. In *Maximum Entropy and Bayesian Methods* (ed G. Heidbreder), pp. 221–234. Dordrecht: Kluwer Academic Press.
[30] Metropolis, N., Rosenbluth, A.W., Rosenbluth, M.N., Teller, A.H. and Teller, E. (1953) Equations of State Calculations by Fast Computing Machines. *J. Chem. Phys.* **21(6)**, 1087–1092.
[31] Moler, E. J., Chow, M. L., and Mian, I. S. (2000) Analysis of molecular profile data using generative and discriminative methods. *Physiological Genetics*, **4**, 109–126.
[32] Neal, R. (1996) *Bayesian Learning for Neural Networks*. New York: Springer-Verlag.
[33] Parzen, E. (1970) Statistical inference on time series by rkhs methods. In *Proc. 12th Biennial seminar* (ed. R. Pyke), pp. 1–37. Canadian Mathematical Congress: Montreal.
[34] Pontil, M., Evgeniou, T., and Poggio, T. (2000). Regularization networks and support vector machines. *Advances in Computational Mathematics*, **13**, 1–50.
[35] Ripley, B.D. (1996) *Pattern Recognition and Neural Networks*. Cambridge: Cambridge University Press.
[36] Rosenblatt F. (1962) *Principles of Neurodynamics*. New York: Spartan Books.
[37] Roth, V. (2002) The generalized LASSO: a wrapper approach to gene selection for microarray data. Department of Computer Science, University of Bonn.
[38] Schena, M., Shalon, D., Davis, R., and Brown, P. (1995) Quantitative monitoring of gene expression patterns with a complementary DNA microarray. *Science*, **270**, 467–470.
[39] Schölkopf, B. and Smola, A. (2002) *Learning with Kernels*. Cambridge, MA: MIT Press.
[40] Sollich, P. (2001) Bayesian methods for support vector machines: evidence and predictive class probabilities. *Machine Learning*, **46**, 21–52.
[41] Specht, D. F. (1990) Probabilistic neural networks. *Neural Networks*, **3**, 109–118.
[42] Tipping, M. (2000) The relevance vector machine. In *Neural Information Processing Systems 12* (eds S. Solla, T. Leen, and K. Muller), pp. 652–658. Cambridge, MA: MIT Press.
[43] Tipping, M. (2001) Sparse Bayesian learning and the relevance vector machine. *J. Mach. Learn. Res.*, **1**, 211–244.

[44] Tseng, G.C., Oh, M.K., Rohlin, L., Liao, J.C., and Wong, W.H. (2001) Issues in cDNA microarray analysis: quality filtering, channel normalization, models of variations and assessment of gene effects. *Nucleic Acids Res.*, **29**, 2549–2557.

[45] Vapnik, V.N. (2000) *The Nature of Statistical Learning Theory*, 2nd edn. New York: Springer.

[46] Wahba, G. (1990) *Spline Models for Observational Data.* Philadelphia: SIAM.

[47] Wahba, G. (1999) Support vector machines, reproducing kernel hilbert spaces and the randomized GACV. In *Advances in Kernel Methods* (eds B. Schölkopf, C. Burges, and A. Smola), pp. 69–88. Cambridge, MA: MIT Press.

[48] Wahba, G., Lin, Y., Lee, Y., and Zhang, H. (2002) Optimal properties and adaptive tuning of standard and nonstandard support vector machines. In *Nonlinear Estimation and Classification* (eds D. Denison, M. Hansen, C. Holmes, B. Mallick, and B. Yu), pp. 125 – 143. New York: Springer.

[49] West, M., Blanchette, C., Dressman, H., Huang, E., Ishida, S., Spang, R., Zuzan, H., Olson Jr, J.A., Marks, J.R. and Nevins, J.R. (2001) Predicting the clinical status of human breast cancer by using gene expression profiles. *Proc. Nat. Acad. Sci.*, **98(20)**, 11462–11467.

[50] West, M. (2003) Bayesian factor regression models in the "large p, small n" paradigm. Technical Report, Institute for Statistics and Decision Sciences, Duke University.

[51] Williams, C. and Barber, D. (1998) Bayesian classification with Gaussian priors. *IEEE Trans. Patt. Anal. Mach. Intell.*, **20**, 1342–1351.

[52] Xiong, M.M., Jin, L., Li, W. and Boerwinkle, E. (2000) Tumor classification using gene expression profiles. *Biotechiques*, **29**, 1264–1270.

[53] Yang, Y.H., Dudoit, S., Luu, P., Lin, D.M., Peng, V., Ngai, J. and Speed, T.P. (2002) Normalization for cDNA microarray data: a robust composite method addressing single and multiple slide systematic variation. *Nucleic Acids Res.* **30(4)**, e15.

[54] Zhu, J. and Hastie, T. (2002) Kernel logistic regression and the import vector machine. In *Neural Information Processing Systems 14* (eds T. Dietterich, S. Becker, and Z. Ghahramani), pp. 1081–1088. Cambridge, MA: MIT Press.

CHAPTER 11

Modeling and Analysis of Quantitative Proteomics Data Obtained from iTRAQ Experiments

George Michailidis and Philip Andrews
University of Michigan, Ann Arbor

11.1 Introduction

The advent of high throughput technologies over the last two decades (e.g., DNA sequencing, gene microarrays, etc) have led to fundamental insights into the workings of various biological processes. In addition they have provided inferences about protein interactions and regulation. However, they have limitations compared to the direct observations of protein expression levels. High throughput technologies that facilitate this task would enable researchers to build more accurate and comprehensive protein pathways. However, until recently most methods used for inferring protein expression levels, such as Western blots, 2D gel electrophoresis and more recently antibody protein microarrays, are inherently low-throughput.

The recent development of mass spectrometry based techniques for assessing the *relative* abundances of proteins in samples has provided a high-throughput alternative to these traditional techniques. For example, isotope-coded affinity tags (iCAT) relies on mass-difference labeling, but is limited to 'duplex experiments' (two samples/experimental conditions), and so is the case with stable isotope labeling of amino acids in culture (SILAC) [11, 15, 20]. In both technologies, protein quantification is achieved by comparing the mass spectrum intensity of the peptides (protein fragments) derived from two samples. On the other hand, the recent development of the isobaric tagging reagent (iTRAQ) [27] allows one to compare up to four samples, thus expanding the range of potentially interesting experiments. Further, a new iTRAQ reagent that allows the simultaneous processing of up to eight samples would further expand its range of applications.

In this paper, we present statistical methodology for analyzing quantitative data obtained from iTRAQ experiments in order to infer differential expression of proteins. Specifically, we develop a model that takes accounts for the various sources of variability present in the data and also allows for formal testing of the hypothesis of differential expression of proteins. Some other technical issues pertinent to the processing of the data, such as normalization and outlier elimination, are briefly discussed. The remainder of the paper is organized as follows: a brief overview of the iTRAQ technology is provided next. Section 2 presents the statistical modeling framework and briefly discusses estimation and inference issues. In Section 3, the methodology is illustrated on a small data set, while an overview of the range of studies that the technology has been used for is briefly reviewed in Section 4.

11.1.1 Overview of the iTRAQ technology

We provide next a brief overview of the iTRAQ technology (available from Applied Biosystems, Inc. CA) that sets the stage for the statistical model discussed in the next section. A detailed explanation of the technology and its underlying chemistry is provided in [27]. The iTRAQ reagent consists of four isobaric tags for labeling peptides, which are indistinguishable with respect to mass spectrometry (MS), but which yield signature peaks at m/z 114.1, 115.1, 116.1 and 117.1 in MS/MS spectra (see Figure 2, page 1157 in [27]). These peak areas can be used in turn for relative quantification of proteins. The availability of four tags allows one to compare up to four samples (e.g., normal vs different treatments) in a single experiment. More specifically, the four distinct iTRAQ tags can be used with four different complex protein samples, resulting in four labeled peptide pools. The peptide pools are then mixed together and subjected to 2-dimensional liquid chromatography and processed by tandem MS, resulting in a set of MS/MS spectra. The MS/MS spectra observations are then searched against a protein sequence database, using a database search program, such as Mascot [19], X-tandem [35], Paragon [24] or Sequest [31] to identify peptides and proteins in the sample. In addition, the peak areas for the 114.1, 115.1, 116.1 and 117.1 signature iTRAQ masses are extracted.

11.2 Statistical Modeling of iTRAQ Data

As outlined above, every MS/MS spectrum provides up to four measurements of peak areas, together with peptide and protein identification information, as well as a confidence score regarding the quality of the identification. The important thing to note is that the measurements for the different conditions (tags) are *paired*. Furthermore, due to the nature of mass spectrometry technology, the data have an inherent hierarchical structure; namely, there are multiple observations (spectra) of a particular peptide, and each full-length protein is associated with a number of peptides from which it was derived. In order to assess whether a particular protein is differentially expressed across two different experiment conditions, this hierarchical nature of the data structure needs to be considered to account for the different sources of variability.

The following model introduced in [14] achieves the objective. Let $y_{ijk\ell}$ denote the peak area measurement of the k-th spectrum of the j-th peptide, corresponding to the i-th protein, under condition (tag) ℓ, where $i = 1, \cdots, I$, $j = 1, \cdots, J(i)$, $k = 1, \cdots, K(j)$ and $\ell = 114, 115, 116$ and 117, the four possible tags available in an iTRAQ experiment. The notation captures the hierarchical nature of the data; suppose protein i has been identified with $J(i) = 3$ peptides, which in turn are associated with $K(1) = 2$, $K(2) = 3$ and $K(3) = 5$ spectra, respectively. For ease of notation we drop from the subsequent discussion index i, since the proposed model is used for each protein separately. Then, the ratios of two different conditions ℓ and ℓ' in the log-scale (in order to overcome the fact that the ratio scale is bounded from below by zero) can be modeled as follows:

$$\log\left(\frac{y_{jk\ell}}{y_{jk\ell'}}\right) = r + p_j + \epsilon_{jk}, \tag{11.1}$$

where r represents the *average* relative quantification of the protein in the two experimental conditions, p_j is the specific effect of peptide j on the observed ratio and ϵ_{jk} is a MS/MS spectrum specific effect, which is assumed to be normally distributed with mean zero and constant variance σ^2. Due to the fact that the protein fragmentation process into peptides and the peptide identification process are subject to random variability, it is further assumed that the peptide specific effects p_j are normally distributed with mean zero and constant variance τ^2.

The postulated model is known in the literature [18] as a random effects one-way analysis of variance and the unknown parameters σ^2 and τ^2 as random components. The proposed model takes into consideration the variability of the observed ratios both at the MS/MS spectrum level and at the peptide level, or in other words it says that every observed relative peak area ratio can be accounted for, by the overall protein ratio r, by a peptide specific value and by an MS/MS spectrum specific value. Finally, notice that in case a protein is identified by a single peptide the above model reduces to calculating r by simple averaging of the observed peak area ratios.

11.2.1 Estimation and inference

The parameters of the model can be estimated using a restricted maximum likelihood procedure [18], which ensures that the variance components are nonnegative.

The above model offers two advantages over the current treatment of iTRAQ data in the literature (see for example [1,12,15]), where a relative quantification ratio is calculated as a simple average of the observed values $y_{jk\ell}/y_{jk\ell'}$.

Firstly, it takes into consideration the underlying data structure and accounts for different sources of variability (peptide level, MS/MS spectrum level). As previously noted [1], there are cases in which the relative expression levels (observed ratios) exhibit a fairly high degree of variability amongst multiple measurements of the same peptide for a specific protein and even more importantly across peptides for the same protein. Therefore, a simple averaging of those ratios ignores (or bypasses) this issue, unlike the proposed model. Secondly and more importantly, it allows one to formally test for differential expression of proteins as follows:

the null hypothesis of no change between the two conditions ℓ and ℓ' is given by $H_0 : r = 0$, whereas the alternative hypothesis of up/down-regulation of the protein by $H_A : r \neq 0$. The variance of the estimate $\hat{r}$ for a protein associated with J peptides and K MS/MS spectra is given by

$$\text{Var}(\hat{r}) = \frac{\tau^2}{J} + \frac{\sigma^2}{N}.$$

Then, the test statistic $\frac{\hat{r}}{\text{Var}(\hat{r})} \sim t_{K-1}$ follows a t-distribution with $K - 1$ degrees of freedom, which can be used for calculating p-values or the rejection region of the above hypothesis.

However, in many situations investigators are only interested in statistically significant changes of a certain magnitude, since they believe that small magnitude ones do not represent genuine biological activity. This is easily incorporated in the proposed framework by modifying the null and alternative hypotheses as follows: $H_0 : \theta_1 \leq r \leq \theta_2$ vs$H_A : r \leq \theta_1$ or $r \geq \theta_2$, for some thresholds θ_1 and θ_2. For example, $\theta_1 = -.23$ and $\theta_2 = .20$ correspond to a 15% change in the original scale relative expression levels. Carrying out such a composite test poses some technical challenges, but the basic technical details can be found in [12].

11.3 Data Illustration

We illustrate the proposed methodology on two small experiments that were designed to test the reproducibility of the technology. In both cases, the sample consisted of 23 purified proteins that were mixed together and tagged with the iTRAQ reagent. In the first experiment, two equal proportions of the sample were aliquotted; hence, the expected relative ratios between the ratios of each identified protein should be one. In the second experiment, three samples were processed in a 4:2:1 ratio.

For the first experiment, the obtained MS/MS spectra were searched with the Mascot search engine and 286 produced high confidence identifications about the peptides and their corresponding proteins. Eighteen out of the twenty-three proteins in the mixture were identified from 64 unique peptides. A quantile-quantile plot of the data in $\log_2$ scale is shown in Figure 11.1,

which exhibits a very good agreement between the empirical distributions of
the peak areas of the two samples. However, in many cases because of pipeting
errors or other experimental factors, not such good agreement is obtained. In
such a case, normalization of the samples is required (see [32] for a discussion
of the relevant issues in gene microarrays). Our experience has shown that a
quantile based normalization [14] proves adequate. In some instances, prior
biological knowledge dictates that only spectra corresponding to a particular
subset of proteins should be used for normalization purposes. Further, in some
experiments one can include in the sample a number of proteins that act as
internal standards and could be used for such purposes. However, it may happen that some of these proteins may not be identified by the search engine
and hence one may have to resort to a data driven normalization process.
The mean ratio of the proteins in the two samples together with error bars

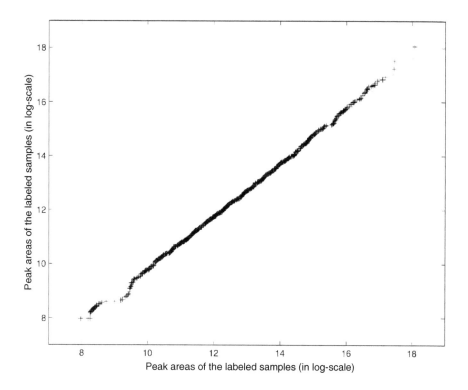

Figure 11.1 *Quantile-quantile plot of peak areas of the two tagged samples.*

corresponding to two standard errors are shown in Figure 11.2. All the ratios
are very close to their expected level of one. The different lengths of the error
bars reflect the inherent variability in the quantification of the relative protein
expression in the two samples, which is due to two factors: the number of peptides associated with each protein and the number of spectra associated with

each peptide. As noted above, the proposed model accounts for both sources of variability appropriately.

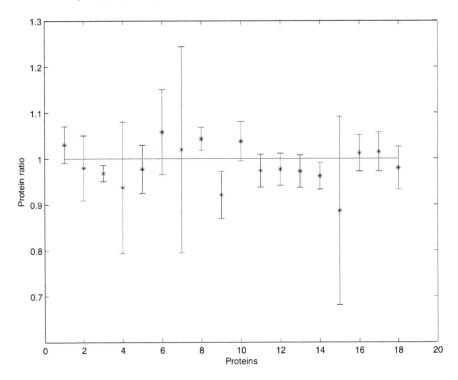

Figure 11.2 *Protein ratios with error bars.*

In the second experiment, much more material was loaded in order to achieve the desired ratios between the samples. This led to obtain 673 spectra that produced identifications. In total, 17 proteins were identified from 63 unique peptides. However, one protein was identified from a single spectrum and was removed from any further analysis. The protein ratios together with error bars are shown in Figure 11.3. It can be seen that to a large extent the protein ratios are close to their expected levels, which illustrates the potential of the iTRAQ technology.

11.4 Discussion and Concluding Remarks

The so-called shotgun proteomic methods [2], coupled with isobaric tagging, enable high throughput proteomic analysis. The iTRAQ technology has been applied to date to a number of diverse areas. For example, a number of studies have looked at cancer biomarkers [16, 8, 6, 14, 26]. Scaling up the technology to deal with large clinical studies remains an issue, although the recent development of an iTRAQ reagent comprised of 8 tags (the four original together

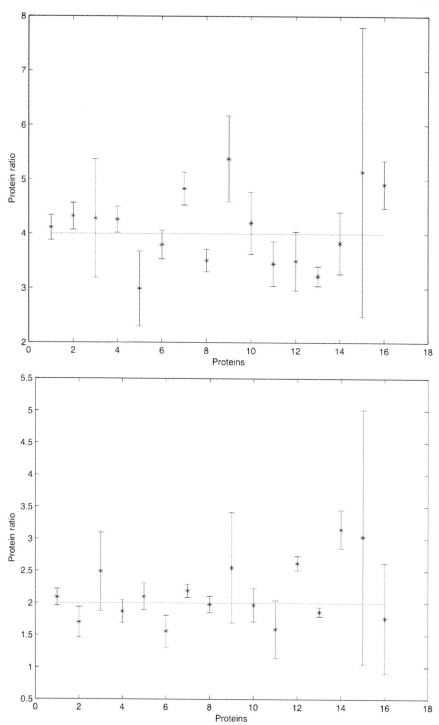

Figure 11.3 *Protein ratios with error bars. Top panel, corresponding to the samples in a 4:1 ratio; bottom panel, samples with a 2:1 ratio.*

with the 113, 118, 119 and 121 m/z peaks of MS/MS spectra) [5] will allow researchers to conduct larger scale studies. The 8-plex iTRAQ reagent is experiencing fast adoption as a number of recent publications attests (see for example [22, 23]).

Another important area of applications is cell biology and the identification of biological processes (see for example, [4, 13, 21, 17, 7, 30, 25]). Emerging areas include neuroscience [33, 1], toxicology [10], organelle profiling [36], sphosphoproteomics [9, 28], identification of signaling pathways [39] and stem cell research [34, 29, 37, 38].

To conclude, we presented a statistical model for processing high throughput protein expression data obtained from a novel isobaric tagging technology. The main advantages of the proposed modeling framework, based on a one-way random effects analysis of variance model, are: (i) that it takes into consideration the underlying structure of the data and accounts for the various sources of variability and (ii) that it allows one to set up a rigorous hypothesis testing procedure regarding differential expression of proteins.

The availability of only up to eight tags per iTRAQ experiment acts as a limiting factor in setting up experiments with multiple conditions. In that case one needs to appropriately combine the data from several experiments, which introduces an additional source of variability that needs to be taken into consideration by the model. Ideas from experimental design can be employed to resolve this issue and this constitutes a topic of current work.

11.5 Acknowledgments

This work is supported in part by NIH grant P41-18627.

References

[1] Abdi, F., Quinn, J.F., Jankovic, J., McIntosh, M., Leverenz, J.B., Peskind, E., Nixon, R., Nutt, J., Chung, K., Zabetian, C., Samii, A., Lin, M., Hattan, S., Pan, C., Wang, Y., Jin, J., Zhu, D., Li, G.J., Liu, Y., Waichunas, D., Montine, T.J. and Zhang, J. (2006) Detection of biomarkers with a multiplex quantitative proteomic platform in cerebrospinal fluid of patients with neurodegenerative disorders. *J. Alzheimers Dis.*, **9**, 293–348.

[2] Aebersold, R. and Mann, M. (2003) Mass spectrometry based proteomics. *Nature*, **422**, 198–207.

[3] Aggarwal, K., Choe, L.H., and Lee, K.H. (2006) Shotgun proteomics using the iTRAQ isobaric tags. *Proteomics*, **5**, 2297.

[4] Bantscheff, M., Eberhard, D., Abraham, Y., Bastuck, S., Boesche, M., Hobson, S., Mathieson, T., Perrin, J., Raida, M., Rau, C., Reader, V., Sweetman, G., Bauer, A., Bouwmeester, T., Hopf, C., Kruse, U., Neubauer, G., Ramsden, N., Rick, J., Kuster, B. and Drewes, G. (2007) Quantitative chemical proteomics

reveals mechanisms of action of clinical ABL kinase inhibitors. *Nat. Biotechnol.*, **25**, 1035–1044.

[5] Choe, L., D'Ascenzo, M., Relkin, N.R., Pappin, D., Ross, P., Williamson, B., Guertin, S., Pribil, P. and Lee, K.H. (2007) 8-Plex quantitation of changes in cerebrospinal fluid protein expression in subjects undergoing intravenous immunoglobulin treatment for Alzheimer's disease. *Proteomics*, **7**, 3651–3660.

[6] Comuzzi, B. and Sadar, M.D. (2006) Proteomic analyses to identify novel therapeutic targets for the treatment of advanced prostate cancer. *Cell Sci*, **3**, 61–81.

[7] Dean, R.A. and Overall, C.M. (2007) Proteomics discovery of metalloproteinase substrates in the cellular context by iTRAQ labeling reveals a diverse MMP-2 substrate degradome. *Mol. Cell. Prot.* **6**, 611–623.

[8] Desouza, L.V., Grigull, J., Ghanny, S., Dube, V., Romaschin, A.D., Colgan, T.J. and Siu, K.W. (2007) Endometrial carcinoma biomarker discovery and verification using differentially tagged clinical samples with multidimensional liquid chromatography and tandem mass spectrometry. *Mol. Cell. Prot.*, **6**, 1170.

[9] Gafken, P.R. and Lampe, P.D. (2006) Methodologies for characterizing phosphoproteins by mass spectrometry. *Cell Commun. Adhes.* **13**, 249–262.

[10] Gluckmann, M., Fella, K., Waidelich, D., Merkel, D., Kruft, V., Kramer, P.J., Walter, Y., Hellmann, J., Karas, M. and Kroger, M. (2007) Prevalidation of potential protein biomarkers in toxicology using iTRAQ reagent technology. *Proteomics*, **7**, 1564–1574.

[11] Gygi, S.P., Rist, B., Gerber, S.A., Turecek, F., Gelb, M.H. and Aebersold, R. (1999) Quantitative analysis of complex protein mixtures using isotope-coded affinity tags. *Nat. Biotechnol.*, **17**, 994–999.

[12] Hodges, J.L. and Lehmann, E.L. (1954) Testing the approximate validity of statistical hypotheses. *J. Roy. Stat. Soc. B*, **16**, 261–268.

[13] Jagtap, P., Michailidis, G., Zielke, R., Walker, A.K., Patel, N., Strahler, J.R., Driks, A., Andrews, P.C. and Maddock, J.R. (2006) Early events of *Bacillus anthracis* germination identified by time-course quantitative proteomics. *Proteomics*, **6**, 5199–5211.

[14] Keshamouni, V.G., Michailidis, G., Grasso, C.S., Anthwal, S., Strahler, J.R., Walker, A., Arenberg, D.A., Reddy, R.C., Akulapalli, S., Thannickal, V.J., Standiford, T.J., Andrews, P.C. and Omenn, G.S. (2006) Differential protein expression profiling by iTRAQ-2DLC-MS/MS of lung cancer cells undergoing epithelial-mesenchymal transition reveals a migratory/invasive phenotype. *J. Proteome Res.*, **5**, 1143–1154.

[15] Krijgsveld, J., Ketting, R.F., Mahmoudi, T., Johansen, J., Artal-Sanz, M., Verrijzer, C.P., Plasterk, R.H.A. and Heck, A.J.R. (2003) Metabolic labeling of *C. elegans* and *D. melanogaster* for quantitative proteomics. *Nat Biotechnol.*, **21**, 927–931.

[16] Kristiansson, M.H., Bhat, V.B., Babu, I.R., Wishnok, J.S. and Tannenbaum, S.R. (2007) Comparative time-dependent analysis of potential inflammation biomarkers in lymphoma-bearing SJL mice. *J. Proteome Res.*, **4**, 1735–1744.

[17] Lund, T.C., Anderson, L.B., McCullar, V., Higgins, L., Yun, G.H., Grzywacz, B., Verneris, M.R. and Miller, J.S. (2007) iTRAQ is a useful method to screen for membrane-bound proteins differentially expressed in human natural killer cell types. *J. Proteome Res.*, **6**, 644–653.

[18] McCulloch, C.E. and Searle, S.R. *Generalized, Linear and Mixed Models*, Wiley (2001).

[19] Mascot, Matrix Science.

[20] Ong, S.E., Blagoev, B., Kratchmarova, I., Kristensen, D.B., Steen, H., Pandey, A. and Mann, M. (2002) Stable isotope labeling by amino acids in cell culture, SILAC, as a simple and accurate approach to expression proteomics. *Mol. Cell. Prot.*, **1**, 376–386.

[21] Ono, Y., Hayashi, C., Doi, N., Kitamura, F., Shindo, M., Kudo, K., Tsubata, T., Yanagida, M. and Sorimachi, H. (2007) Comprehensive survey of p94/calpain 3 substrates by comparative proteomics - Possible regulation of protein synthesis by p94. *Biotechnol J.*, **2**, 565–576.

[22] Ow, S.Y., Cardona, T., Taton, A., Magnuson, A., Lindblad, P., Stensjo, K. and Wright, P.C. (2008) Quantitative shotgun proteomics of enriched heterocysts from Nostoc sp. PCC 7120 using 8-Plex isobaric peptide tags. *J. Proteome Res.*, **7**, 1615–1628.

[23] Pierce, A., Unwin, R.D., Evans, C.A., Griffiths, S., Carney, L., Zhang, L., Jaworska, E., Lee, C.F., Blinco, D., Okoniewski, M.J., Miller, C.J., Bitton, D.A., Spooncer, E. and Whetton, A.D. (2007) Eight-channel iTRAQ enables comparison of the activity of 6 leukaemogenic tyrosine kinases. To appear in *Mol. Cell. Prot.*

[24] ProteinPilot Software, Applied Biosystems.

[25] Radosevich, T.J., Reinhardt, T.A., Lippolis, J.D., Bannantine, J.P. and Stabel, J.R. (2007) Proteome and differential expression analysis of membrane and cytosolic proteins from *Mycobacterium avium* subsp. paratuberculosis strains K-10 and 187. J. Bacteriol., **189**, 1109–1117.

[26] Ralhan, R., Desouza, L.V., Matta, A., Chandra Tripathi, S., Ghanny, S., Datta Gupta, S., Bahadur, S. and Siu, K.W. (2008) Discovery and verification of head-and-neck cancer biomarkers by differential protein expression analysis using iTRAQ-labeling and multidimensional liquid chromatography and tandem mass spectrometry. To appear in *Mol. Cell. Prot.*

[27] Ross, P.L., Huang, Y.L.N., Marchese, J.N., Williamson, B., Parker, K., Hattan, S., Khainovski, N., Pillai, S., Dey, S., Daniels, S., Purkayastha, S., Juhasz, P., Martin, S., Bartlet-Jones, M., He, F., Jacobson, A. and Pappin, D.J. (2004) Multiplexed protein quantitation in *Saccharomyces cerevisiae* using amine-reactive isobaric tagging reagents. *Mol. Cell. Prot.*, **3**, 1154–1169.

[28] Sachon, E., Mohammed, S., Bache, N. and Jensen, O.N. (2006) Phosphopeptide quantitation using amine-reactive isobaric tagging reagents and tandem mass spectrometry: application to proteins isolated by gel electrophoresis. *Rapid Commun. Mass Sp.*, **20**, 1127–1134.

[29] Salim, K., Kehoe, L., Minkoff, M.S., Bilsland, J.G., Munoz-Sanjuan, I. and Guest, P.C. (2006) Identification of differentiating neural progenitor cell markers using shotgun isobaric tagging mass spectrometry. *Stem Cells Dev.*, **15**, 461–470.

[30] Schilling, O. and Overall, C.M. (2007) Proteomic discovery of protease substrates. *Curr. Opin. Chem. Biol.*, **11**, 36–45.

[31] Sequest Sorcerer, Thermo Scientific.

[32] Speed, T. (ed.) *Statistical Analysis of Gene Expression Microarray Data*, Chapman & Hall (2003).

[33] Tannu, N.S. and Hemby, S.E. (2006) Methods for proteomics in neuroscience. *Prog. Brain Res.*, **158**, 41–82.

[34] Unwin, R.D., Smith, D.L., Blinco, D., Wilson, C.L., Miller, C.J., Evans, C.A., Jaworska, E., Baldwin, S.A., Barnes, K., Pierce, A., Spooncer, E. and Whet-

ton, A. (2006) Quantitative proteomics reveals post-translational control as a regulatory factor in primary hematopoietic stem cells. *Blood*, **107**, 4687–4694.

[35] X–tandem search algorithm, The Global Proteome Machine Organization.

[36] Yan, W., Hwang, D. and Aebersold, R. *Quantitative proteomic analysis to profile dynamic changes in the spatial distribution of cellular proteins.* Humana Press (2008).

[37] Yocum, A.K., Gratsch, T.E., Leff, N., Strahler, J.R., Hunter, C.L., Walker, A.K., Michailidis, G., Omenn, G.S., OShea, S.K. and Andrews, P.C. (2008) Coupled global and targeted proteomics of human embryonic stem cells during induced differentiation. *Mol. Cell. Prot.*, **7**, 750–767.

[38] Zhang, J., Sui, J., Ching, C.B. and Chen, W.N. (2008) Protein profile in neuroblastoma cells incubated with S- and R-enantiomers of ibuprofen by iTRAQ-coupled 2-D LC-MS/MS analysis: Possible action of induced proteins on Alzheimer's disease. To appear in *Proteomics*.

[39] Zhang, Y., Wolf-Yadlin, A., Ross, P.L., Pappin, D.J., Rush, J., Lauffenburger, D.A. and White, F.M. (2005) Time-resolved mass spectrometry of tyrosine phosphorylation sites in the epidermal growth factor receptor signaling network reveals dynamic modules. *Mol. Cell. Prot.*, **4**, 1240–1250.

CHAPTER 12

Statistical Methods for Classifying Mass Spectrometry Database Search Results

Zhaohui S. Qin, Peter J. Ulintz, Ji Zhu and Philip Andrews
University of Michigan, Ann Arbor

12.1 Introduction

Proteomics, the large scale analysis of proteins, holds great promise to enhance our understanding of cellular biological process in normal and disease tissue [32]. The proteome is defined as the complete set of proteins expressed by a cell, tissue or organism. Transcript level analysis, as typically measured by DNA microarray technologies [35, 26], does not provide complete information on the proteome in that the DNA transcriptional template of an organism is static, whereas the proteome is constantly changing in response to environmental signals and stress. Recent studies have been unable to establish a consistent relationship between the transcriptome and proteome [3, 30]. The main reason is that transcript profiles do not provide any information on posttranslational modifications of proteins (such as phosphorylation and glycosylation) that are crucial for protein transport, localization and function. Studies have shown that the correlation between mRNA levels and protein abundance was poor for low expression proteins [17]. Posttranslational modifications such as glycosylation, phosphorylation, ubiquitination and proteolysis produce further modifications, which lead to changes in protein abundance.

The ultimate goal of proteomics is to identify and characterize all proteins expressed in cells collected from a variety of conditions to facilitate comparisons [30, 1]. Although transcriptome data such as those from microarray studies reflect the genome's objectives for protein synthesis, they do not provide information about the final results of such objectives. Proteomics analysis provides a more accurate view of biological porcesses, thus offering a better understanding of the physiologic and pathologic states of an organism, and serving as an important step in the development and validation of diagnostic and therapeutics.

In the fast-advancing proteomics field, increasingly high-throughput assays for the identification and characterization of proteins are being developed. Mass spectrometry is the primary experimental method for protein identification; tandem mass spectrometry (MS/MS) in particular is now the *de facto* standard identification technology, providing the ability to rapidly characterize thousands of peptides in a complex mixture. In such assays, proteins are first digested into smaller peptides, and subjected to reverse-phase chromatography. Pepetides are then ionized and fragmented to produce signature MS/MS spectra that are used for identification. More details about this technique can be found in Link et al. [25] and Washburn et al. [41].

Instrument development continues to improve the sensitivity, accuracy and throughput of analysis. Current instruments are capable of routinely generating several thousand spectra per day, detecting sub-femtomolar levels of protein at better than 10 ppm mass accuracy. Such an increase in instrument performance is limited, however, without effective tools for automated data analysis. In fact, the primary bottleneck in high-throughput proteomics production "pipelines" is in many cases the quality analysis and interpretation of the results to generate confident protein assignments. This bottleneck is primarily due to the fact that it is often difficult to distinguish true hits from false positives in the results generated by automated mass spectrometry database search algorithms. In most cases, peptide identifications are made by searching MS/MS spectra against a sequence database to find the best matching database peptide [21]. All MS database search approaches produce scores describing how well a peptide sequence matches experimental data, yet classifying hits as "correct" or "incorrect" based on a simple score threshold frequently produces unacceptable false positive/false negative rates. Consequently, manual validation is often required to be truly confident in the assignment of a database protein to a spectrum. Design and implementation of an efficient algorithm to automate the peptide identification process is of great interest and presents a grand challenge for the statistician and bioinformatician in the post-genomics era.

An array of strategies for automated and accurate spectral identification ranging from heuristics to probabilistic models has been proposed [11, 33, 8, 4, 19, 9, 16]. Discrimination of correct and incorrect hits is thus an ongoing effort in the proteomics community [28, 27, 43, 23, 34, 13, 2, 12, 37, 39], with the ultimate goal being completely automated MS data interpretation. In this chapter, we intend to provide a concise and up to date review of automated techniques for classification of database search results, with more details on a probabilistic approach and a recent machine learning-based approach. The remaining part of this chapter is organized as follows. In Section 12.2, we provide a concise overview of the current proteomics technologies. In Section 12.3, we summarize all the methods proposed for classifying database search results. In Section 12.4, we describe a dataset and the detailed procedure of

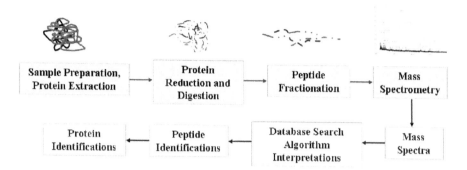

Figure 12.1 *Schematic Diagram of a typical proteomics experiment. Proteins are extracted from a biological sample, digested and fractionated (separated) prior to introduction into the mass spectrometer. Spectra generated by the mass spectrometer are interpreted using algorithms which compare them to amino acid sequences from a database, returning a list of database peptides which best match each spectrum. Peptide predictions are validated and combined to produce a list of proteins predicted to be present in the biological sample.*

analyzing it. In Sections 12.5 and 12.6, we close with concluding remarks and discussions.

12.2 Background on Proteomics

A typical proteomics experiment involves analyzing a biological sample to identify the proteins present. Generating a list of protein identifications follows a somewhat standard set of steps that are now briefly outlined in Figure 12.1 (please see Steen and Mann [36] for a general review).

An experiment begins by extracting proteins from a biological sample. This complex mixture of proteins is reduced to disrupt the tertiary structure of the individual proteins, the cysteines are blocked to prevent their refolding and the proteins are digested with a proteolytic enzyme— typically trypsin— into shorter peptide sequences to render the mixture more amenable to mass spectrometry. This complex mixture of peptides must be fractionated into simpler mixtures for introduction into the mass spectrometer. There is a variety of methods for resolving complex peptide mixtures, all exploiting different physical properties of individual peptides, e.g., size, charge, hydrophobicity, or isoelectric point (pI). Two-dimensional separation of peptide mixtures via liquid chromatography (2D-LC) is now standard. In such an experiment, the multiple simplified fractionated mixtures are analyzed individually in the mass spectrometer and the results combined to produce a final dataset of evidence for the presence of all proteins in the mixture.

Once a peptide mixture from 2D-LC is introduced, the mass spectrometer measures the masses of the individual peptides. A measurement of the intact mass of an individual peptide is rarely sufficient to identify it unambiguously, therefore tandem mass spectrometry (MS/MS) is typically employed. Tandem mass spectrometry involves two sequential measurements. In the first stage, the masses of all peptides in the mixture are measured as described above. From this measurement, very narrow mass ranges corresponding to these individual "precursor" masses may be selected for further analysis. These selections, ideally consisting of a homogeneous mixture of identical peptides, undergo further fragmentation. The fragmentation is induced by adding energy to the isolated peptides, typically by collision with an inert gas in a process known as collision-induced dissociation (CID). The additional energy causes these peptides to fragment in a predictable manner. The masses of these fragments are detected, producing the tandem mass spectrum. This spectrum may be interpreted to infer the amino acid sequence of the peptide from which it was produced. A raw tandem mass spectrum may be interpreted manually, but this process is time-consuming; a typical mass spectrum requires 15-45 minutes to interpret by a trained expert. Consequently, the aforementioned algorithms such as Sequest [11] and Mascot [33] are typically used to interpret the spectra automatically. These algorithms make use of a sequence database, a list of protein amino acid sequences, although it must be noted that a number of algorithms exist which attempt to infer the peptide sequence directly from spectra "de novo" without the use of a sequence database (see Johnson et al. [21] for a general review of both standard "uninterpreted" and "de novo" methods). The primary goal of the database search algorithms is to return the peptide sequence in the database which best explains the experimental spectrum. A significant amount of attention has been paid over the past fourteen years to algorithms and heuristics for ranking and scoring sequence/spectrum matches, but most follow the same underlying principle: for each peptide in the database, the algorithms generate a theoretical tandem mass spectrum and match this theoretical spectrum to the experimental spectrum, producing a score for the match. The result of performing this matching process is a ranked list of amino acid sequences which best explain the experimental spectrum (see Figure 12.2). Each experimental spectrum generated by the instrument will have this associated list of search results. If successful, the top-scoring result or 'hit' will be the correct one. However, the top-scoring hit is frequently incorrect. Therefore, the fundamental problem becomes one of identifying which hits are correct given the data returned by the search algorithm for each experimental mass spectrum.

12.3 Classification Methods

The most straightforward approach to automated analysis of the results of mass spectrometry database search result is to define specific score-based filtering thresholds as discriminators of correctness, e.g., accepting Sequest scores of doubly-charged fully tryptic peptides with XCorr greater than 2.2

```
denSilac-150mM.2684.2684.2.out
TurboSEQUEST v.27 (rev. 12), (c) 1999-2005
Molecular Biotechnology, Univ. of Washington, J.Eng/S.Morgan/J.Yates
Licensed to ThermoFinnigan Corp.
05/04/2006, 10:17 AM, 0.1 sec on MPCORBITRAP
(M+H)+ mass = 1688.8477 ~ 10.0000 (+2), fragment tol = 2.0000, MONO/MONO
total inten = 6941.0, lowest Sp = 96.4, # matched peptides = 446
# amino acids = 20076, # proteins = 13698, C:\Xcalibur\database\uniprot_sprot-HUMAN-SILAC.fasta.hdr
ion series nABY ABCDVWXYZ: 0 1 1 0.0 1.0 0.0 0.0 0.0 0.0 0.0 1.0 0.0
display top 10/0, ion % = 0.0, CODE = 101040
(K* +6.02040) (R# +10.01240) C=161.03918  Enzyme:Trypsin(KR) (2)
```

#	Rank/Sp	Id#	(M+H)+	deltCn	XCorr	Sp	Ions	Reference		Peptide
1.	1 / 1	7073	1688.84253	0.0000	2.8175	758.1	17/28	MCM5_HUMAN		K.LQELPDAVPHGEMPR.H
2.	2 / 44	13268	1688.83508	0.5669	1.2201	272.6	14/28	ZDH18_HUMAN		K.QIDNTGSSTYRPPPR.T
3.	3 / 63	688	1688.84891	0.5754	1.1964	221.9	10/22	APEX1_HUMAN		R.LEYR#QR#WDEAFR.K
4.	4 / 70	3451	1688.84107	0.6064	1.1089	206.8	12/30	DUOX2_HUMAN		K.EAAKDGVPAMEWPGPK*.E
5.	5 /178	2400	1688.85034	0.6296	1.0437	130.5	10/34	COS5A1_HUMAN		K.GGQGPPGPQGPIGYPGPR.G
6.	6 / 41	4170	1688.84087	0.6398	1.0148	286.6	13/30	FILA_HUMAN	+1	R.HSQVGQGQSSGPR#TSR#.N
7.	7 / 6	9401	1688.83210	0.6405	1.0128	489.6	13/24	PPAL_HUMAN		R.QRYHGFLNTSYHR#.Q
8.	8 /144	12630	1688.85571	0.6537	0.9757	152.0	11/28	UBAP1_HUMAN		K.PNGFITLPQLGNCEK.M
9.	9 /152	5742	1688.86352	0.6774	0.9089	145.5	9/26	INVS_HUMAN		R.HMKQLGAGDVDR#WR#.Q
10.	10 / 86	3086	1688.83641	0.6902	0.8730	193.3	11/28	DCP2_HUMAN		K.GK*VNKEEAPHDCAAR.E

Figure 12.2 *Example of Sequest search result (*.out file): A Sequest report resulting from the search of an MS/MS peaklist against a sequence database. In addition to search parameter information and some background statistics of the search (the top portion of the report), a ranked list of matching peptides is provided. Each row in this list is a potential identification of a peptide that explains the mass spectrum, the top match being the peptide that the algorithm believes is the best match. Various scores or attributes are provided for each potential "hit": Sp, XCorr, deltCn, etc. These attributes, or additional attributes calculated based on the information provided for each hit, are the primary data being used in this work.*

and delta Cn values at least 0.1 [41]; these thresholds are typically published as the criteria for which correctness is defined. Other efforts have focused on establishing statistical methods for inferring the likelihood that a given hit is a random event. A well known example of this is the significance threshold calculated by the Mascot search algorithm, which by default displays a threshold indicating the predicted probability of an assignment being greater than 5% likely to be a false positive based on the size of the database. Use of a reverse database search to provide a measure of false positive rate is another method frequently used [28, 31]. More formally, Sadygov and Yates [34] model the frequency of fragment ion matches from a peptide sequence database matching a spectrum as a hypergeometric distribution, a model also incorporated into the openly available X!Tandem algorithm [9, 13]; while Geer et al. model this distribution as a Poisson distribution [16].

Several of these approaches have been implemented directly in the scoring calculation of new search algorithms [9, 16, 12]. Alternatively, external algorithms may be developed that process the output of the more standard search platforms such as Mascot or Sequest, classifying results as either correct or incorrect with an associated probability. Examples of the latter type include

PeptideProphet [23] and QScore [28]. These tools have the advantage of being able to accommodate results from these existing, well-established search engines that may already be in place in a production lab; conversely, approaches in which the quality measures are built into the search algorithm scoring are arguably more user-friendly in that they eliminate the extra post-search processing step of having to run a second algorithm.

Keller et al. were among the first to implement a generic tool for classifying the results of common search algorithms as either correct or incorrect [23]. Their PeptideProphet tool represents arguably the most widely used openly-available package implementing a probabilistic approach to assess the validity of peptide assignments generated by MS database search algorithms. Their approach contains elements of both supervised and unsupervised learning, achieves a much higher sensitivity than conventional methods based on simple scoring thresholds. One concern with PeptideProphet, however, is the degree to which the supervised component of the model can be generalized to new types of data and the ease with which new potentially useful information can be added to the algorithm.

Ulintz et al. attempt to address these difficulties by applying a set of standard "over the counter" methods to the challenging peptide identification problem [39]. Anderson et al. demonstrated that support vector machines could perform well on ion-trap spectra searches using the Sequest algorithm [2]. Our approaches, based on the latest machine learning techniques, extend this idea, providing further support for the flexibility and generality of using machine learning tools with these data. We focus on establishing the effectiveness of two statistical pattern classification approaches—boosting and random forests—at improving peptide identification, even in cases in which individual pieces of information obtained from a particular search result are not completely independent or strongly discriminatory (but are easily obtained). Such work will hopefully result in development of software tools that are easily installed in a production laboratory setting that would allow convenient filtering of false identifications with an acceptably high accuracy, either as new tools or as a complement to currently existing software. The problem of classification of mass spectrometry-based peptide identification, a binary classification problem, seems well suited to these algorithms and could lead to more readily-usable software for automated analysis of the results of mass spectrometry experiments.

12.3.1 Mixture model approach in peptide prophet

Among all the methods that have been proposed in the literature for the peptide identification problem, the mixture model approach implemented in the PeptideProphet algorithm [23] is perhaps the best known. In this method, a discriminant score function $F(x_1, x_2, \ldots, x_S) = c_0 + c_1 x_1 + \cdots + c_S x_S$ is defined to combine database search scores x_1, x_2, ..., x_S where c_i's are weights. Based on a training dataset, a Gaussian distribution is chosen to model the discriminant scores corresponding to correct peptide assignments, and a Gamma distribution is selected to model the asymmetric discriminant scores corresponding to incorrect peptide assignments. All the scores are therefore represented by a mixture model

$$p(x) = rf_1(x) + (1-r)f_2(x),$$

where $f_1(x)$ and $f_2(x)$ represent the density functions of the two types of discriminant scores, and r is the proportion of correct peptide identification. For each new test dataset, the EM algorithm [10] is used to estimate the probability that the peptide identified is correct. A decision can be made by comparing the probability to a pre-specified threshold. When compared to conventional means of filtering data based on Sequest scores and other criteria, the mixture model approach achieves much higher sensitivity.

A crucial part of the above approach is the choice of discriminant score function F. In Keller et al. (2002a), the c_i's are derived in order to maximize the between- versus within-class variation under the multivariate normal assumption using training data with peptides assignments of known validity. To make this method work, one has to assume that the training data and the test data are generated from the same source. In other words, when a new set of discriminant scores is generated requiring classification, one has to retrain the weight parameters c_i's using a new corresponding training set; the discriminant function F is data dependent. In an area such as proteomics in which there is a good amount of heterogeneity in instrumentation, protocol, database and database searching software, it is fairly common to come across data which display significant differences. Its unclear to what degree the results of a classification algorithm are sensitive to these differences, hence it is desirable to automate the discriminant function training step. Another potential issue is the normal and Gamma distributions used to model the two types of discriminant scores. There is no theoretical explanation why the discriminant scores should follow these two distributions; in fact, a Gamma distribution rather than a normal distribution may be appropriate for both positive and negative scores when using the Mascot algorithm [33]. It is likely that for a new set of data generated by different mass spectrometers and/or different search algorithms, the two distributions may not fit the discriminant scores well. If they are generically applied, significant higher classification errors may be produced.

12.3.2 Machine learning techniques

Distinguishing correct from incorrect peptide assignments can be regarded as a classification problem, or supervised learning, a major topic in the statistical learning field. Many powerful methods have been developed such as CART,

SVM, random forest, boosting and bagging [18]. Each of these approaches has some unique features that enable them to perform well in certain scenarios; SVMs, for example, are an ideal tool for small sample size, large feature space situations. On the other hand, all approaches are quite flexible and have been applied to an array of biomedical problems, from classifying tissue types using microarray data [7] to predicting functions of single nucleotide polymorphisms [5]. In this chapter, we apply state-of-the-art machine learning approaches to the peptide assignment problem.

Boosting
The boosting idea, first introduced by Freund and Schapire with their AdaBoost algorithm [14], is one of the most powerful learning techniques introduced during the past decade. It is a procedure that combines many "weak" classifiers to achieve a final powerful classifier. Here we give a concise description of boosting in the two-class classification setting. Suppose we have a set of training samples, where x_i is a vector of input variables (in this case, various scores and features of an individual MS database search result produced from an algorithm such as Sequest—see Figure 12.2) and y_i is the output variable coded as -1 or 1, indicating whether the sample is an incorrect or correct assignment of a database peptide to a spectrum. Assume we have an algorithm that can build a classifier $T(x)$ using weighted training samples so that, when given a new input x, $T(x)$ produces a prediction taking one of the two values $-1, 1$. Then boosting proceeds as follows: start with equal weighted training samples and build a classifier $T_1(x)$; if a training sample is misclassified, i.e., an incorrect peptide is assigned to the spectrum, the weight of that sample is increased (boosted). A second classifier $T_2(x)$ is then built with the training samples, but using the new weights, no longer equal. Again, misclassified samples have their weights boosted and the procedure is repeated M times. Typically, one may build hundreds or thousands of classifiers this way. A final score is then assigned to any input x, defined to be a linear (weighted) combination of the classifiers, where w indicates the relative importance of each classifier. A high score indicates that the sample is most likely a correctly assigned protein with a low score indicating that it is most likely an incorrect hit. By choosing a particular value of the score as a threshold, one can select a desired specificity or a desired ratio of correct to incorrect assignments. In this work, we use decision trees with 40 leaves for the 'weak' classifier, and fix M equal to 1000.

Random Forests
Similar to boosting, the random forest [6] is also an ensemble method that combines many decision trees. However, there are three primary differences in how the trees are grown: 1. Instead of assigning different weights to the training samples, the method randomly selects, with replacement, n samples from the original training data; 2. Instead of considering all input variables at each split of the decision tree, a small group of input variables on which to split are randomly selected; 3. Each tree is grown to the largest extent

possible. To classify a new sample from an input, one runs the input down each of the trees in the forest. Each tree gives a classification (vote). The forest chooses the classification having the most votes over all the trees in the forest. The random forest enjoys several nice features: it is robust with respect to noise and overfitting, and it gives estimates of what variables are important in the classification. A discussion of the relative importance of the different parameters used in our analysis of MS search results is given in the results section. The performance of the random forest depends on the strength of the individual tree classifiers in the forest and the correlation between them. Reducing the number of randomly selected input variables at each split reduces both the strength and the correlation; increasing it increases both. Somewhere in between is an optimal range–usually quite wide.

Support vector machines

The support vector machine (SVM) is another successful learning technique introduced in the past decade (Vapnik 1999). It typically produces a non-linear classification boundary in the original input space by constructing a linear boundary in a transformed version of the original input space. The dimension of the transformed space can be very large, even infinite in some cases. This seemingly prohibitive computation is achieved through a positive definite reproducing kernel, which gives the inner product in the transformed space. The SVM also has a nice geometrical interpretation in the finding of a hyperplane in the transformed space that separates two classes by the biggest margin in the training samples, although this is usually only an approximate statement due to a cost parameter. The SVM has been successfully applied to diverse scientific and engineering problems, including bioinformatics [20, 7, 15]. Anderson et al. [2] introduced the SVM to MS/MS spectra analysis, classifying Sequest results as correct and incorrect peptide assignments. Their result indicates that the SVM yields less false positives and false negatives compared to other cutoff approaches.

However, one weakness of the SVM is that it only estimates the category of the classification, while the probability $p(x)$ is often of interest itself, where $p(x) = P(Y = 1|X = x)$ is the conditional probability of a sample being in class 1 (i.e., correctly identified peptide) given the input x. Another problem with the SVM is that it is not trivial to select the best tuning parameters for the kernel and the cost. Often a grid search scheme has to be employed, which can be time consuming. In comparison, boosting and the random forest are very robust, and the amount of tuning needed is rather modest compared with the SVM.

12.4 Data and Implementation

12.4.1 Reference datasets

In the remainder of this chapter, we present performance comparison results from the aforementioned algorithms on a published dataset of Keller et al.

Table 12.1 *Number of spectra examples for the ESI-SEQUEST dataset.*

	Training	Testing
Correct	1930	827
Incorrect	24001	10286

(2002b) where the protein content is known. In particular, we intend to benchmark the performance of boosting and random forest methods in comparison
with other approaches using a "gold-standard" dataset. The assay used to
generate this dataset represents one of the two most common protein MS
ionization approaches: electrospray ionization (ESI) (the other being Matrix-
Assisted Laser Desorption ionization (MALDI)). The results are selected and
summarized from our previous publication [39]. This dataset is referred to as
the ESI-Sequest dataset from here on.

The electrospray dataset was kindly provided by Andy Keller, as described in
Keller et al. [23, 24]. These data are combined MS/MS spectra generated from
22 different LC/MS/MS runs on a control sample of 18 known (nonhuman)
proteins mixed in varying concentrations. A ThermoFinnigan ion trap mass
spectrometer was used to generate the dataset. In total, the data consists of
37,044 spectra of three parent ion charge states: $[M + H]^+$, $[M + 2H]^{2+}$ and
$[M + 3H]^{3+}$. Each spectrum was searched by Sequest against a human protein
database with the known protein sequences appended. The top-scoring peptide hit was retained for each spectrum; top hits against the known eighteen
proteins were labeled as "correct," and manually verified by Keller et al. All
peptide assignments corresponding to proteins other than the eighteen in the
standard sample mixture and common contaminants were labeled as "incorrect." In all, 2698 (7.28%) peptide assignments were determined to be correct.
The distribution of hits is summarized in Table 12.1.

12.4.2 Dataset parameters

The attributes extracted from SEQUEST assignments are listed in Table 12.2.
Attributes include typical scores generated by the SEQUEST algorithm (preliminary score (Sp), Sp rank, deltaCn, XCorr), as well as other statistics included in a SEQUEST report (total intensity, number of matching peaks,
fragment ions ratio). Number of tryptic termini (NTT) is a useful measure
for search results obtained by specifying no proteolytic enzyme, and is used
extensively in Keller et al. [23]. Other parameters include features readily obtainable from the candidate peptide sequence: C-term residue (K='1', R='2',
others='0'), number of prolines and number of arginines. A new statistic, the
Mobile Proton Factor (MPF), is calculated, which attempts to provide a simple measure of the mobility of protons in a peptide; it is a theoretical measure
of the ease of which a peptide may be fragmented in the gas phase [42, 22, 38].

Table 12.2 *SEQUEST attribute descriptions. Attribute names in bold are treated as discrete categorical variables.*

Attribute Group	Attribute Name	SEQUEST Name	Description
PeptideProphet (**I**)			
	Delta MH+	(M+H)+	Parent ion mass error between observed and theoretical
	Sp Rank	Rank/Sp	Initial peptide rank based on preliminary score
	Delta Cn	deltCn	1 - Cn: difference in normalized correlation scores between next-best and best hits
	XCorr	XCorr	Cross-correlation score between experimental and theoretical spectra
	Length	*Inferred from Peptide*	Length of the peptide sequence
NTT (**II**)			
	NTT	*Inferred from Peptide*	Measures whether the peptide is fully tryptic (2), partially tryptic (1), or non-tryptic (0)
Additional (**III**)			
	Parent Charge	(+1), (+2), (+3)	Charge of the parent ion
	Total Intensity	total inten	Normalized summed intensity of peaks
	DB peptides within mass window	# matched peptides	Number of database peptides matching the parent peak mass within the specified mass tolerance
	Sp	Sp	Preliminary score for a peptide match
	Ion Ratio	Ions	Fraction of theoretical peaks matched in the preliminary score
	C-term Residue	*Inferred from Peptide*	Amino acid residue at the C-term of the peptide (1 = R, 2 = 'K', 0 = 'other')
	Number of Prolines	*Inferred from Peptide*	Number of prolines in the peptide
	Number of Arginines	*Inferred from Peptide*	Number of arginines in the peptide
Calculated (**IV**)			
	Proton Mobility Factor	*calculated*	A measure of the ratio of basic amino acids to free protons for a peptide

We include MPF and the other derived parameters to demonstrate the ease of accommodation of additional information into the classification algorithms.

12.4.3 Implementation

For each of the ESI-Sequest datasets, we construct one balanced training and testing dataset by random selection. To be specific, correct-labeled and incorrect-labeled spectra were sampled separately so that both the training and the testing datasets contain the same proportion of correctly identified spectra. For all results, evaluation was performed on a test set that does not overlap the training set. Two-thirds of all data were used for training and one-third was used for testing.

The PeptideProphet application used in this analysis was downloaded from peptideprophet.sourceforge.net. The *.out search result files from Sequest were parsed; only the top hit for each spectrum was kept. PeptideProphet was run by executing the runPeptideProphet script using default parameters.

For the boosting and random forest approaches, we used contributed packages for the R programming language (`http://www.r-project.org`). In general, we did not fine-tune the parameters (i.e., tree size, number of trees, etc.) of the random forest and boosting for two reasons: classification performances of both the random forest and boosting are fairly robust to these parameters, and also because our ultimate goal is to provide a portable software tool that can be easily used in a production laboratory setting. Therefore, we wanted to demonstrate the superior performances of these methods in a user-friendly way.

For the AdaBoost [14] analysis, we used decision tree with 40 leaves for the weak classifier and fixed the number of boosting iterations to 1000. For random forest, the default number of attributes for each tree (one-third of total number of attributes) was used except for the five-variable case in which the number of attributes was fixed at two. The default number of trees in the forest was 500, and each tree in the forest was grown until the leaf was either pure or had only fives samples. We didn't use cross-validation to fine tune the random forest (as we do below for the SVM). All settings reflect the defaults of the randomForest v4.5-8 package available in the R statistical programming language.

With the support vector machine, we chose a radial kernel to classify the spectra as implemented in the libSVM package (version 2.7)*. The radial kernel is flexible and performed well in preliminary studies. In order to select the optimal set of tuning parameters for radial kernel, a grid search scheme was adopted using a modified version of the grid.py python script distributed with the libSVM package. The training set was randomly decomposed into two subsets, and we use cross-validation to find the best combination of tuning parameters. The optimal parameters for these data found via course grid search were $\gamma = 3.052e - 05$ and $cost = 32768.0$.

12.5 Results and Discussion

12.5.1 Classification performance on the ESI-SEQUEST dataset

Each mass spectrometry database search result can be sorted based on the resulting scoring statistics, from most confident to least confident. For Peptide-Prophet, the samples are ordered highest to lowest on the basis of a posterior probability as described in Keller et al. [23]. In the case of the machine learning algorithms discussed here, in addition to a correct or incorrect label, the algorithms also return an additional "fitness" term. For random forest, the fitness term can be interpreted as a probability of the identification being correct. A probability score can be generated from the boosting fitness measure as well using a simple transformation. The SVM returns a classification and a measure of the distance to a distinguishing hyperplane in attribute space that

* `http://www.csie.ntu.tw/~cjin/libsvm/`

can be considered a confidence measure. When samples are ordered in this way, results of such classification and scoring can be represented as a Receiver Operating Characteristic (ROC) plot, which provides a way of displaying the ratio of true positive classifications (sensitivity) to the fraction of false positives (1 specificity) as a function of a variable test threshold, chosen on the ranked ordering of results produced by the classifier. Decision problems such as this are always a tradeoff between being able to select the true positives without selecting too many false positives. If we set our scoring threshold very high, we can minimize or eliminate the number of false positives, but at the expense of missing a number of true positives; conversely, as we lower the scoring threshold, we select more true positives but more false positives will be selected as well. The slopes of the ROC plot are a measure of the rate at which one group is included at the expense of the other.

The ESI-Sequest dataset allows us to compare all four classification approaches: boosting, random forests, PeptideProphet and the SVM. ROC plots showing the results of classifying correct vs incorrect peptide assignments of the ESI-Sequest dataset using these methods are shown in Figure 12.3. All methods perform well on the data. As can be seen, the boosting and random forest methods provide a slight performance improvement over PeptideProphet and the SVM classification using the same six attributes. At a false positive rate of roughly 0.05%, the boosting and random forest achieves a sensitivity of 99% while PeptideProphet and SVM provide a 97-98% sensitivity. We note that, although a systematic difference of 1-2% can be seen in these results, this corresponds to a relatively small number of total spectra. Also indicated in Figure 12.3 are points corresponding to well-known thresholds from several literature citations. Each point shows the sensitivity and specificity that would be obtained on the test dataset by applying these published thresholds to the Sequest attributes Charge, Xcorr, Delta Cn and NTT.

Panels 1B and 1C compare the performance of the boosting and random forest methods using different sets of input attributes, as shown in Table 12.2. The panels contain the results of these algorithms using three combinations of features: 1) attribute groups I and II: the six attributes used by the PeptideProphet algorithm (SEQUEST XCorr, Delta Cn, SpRank, Delta Parent Mass, Length and NTT); 2) attribute groups I, III and IV (all attributes except NTT); and 3) attribute group I-IV (all fifteen variables shown in Table 12.2). Overall, it can be seen that both machine learning approaches provide improvement over the scoring thresholds described in the literature. The best performance was obtained by including all fifteen variables, indicating that accommodation of additional information is beneficial. The random forest appears to be slightly more sensitive to the presence of the NTT variable than boosting. Of note is the fact that effective classification is attained by the boosting and random forest tools even in the explicit absence of the NTT variable, as demonstrated by feature combination 2), despite the fact that the ESI dataset was generated using the no enzyme feature of Sequest. No enzyme

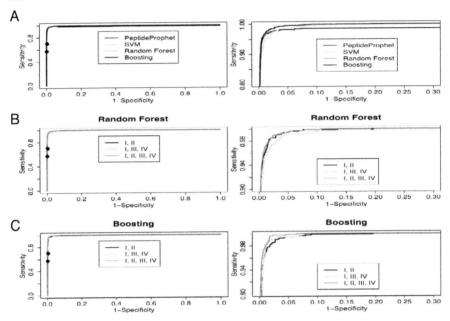

Figure 12.3 *Performance of boosting and random forest methods on the ESI-SEQUEST dataset. A. The top two plots are ROC plots of classification of the test set by PeptideProphet, SVM boosting, and random forest methods using attribute groups I and II. The plot on the right is a blowup of the upper left region of the figure on the left. Also displayed are points corresponding to several sets of SEQUEST scoring statistics used as linear threshold values in published studies. The following criteria were applied for choosing correct hits, the +1, +2, +3 numbers indicating peptide charge: a) +1: XCorr $\geq$ 1.5, NTT = 2; +2, +3: XCorr $\geq$ 2.0, NTT = 2; b) ΔCn > 0.1, +1: XCorr $\geq$ 1.9, NTT = 2; +2: XCorr $\geq$ 3 OR 2.2 $\leq$ XCorr $\leq$ 3.0, NTT $\geq$ 1; +3: XCorr $\geq$ 3.75, NTT $\geq$ 1; c) ΔCn $\geq$ 0.08, +1: XCorr $\geq$ 1.8; +2: XCorr $\geq$ 2.5; +3: XCorr $\geq$ 3.5; d) ΔCn $\geq$ 0.1, +1: XCorr $\geq$ 1.9, NTT = 2; +2: XCorr $\geq$ 2.2, NTT $\geq$ 1; +3: XCorr $\geq$ 3.75, NTT $\geq$ 1; e) ΔCn $\geq$ 0.1, Sp Rank $\leq$ 50, NTT $\geq$ 1, +1: not included; +2: XCorr $\geq$ 2.0; +3: XCorr $\geq$ 2.5. It can be seen that all machine learning approaches provide a significant improvement over linear scoring thresholds. B. The middle two plots show results of the random forest method using various sets of attributes. The black line represents the result of the random forest using six attributes defined in Table 12.2 as groups I and II: the SEQUEST XCorr, Sp Rank, ΔCn, Delta Parent Mass, Length and Number of Tryptic Termini (NTT). The lighter gray line is the result using fourteen attributes, groups I, III, and IV (no NTT). The darker gray line represents the result using all attribute groups I-IV, all fifteen variables. C. The bottom two plots are ROC plots of the boosting method using attribute groups I and II (black); I, III, and IV (lighter gray); and I-IV (darker gray).*

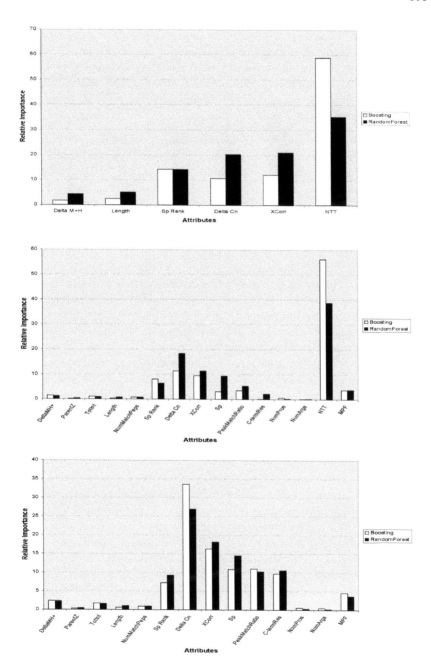

Figure 12.4 *Relative importance of data attributes used for classification by boosting and random forest methods. The top figure shows SEQUEST attribute importance from boosting and random forest classification using attribute groups I and II. The middle figure shows SEQUEST attribute importance using attribute groups I, III and IV. The bottom figure shows SEQUEST attribute importance using all attributes in random forest and boosting methods.*

specificity in the database search is often time-prohibitive in routine production work; it is much more common to restrict searches to tryptic peptides (or any other proteolytic enzyme used to digest the protein sample). Restricting to trypsin restricts results to having an NTT=2, rendering the attribute non-discriminatory. It must be noted, however, that in this analysis the C-term Residue attribute is not completely independent of NTT in that it contains residue information on one of the termini.

Figure 12.3 contains the results of these algorithms using three combinations of features: 1) the five features used by the PeptideProphet algorithm-Sequest XCorr, Delta Cn, SpRank, Delta parent mass and NTT; 2) all fourteen variables discussed above except NTT; and 3) all eleven variables discussed above including NTT. Overall, it can be seen that all machine learning approaches provide improvement over linear scoring thresholds. A modest performance improvement is achieved by these approaches over PeptideProphet using the same five variables. The best performance was obtained by including all fourteen variables, indicating that accommodation of additional information is beneficial. It is interesting that effective classification is attained by the boosting and random forest tools even in the absence of the NTT variable, as demonstrated by feature combination 2), despite the fact that the ESI dataset was generated using the 'no enzyme' feature of Sequest. This type of search is often time-prohibitive in routine production work; it is much more common to restrict searching to typtic peptides (or whatever other proteolytic enzyme used to digest the protein sample). SVM classification performed well but showed the most error of all methods; this performance could likely be improved by more precise parameter tuning.

12.5.2 The impact of individual attributes on the final prediction accuracy

It is interesting to examine the relative usefulness of the various parameters used by the machine learning algorithms to classify search results. One can gain a feeling for the relative importance of each variable by randomly scrambling the values of any one variable amongst all search results and re-running each search result (with the 'noised up' variable) through the classifier. The increase in misclassification that results is a measure of the relative importance of the randomized variable. This is valuable information in that the development of new scoring measures is an active area of investigation in proteomics research.

Figure 12.4 displays the relative importance each of the features for the SE-QUEST approaches using the various variables. Results for classification of the ESI-SEQUEST dataset incorporating the five features used by Peptide-Prophet for the ESI-SEQUEST dataset are shown in A. All five features show a contribution to the discrimination, with the most important contribution from the Number of Tryptic Terminii (NTT) categorical variable. PeptideProphet incorporates only the first four variables in the calculation of the discriminant function, introducing NTT distributions using a separate joint probability calculation. The coefficients for their discriminant score weight XCorr highest, followed by Delta Cn, with much lower contributions due to Delta M+H and Sp Rank. Our results indicate a larger contribution from Delta Cn, followed by XCorr, with moderate contributions from Sp Rank and Delta M+H. These results agree with those from Anderson et al., with the exception of Delta M+H which showed very little contribution using their SVM approach. The five features display a high importance when used in conjunction with the other six variables, as indicated in Figure 12.4. Of these additional six, Sp score shows a surprising contribution, as this scoring measure is rarely used for discrimination in popular usage. Also significant is the Ions Ratio measure. These results are in agreement with the Fisher Scores calculated by Anderson et al. The number of arginine and proline measures, as well as parent charge, appear to provide very little discriminative value.

The NTT variable provides by far the most important contribution, particularly for the boosting approach, but is only obtainable in nonenzyme-specific searches. The results above indicate, however, that the machine learning approaches perform quite well even in the absence of this variable. The relative importances of the other measures in the absence of this variable are shown in Figure 12.4. In this scenario, the Delta Cn measure provides the most importance contribution.

12.5.3 Unsupervised learning and generalization/comparison to PeptideProphet

In general, there are two primary classes of algorithms for addressing the pattern recognition problem, the rule-based approach and the model-based approach. Algorithms such as boosting and RFs are rule-based (see Vapnik [40] for a general description). Rule-based approaches are characterized by choosing one particular classification rule out of a set of rules that minimizes the number of mistakes on the training data. Model-based approaches are based on estimation of distributions for each of the classes being modeled, and use these distributions to calculate the probability that a data point belongs to each class. PeptideProphet is an example of this latter type, modeling correct identifications as a Gaussian distribution and incorrect identifications as a Gamma distribution. If the distributions describing the different classes in the problem accurately reflect the physical processes by which the data are generated, the modeling approach works well even for a small amount of training data. On the other hand, if the data diverge from the modeled distributions in a significant way, classification errors proportional to the degree of divergence result. Rule-based approaches are a less risky option if there is little knowledge of the distributions of classes of the data, and become increasingly safe, approaching optimality, as data size increases. Keller et al. [23] and

Nesvizhskii et al. [29] demonstrate that, for their data, the distributions described in their approach model the data well. Whether these distributions are appropriate for all types of instruments and MS search engines, and whether they are optimal, are two research questions. Given the large amount of mass spectrometry data available, rule-based approaches may generalize well to all types of data.

We note that both PeptideProphet and the boosting/Random Forest methods are supervised approaches, relying on training data for their functionality. PeptideProphet uses training data to learn coefficients in the calculation of the discriminate score; it subsequently uses these scores to establish the basic shape of the probability distributions modeling correct and incorrect search hits as a function of parent peptide charge. For each unique dataset, the distribution parameters are refined using an EM algorithm. Our approach provides a framework for performing the supervised aspect of the problem in a more general way, using established out-of-the-box functionality. This approach can be coupled with an unsupervised component to provide the same functionality, assuming appropriate training datasets are available that match the input data. The degree to which an individual training dataset provides adequate parameterization for a particular test set is an open question. Certainly, training sets will need to be search algorithm specific, but whether instrument-specific datasets are necessary is an area of investigation.

12.6 Conclusions

A serious challenge faced by researchers in a proteomics lab is curating large lists of protein identifications based on various confidence statistics generated by the search engines. The common methodology for selecting true hits from false positives is based on linear thresholding. These approaches can lead to a large number of false positives (using a more promiscuous threshold), or a large number of false negatives (using a relatively stringent threshold). Machine learning approaches such as boosting and random forest provide a more accurate method for classification of the results of MS/MS search engines as either correct or incorrect. Additionally, algorithmic development continues to improve the ability of automated tools to better discriminate true search results, and can complement the standard scoring measures generated by popular search engines. Flexible methods that allow for accommodation of these new scoring measures are necessary to allow them to be easily incorporated into production use. Tools such as PeptideProphet require significant work to accommodate any new features, and are based on statistical models which may not generalize well to all situations. Modern machine learning approaches can perform equally well, if not better, out-of-the box with very little tuning. Improved results could very likely be obtained by tuning these tools to particular data sets, i.e., by making use of class prior probabilities to accommodate the imbalanced sizes of the correct and incorrect datasets. These approaches can additionally be used to generate measures of relative importance of scoring variables, and may be useful in the development of new scoring approaches.

Peptide identification is not the ultimate goal in proteomics; accurately identifying the presence of proteins will allow the comparison between different samples and is the end result of biological analysis.

12.7 Acknowledgments

We thank Dr. Philip Andrews and colleagues in National Resources for Proteomics and Pathways for fruitful discussion on proteomics research and Ms. Rhiannon Popa for editorial assistance.

References

[1] Aebersold, R. and Goodlett, D. R. (2001) Mass spectrometry in proteomics. *Chem. Rev.* **101**, 269-95.

[2] Anderson, D. C., Li, W., Payan, D. G. and Noble, W. S. (2003) A new algorithm for the evaluation of shotgun peptide sequencing in proteomics: support vector machine classification of peptide MS/MS spectra and SEQUEST scores. *J. Proteome Res.* **2**, 137-46.

[3] Anderson, L. and Seilhamer, J. (1997) A comparison of selected mRNA and protein abundances in human liver. *Electrophoresis* **18**, 533-7.

[4] Bafna, V. and Edwards, N. (2001) SCOPE: a probabilistic model for scoring tandem mass spectra against a peptide database. *Bioinformatics* **17**, Suppl. 1, S13-21.

[5] Bao, L. and Cui, Y. (2005) Prediction of the phenotypic effects of nonsynonymous single nucleotide polymorphisms using structural and evolutionary information. *Bioinformatics* **21**, 2185-90.

[6] Breiman, L. (2001) Random Forests. *Machine Learning* **45**, 5-32.

[7] Brown M. P., Grundy, W. N., Lin, D., Cristianini, N., Sugnet, C. W., Furey, T. S., Ares, M. Jr. and Haussler, D. (2000) Knowledge-based analysis of microarray gene expression data by using support vector machines. *Proc. Natl. Acad. Sci. U. S. A.* **97**, 262-7.

[8] Clauser, K. R., Baker, P. and Burlingame A. L. (1999) Role of accurate mass measurement (+/- 10 ppm) in protein identification strategies employing MS or MS/MS and database searching. *Anal. Chem.* **71**, 2871-82.

[9] Craig, R. and Beavis, R. C. (2004), TANDEM: matching proteins with tandem mass spectra. *Bioinformatics* **20**, 1466-7.

[10] Dempster, A. P., Laird, N. M. and Rubin, D. B. (1977) Maximum likelihood from incomplete data via EM algorithm. *J Roy Stat Soc Series B* /bf 39, 1-38.

[11] Eng, J., McCormack, A. and Yates, J. (1994) An approach to correlate tandem mass spectral data of peptides with amino acid sequences in a protein database. *J. Am. Soc. Mass Spec.* **5**, 976-989.

[12] Eriksson, J. and Fenyo, D. (2004) Probity: a protein identification algorithm with accurate assignment of the statistical significance of the results. *J Proteome Res.* **3**, 32-6.

[13] Fenyo, D. and Beavis, R. C. (2003) A method for assessing the statistical significance of mass spectrometry-based protein identifications using general scoring schemes. *Anal Chem.* **75**, 768-74.

[14] Freund, Y. and Schapire, R. (1995) A decision theoretic generalization of on-line learning and an application to boosting. *Proceedings of the 2nd European Conference on Computational Learning Theory.* 23-37, Springer, New York.

[15] Furey, T. S., Cristianini, N., Duffy, N., Bednarski, D. W., Schummer, M. and Haussler, D. (2000) Support vector machine classification and validation of cancer tissue samples using microarray expression data. *Bioinformatics* **16**, 906-914.

[16] Geer, L. Y. , Markey, S. P., Kowalak, J. A., Wagner, L., Xu, M., Maynard, D. M., Yang, X., Shi, W. and Bryant, S. H. (2004) Open mass spectrometry search algorithm. *J Proteome Res.* **3**, 958-64.

[17] Gygi, S. P., Rochon. Y., Franza, B.R. and Aebersold, R. (1999) Correlation between protein and mRNA abundance in yeast. *Mol. Cell Biol.* **19**, 1720-30.

[18] Hastie, T., Tibshirani, R. and Friedman, J. H. (2001) *Elements of Statistical Learning.* Springer, New York.

[19] Havilio, M., Haddad, Y. and Smilansky, Z. (2003) Intensity-based statistical scorer for tandem mass spectrometry. *Anal. Chem.* **75**, 435-44.

[20] Jaakkola, T., Diekhans, M. and Haussler, D. (1999) Using the Fisher kernel method to detect remote protein homologies. *Proc. Int. Conf. Intell. Syst. Mol. Bio.,* 149-158. AAAI Press, Menlo Park, CA.

[21] Johnson, R. S., Davis, M. T., Taylor, J. A. and Patterson, S. D. (2005) Informatics for protein identification by mass spectrometry. *Methods.* **35**, 223-236.

[22] Kapp, E. A., Schutz, F., Reid, G. E., Eddes, J. S., Moritz, R. L, O'Hair, R. A., Speed, T. P. and Simpson, R. J. (2003) Mining a tandem mass spectrometry database to determine the trends and global factors influencing peptide fragmentation. *Anal. Chem.* **75**, 6251-64.

[23] Keller, A., Nesvizhskii, A. I., Kolker, E. and Aebersold, R. (2002) Empirical statistical model to estimate the accuracy of peptide identifications made by MS/MS and database search. *Anal. Chem.* **74**, 5383-5392.

[24] Keller, A., Purvine, S., Nesvizhskii, A. I., Stolyar, S., Goodlett, D. R. and Kolker, E. (2002) Experimental protein mixture for validating tandem mass spectral analysis. *OMICS.* **6**, 207- 12.

[25] Link, A. J., Eng, J., Schieltz, D. M., Carmack, E., Mize, G. J., Morris, D. R., Garvik, B. M. and Yates, J. R. III (1999) Direct analysis of protein complexes using mass spectrometry. *Nat. Biotechnol.* **17**, 676- 682.

[26] Lockhart, D. J., Dong, H., Byrne, M. C., Follettie, M. T., Gallo, M. V., Chee, M. S., Mittmann, M., Wang, C., Kobayashi, M., Horton, H. *et al.* (1996) Expression monitoring by hybridization to high-density oligonucleotide arrays. *Nat. Biotechnol.* **14** 1675-1680.

[27] MacCoss, M. J., Wu, C. C. and Yates, J. R. 3rd. (2002) Probability-based validation of protein identifications using a modified SEQUEST algorithm. *Anal. Chem.* **74**, 5593-9.

[28] Moore, R. E., Young, M. K. and Lee, T. D. (2002) Qscore: an algorithm for evaluating SEQUEST database search results. *J. Am. Soc. Mass Spectrom.* **13**, 378-86.

[29] Nesvizhskii, A. I., Keller, A., Kolker, E. and Aebersold, R. (2003) A statistical model for identifying proteins by tandem mass spectrometry. *Anal. Chem.* **75**, 4646-58.

[30] Pandey, A. and Mann, M. (2000) Proteomics to study genes and genomes. *Nature* **405**, 837-46.

[31] Peng, J., Elias, J. E., Thoreen, C. C., Licklider, L. J. and Gygi, S. P. (2003) Evaluation of multidimensional chromatography coupled with tandem mass spectrometry (LC/LC-MS/MS) for large-scale protein analysis: the yeast proteome. *J. Proteome Res.* **2**, 43-50.

[32] Pennington, K., Cotter, D. and Dunn, M. J. (2005) The role of proteomics in investigating psychiatric disorders. *Br. J. Psychiatry* **187**, 4-6.

[33] Perkins, D., Pappin, D., Creasy, D. and Cottrell, J. (1999) Probability-based protein identification by searching sequence databases using mass-spectromety data. *Electorphoresis* **20**, 3551-3567.

[34] Sadygov, R. G. and Yates, J. R. 3rd. (2003) A hypergeometric probability model for protein identification and validation using tandem mass spectral data and protein sequence databases. *Anal. Chem.* **75**, 3792-8. 30

[35] Schena, M., Shalon, D., Davis, R. W. and Brown, P. O. (1995) Quantitative monitoring of gene expression patterns with a complementary DNA microarray. *Science* **270**, 467-470.

[36] Steen, H. and Mann, M. (2004) The ABC's (and XYZ's) of peptide sequencing. *Nat. Rev. Mol. Cell Biol.* **5**, 699-711.

[37] Sun, W., Li, F., Wang, J., Zheng, D. and Gao, Y. (2004) AMASS: software for automatically validating the quality of MS/MS spectrum from SEQUEST results. *Mol. Cell Proteomics.* **3**, 1194-9.

[38] Tabb, D. L., Huang, Y., Wysocki, V. H. and Yates, J. R. 3rd. (2004) Influence of basic residue content on fragment ion peak intensities in low-energy collision-induced dissociation spectra of peptides. *Anal. Chem.* **76**, 1243-8.

[39] Ulintz, P. J., Zhu J., Qin, Z. S. and Andrews P. C. (2006) Improved classification of mass spectrometry database search results using newer machine learning approaches. *Mol. Cell Proteomics* **5**, 497-509.

[40] Vapnik, V. N. (1999) *The Nature of Statistical Learning Theory.* **30**, 138-167. Springer, New York.

[41] Washburn, M. P., Wolters, D. and Yates, J. R. 3rd. (2001) Large-scale analysis of the yeast proteome by multidimensional protein identification technology. *Nat. Biotechnol.* **19**, 242-7.

[42] Wysocki, V. H., Tsaprailis, G., Smith, L. L. and Breci, L.A. (2000) Mobile and localized protons: a framework for understanding peptide dissociation. *J. Mass Spectrom.* **35**, 1399-406.

[43] Zhang, N., Aebersold, R. and Schwikowski, B. (2002) ProbID: a probabilistic algorithm to identify peptides through sequence database searching using tandem mass spectral data. *Proteomics* **2**, 1406-12.

Index

α-helix, 13
β-sheet, 13
β-strand, 13
0/1 loss function, 105
1-D PAGE, 20
2-D PAGE, 20

ACA, 281
ACMI, 247, 264, 274
AdaBoost, 122, 346
adenine, 8
alternative splicing, 11, 12
amino acid, 7, 10, 12
    residue, 13
analysis of covariance, 69
analysis of variance, 69
angiogenesis, 18
ANN, 161, 245
    LMS error, 164
    LVQ, 146
    MLP, 163
    perceptron, 163
    RBF, 165
ANOVA, 1
antibody, 14
    as a probe, 21
apoptosis, 18
ARP/WARP, 246, 247, 273
Artificial neural networks, 161
attribute, 103
Attribute clustering algorithm, 281
automatic relevance determination, 306
average linkage, 147

Backpropagation, 164
backpropagation, 245
backward algorithm, 63
bacteria, 5
bagging, 120

base pair, 8
    complementary, 8
Baum-Welch algorithm, 65
Bayes factor, 92
Bayes posterior risk, 91
Bayes principle, 90
Bayes risk, 90, 105
Bayes rule, 90, 105
Bayes theorem, 37
Bayesian network, 230
    dynamic, 230
Bayesian paradigm, 2
Bernoulli random variable, 46
Beta distribution, 49
bias, 26, 68
bicluster, 149
biclustering, 148, 277
Binomial random variable, 46
biplot, 135
BLOSUM, 214
bones, 246
boosting, 122, 346
bootstrap, 26
breast cancer, 305
Bregman divergence, 282

cancer, 18
Carolas Linnaeus, 1
case-based reasoning, 244
catalysis, 14
Cauchy-Schwarz inequality, 50
cell
    division, 9
    proliferation, 9, 17
cells, 5
central limit theorem, 48
channel, 14
Charles Darwin, 1
Chebyshev's inequality, 56
chi-square distribution, 49

ChIP-chip, 23
chromatin immunoprecipitation, 23
chromosome, 9
class, 103
class conditional density, 104
classification, 3, 101
    bagging, 120
    boosting, 122
    ensemble methods, 120
    logistic, 110
    nearest neighbor, 112
    nonparametric, 112
    parametric, 108
    support vector machine, 117
    tree, 113
CLUSTAL, 216
cluster, 141
clustering, 3, 129, 141, 277
    hierarchical, 142, 146
    k-means, 142
    model based, 143
    partitional, 141
    SOM, 144
co-immunoprecipitation, 23
coding region, 10
codon, 10
coexpression, 226, 277
coherence index, 286
colon cancer, 305
competitive learning, 146
complete linkage, 147
complex, 13
conditional probability, 35
conjugate distribution, 91
conjugate priors, 305
coregulation, 226, 229, 277
correlation, 50, 69
correlation matrix, 51
covariance, 50
credible region, 92
    highest posterior density, 92
cross entropy, 115
cross validation, 107, 304, 314
cytoplasm, 6
cytosine, 8
cytoskeleton, 14

DBF, 281, 296
decision boundary, 106

decision rule, 103
deletion, 212
differential equation, 228
differentiation, 6
dimensionality reduction, 129
    manifold learning, 140
    MDS, 136
    PCA, 129
Dirichlet prior, 218
discriminant analysis, 108
    linear, 108
    quadratic, 108
dispersion matrix, 50
DNA, 7, 8
    3-D structure, 8
    base, 8
    damaga, 19
    polymerase, 9
    probe, 21
    replication, 9
    sequencing, 19
    sequential structure, 8
    strand, 8
double helix, 2

electron density map, 237
EM, 280, 345
empirical Bayes, 94
endoplasmic reticulum, 6, 7
endosymbiosis, 6
ensemble methods, 120
entropy, 56
enzyme, 13
equilibrium, 16
Escherichia coli, 5, 9, 12, 14
eukaryote, 6
events, 29
    independent, 34
    mutually exclusive, 30
evolution, 1
exon, 11
expectation-maximization, 27, 86, 144,
    280, 345
expected value, 41
exponential distribution, 48

false negative, 36
false positive, 36

feature, 103
FFFEAR, 246
final precision, 90
Fisher information matrix, 57
FLOC, 280, 296
forward algorithm, 63
functional data analysis, 229
Fuzzy logic, 158
Fuzzy sets, 157
    basic operations, 160
    membership, 157
    membership function, 159

GA, 167, 282
Gamma distribution, 48
gap, 213
Gaussian distribution, 48
gel electrophoresis, 20
gene, 8
    coding region, 10
    promoter, 10
    regulatory network, 11
    regulatory region, 10
    transcription, 10
    translation, 10, 11
gene expression
    analysis, 218
    clustering, 220
    differential, 222
    dimensionality reduction, 221
    dynamics, 228
    gene enrichment, 222
gene finding, 211
gene ontology, 222, 281, 293
gene regulatory network, 11
    inference, 223
generalized linear model, 69
Genetic algorithms, 167
genetic network inference, 223
geometric distribution, 46
Gibbs sampler, 90, 96, 97, 312
Gibbs sampling, 27
Gini index, 115
Golgi apparatus, 6, 7
growth factor, 18
growth inhibition factor, 18
guanine, 8

hidden Markov models, 62
hierarchical clustering
    average linkage, 147
    complete linkage, 147
    single linkage, 147
histone, 9
HMM, 211
hormone, 7
hydrophilic, 7
hydrophobic, 7
hypergeometric distribution, 47
hyperparameters, 94, 306
hyperpriors, 94
hypothesis testing, 26

immunostaining, 22
improper prior, 94
in situ hybridization, 22
information theory, 56
initial precision, 90
insertion, 212
intron, 11
isoelectric point, 20
Isomap, 140
iTRAQ, 327

Jeffreys' prior, 94, 305
Jensen's inequality, 96
joint distribution, 49

k-means, 142
k-nearest neighbor, 244
kernel, 112
kernel trick, 119
knowledge-based networks, 162
Kullback-Leibler divergence, 57

labeled probe, 21
Laplacian Eigenmaps, 140
latent variables, 307
learning vector quantization, 146
least squares, 68
leukemia, 305
likelihood, 37
likelihood function, 26, 66
linear discriminant analysis, 108
linear separability, 310

link function, 69
lipid, 7
LLE, 140
local linear embedding, 140
logistic classifier, 304
logistic discrimination, 110, 124
logistic regression, 110
logit, 305
loss function, 104, 307
LVQ, 146
lysosome, 6

machine learning, 3
Mahalanobis distance, 105
MAID, 247
manifold learning, 140
MANOVA, 142
margin, 310
marginal distribution, 49
marker, 22
Markov chain Monte Carlo, 2, 27, 90, 305
Markov chains, 58
    ergodic, 62
    regular, 60
    steady state, 60
Markov network, 230
Mascot, 342
mass spectrometry, 22, 124, 327, 339
maximum likelihood, 67
maximum likelihood estimation, 26
MDS, 136
membrane protein, 7
metabolic pathway, 15
metabolism, 15
metastatis, 18
method of moments, 66
Metropolis-Hastings algorithm, 27, 90, 96, 312
microarray, 21, 124, 129, 277, 304
    clustering, 220
    data analysis, 218
    differential expression, 222
    dimensionality reduction, 221
    gene enrichment, 222
    normalization, 219
microtubule, 14
Minkowski inequality, 56
misclassification error rate, 106

mitochondria, 6
MLP, 163
module, 218
MOEA, 283, 296
moment generating function, 42
moments, 42
    central, 42
multidimensional scaling, 136
multilayer perceptron, 162–163
multinomial distribution, 52
mutagen, 19

naive Bayes classifier, 112
nearest neighbor classification, 112
Needleman-Wunsch algorithm, 214
negative binomial distribution, 47
negative selection, 213
neural networks, 161
neuro-fuzzy computing, 174
neuro-fuzzy hybridization, 174
neurotransmitter, 14
neutral drift, 213
nonparametric classification, 112
normal distribution, 48
    bivariate, 53
    multivariate, 53
Northern blot, 20
NSGA, 286
nucleotide, 7
nucleus, 6

odds ratio
    posterior, 92
    prior, 92
oncogene, 19
organ, 6
organelle, 6
overfitting, 107

p.d.f., 44
p.m.f., 41
PAGE, 20
PAM, 214
parameter estimation, 66
parametric classification, 108
PBC, 296
PCA, 129, 139
    kernel, 140

local, 139
PCM, 287
peptide, 13
peptide mass fingerprint, 22
PeptideProphet, 344
plasmid, 9
plastid, 6
point mutation, 212
Poisson distribution, 47
polymer, 7
position weight matrix, 217
position-specific scoring matrix, 217
positive selection, 213
possibilistic c-means, 287
posterior distribution, 26
posterior probability, 37
precision matrix, 51
principal components analysis, 129
prior distribution, 26
probabilistic model, 243
probabilistic modeling, 25
probability density curve, 44
probability density function, 44
probability distribution, 40
probability mass function, 41
probability theory, 25
probe, 21
probit, 305
prokaryote, 5
promoter, 10
proposal density, 95, 96
protein, 7, 10, 12
    3-D structure, 13, 17
    channel, 14
    complex, 13, 23
    folding, 13, 237
    identification, 124
    interaction, 13, 23
    primary structure, 13
    quantification, 327
    quaternary structure, 13
    secondary structure, 13, 238
    sequential structure, 238
    structural, 14
    structure prediction, 211, 237
    tertiary structure, 13, 238
    transmembrane, 14
proteomics, 339
pseudocount, 218

pseudopod, 14

QScore, 344
quadratic discriminant analysis, 108

Radial basis function network, 165
random forest, 346
random variables, 40
    central moment, 42
    continuous vs. discrete, 43
    correlation, 50
    covariance, 50
    expected value or mean, 41
    joint distribution, 49
    moments, 42
    skewness, 42
    standard deviation, 42
    variance, 42
RBF, 165
recurrent state, 61
regression, 53, 69
    linear, 1
regularization, 108
regulatory module, 218
regulatory region, 10
relative entropy, 57
relevance vector machines, 305
reproducing kernel Hilbert space, 304,
    306
residue, 13
RESOLVE, 246, 252, 273
REVEAL, 227
reverse engineering, 224
ribosome, 6, 7
risk function, 104
RMA, 220
RNA, 7, 8
    3-D structure, 10
    messenger, 10
    polymerase, 10
    probe, 21
    sequential structure, 10
    strand, 10
    transcript, 10

SAGE, 21
    data analysis, 218
SAM, 222

Sammon mapping, 138
sampling, 95
  importance, 95
  rejection, 95
SEBI, 282
self organizing map, 144
sequence alignment, 211
  global, 212
  local, 212
  multiple, 215
  pairwise, global, 213
  pairwise, local, 213
sequential evolutionary biclustering, 282
Sequest, 342
serial analysis of gene expression, 21
signal transduction, 7
signaling, 14
simulated annealing, 229, 283
single linkage, 147
single nucleotide polymorphism, 19
Single-layer perceptron, 163
singular value decomposition, 134
skeletonization, 246
skewness, 42
Smith-Waterman algorithm, 215
SNP, 19
soft computing, 3, 277
SOM, 144
Southern blot, 20
splice variant, 11
splicing, 11, 12
  regulation, 12
SSC, 280
standard deviation, 42
state space, 58
Statistical Subspace Clustering, 280
stem cell, 6
stochastic processes, 58
stress function, 138
strong law of large numbers, 95
substitution matrix, 213, 214
sugar, 7
supervised learning, 3, 101
support vector machine, 117, 347
  kernel trick, 119
support vector machines, 304
  Bayesian, 311
SVM, 117, 347

t-test, 1
testing data, 106
TEXTAL, 247, 258, 273
thymine, 8
total risk, 104
training data, 106
transcript, 10
transcription, 10
  regulation, 11
transcription factor, 11
transcription factor binding site, 11
  identification, 216, 229
transient state, 61
transition matrix, 58
translation, 10, 11
  regulation, 12
trinomial distribution, 52
tumor, 18
tumor classification, 303
tumor suppressor gene, 19

uniform density, 44
universal set, 29
unsupervised learning, 3, 129
uracil, 10

vacuole, 6
variable selection, 123
variance, 42
variance-covariance matrix, 50
vesicle, 8

Ward's method, 147
Western blot, 20

X-AUTOFIT, 246
x-ray crystallography, 239

yeast cell cycle, 220, 291
yeast two-hybrid, 23

z-score, 48